The Invisible Hand in Economics

This is a book about one of the most controversial concepts in economics: the invisible hand. The author explores the unintended social consequences implied by the invisible hand and discusses the mechanisms that bring about these consequences.

The book questions, examines and explicates the strengths and weaknesses of invisible-hand explanations concerning the emergence of institutions and macro-social structures, from a methodological and philosophical perspective. Aydinonat analyses paradigmatic examples of invisible-hand explanations, such as Carl Menger's 'Origin of Money' and Thomas Schelling's famous chequerboard model of residential segregation, in relation to contemporary models of emergence of money and segregation. Based on this analysis, he provides a fresh look at the philosophical literature on models and explanation and develops a philosophical framework for interpreting invisible-hand type of explanations in economics and elsewhere. Finally, the author applies this framework to recent game theoretic models of institutions and outlines the way in which they should be evaluated.

Covering areas such as history, philosophy of economics and game theory, this book will appeal to philosophers of social science and historians of economic thought, as well as to practising economists.

There is a long tradition in the social sciences, going back to Adam Smith, of explaining social phenomena as the unintended consequences of human actions. In this illuminating book, Aydinonat investigates the structure of such explanations and the nature of the claims that can legitimately be derived from them. In the process, he analyses some of the classic ideas in social theory – Smith's invisible hand, Carl Menger's explanation of the emergence of money, Thomas Schelling's analysis of racial segregation, and David Lewis's theory of convention – with acuity and subtlety. This is a significant contribution to the philosophy of social science, which will also engage the interest of reflective economic theorists.

Robert Sugden (Professor of Economics, University of East Anglia)

Conjectural models aim at devising the initial conditions required for individual actions to generate a given social phenomenon as unintended consequence. Social scientists make frequent use of this modelling technique. Indeed the list of those who have applied it – from Smith to Menger, from Schelling to Lewis – reads like a veritable 'who's who' of the last 250 years of social sciences. Yet the question of how these purely speculative models may actually enjoy any explanatory power with respect to real world phenomena has only rarely been tackled. Aydinonat's outstanding work fills this gap by thoroughly investigating the philosophical and methodological challenges posed by conjectural models and by developing a coherent and persuasive framework to account for the role of abstract theorizing in the social sciences. The book is a candidate to become compulsory reading for methodologists and philosophers of science, as well as for those economists who take seriously the issue of their models' epistemological foundations.

Nicola Giocoli (Professor of Economics, University of Pisa)

N. Emrah Aydinonat is Lecturer in Philosophy of Economics and Economic Growth at Ankara University, Turkey.

Routledge INEM Advances in Economic Methodology

Edited by Esther-Mirjam Sent
University of Nijmegen, the Netherlands

The field of economic methodology has expanded rapidly during the last few decades. This expansion has occurred in part because of changes within the discipline of economics, in part because of changes in the prevailing philosophical conception of scientific knowledge, and also because of various transformations within the wider society. Research in economic methodology now reflects not only developments in contemporary economic theory, the history of economic thought, and the philosophy of science; but also developments in science studies, historical epistemology, and social theorising more generally. The field of economic methodology still includes the search for rules for the proper conduct of economic science, but it also covers a vast array of other subjects and accommodates a variety of different approaches to those subjects.

The objective of this series is to provide a forum for the publication of significant works in the growing field of economic methodology. Since the series defines methodology quite broadly, it will publish books on a wide range of different methodological subjects. The series is also open to a variety of different types of works: original research monographs, edited collections, as well as republication of significant earlier contributions to the methodological literature. The International Network for Economic Methodology (INEM) is proud to sponsor this important series of contributions to the methodological literature.

1   **Foundations of Economic Method, 2nd Edition**
     A Popperian perspective
     Lawrence A. Boland

2   **Applied Economics and the Critical Realist Critique**
     Edited by Paul Downward

3   **Dewey, Pragmatism and Economic Methodology**
     Edited by Elias L. Khalil

4   **How Economists Model the World into Numbers**
     Marcel Boumans

5   **McCloskey's Rhetoric**
    Discourse ethics in economics
    Benjamin Balak

6   **The Foundations of Paul Samuelson's Revealed Preference Theory**
    A study by the method of rational reconstruction, revised edition
    Stanley Wong

7   **Economics and the Mind**
    Edited by Barbara Montero and Mark D. White

8   **Error in Economics**
    Towards a more evidence-based methodology
    Julian Reiss

9   **Popper and Economic Methodology**
    Contemporary challenges
    Edited by Thomas A. Boylan and Paschal F. O'Gorman

10 **The Invisible Hand in Economics**
    How economists explain unintended social consequences
    N. Emrah Aydinonat

The Invisible Hand in Economics

How economists explain
unintended social consequences

N. Emrah Aydinonat

Routledge
Taylor & Francis Group

LONDON AND NEW YORK

First published 2008
by Routledge
2 Park Square, Milton Park, Abingdon, Oxon, OX14 4RN

Simultaneously published in the USA and Canada
by Routledge
605 Third Avenue, New York, NY 10017

*Routledge is an imprint of the Taylor & Francis Group, an informa
business*

Typeset in Times New Roman by Prepress Projects Ltd, Perth, UK

British Library Cataloguing in Publication Data
A catalogue record for this book is available from the British Library

Library of Congress Cataloging in Publication Data
Aydinonat, N. Emrah
The invisible hand in economics: how economists explain unintended
social consequences/ N. Emrah Aydinonat
    p. cm
Includes bibliographical reference and index
ISBN 978-0-415-41783-9 (hardcover) — ISBN 978-0-203-93094-6
(e-book) 1. Economics–Philosophy. 2. Free enterprise. 3. Self-interest–
Social aspects. 4. Economics–Methodology. I. Title. II. Title: How
economists explain unintended social consequences.
HB72.A93 2008
330.01–dc22
                    2007034246

ISBN 13: 978-0-415-56954-5 (pbk)
ISBN 13: 978-0-415-41783-9 (hbk)

Annem ve Babam için

(For my mother and father)

Contents

List of figures x
List of tables xi
Foreword xiii
Acknowledgements xv

1 Introduction 1

2 Unintended consequences 11

3 The origin of money 27

4 Segregation 50

5 The invisible hand 68

6 The origin of money reconsidered 93

7 Models and representation 119

8 Game theory and conventions 146

9 Concluding remarks 165

Appendix I: Smith, Jevons and Mises on money 169
Appendix II: Models of emergence of money 173
Appendix III: Explorations of the chequerboard world 178
Appendix IV: Focal points and risk dominance 182
Notes 194
Bibliography 216
Index 243

Figures

3.1 States of the world 35
3.2 Real world vs model world 44
4.1 The chequerboard city 53
4.2 Randomly distributed residents 54
4.3 Segregated city 54
4.4 Model world and the real world 58
6.1 Kiyotaki–Wright model vs Menger's model 96
6.2 Absorbing points in Schotter (1981) 108
6.3 Models of emergence of money in relation to Menger's model 113
6.4 Schotter and Young's models in comparison to Menger's 115
7.1 Model world and the real world 126
7.2 A possible way to conceptualise segregation 131
7.3 The chequerboard model 133
7.4 Conceptual tools for model building 141
AIV.1 Simple games of Bacharach and Bernasconi (1997) 187
AIV.2 A coordination game 188

Tables

2.1	Possible intention–action pairs	17
2.2	Table of possibilities	19
4.1	Mechanisms	56
6.1	Production and consumption in Kiyotaki and Wright (1989)	97
6.2	Description of the economies in Marimon *et al.* (1990)	102
6.3	Money game	107
6.4	Models in comparison to Menger (1892a)	114
8.1	The driving game	148
AIV.1	Coordination game	185
AIV.2	Game	186
AIV.3	Stag hunt game	186
AIV.4	Telephone game	190

Foreword

Why do people abide by rules (think of traffic rules, for example)? Because they know that if they transgress the rules there is a chance they will be caught in the act, and because they also know that if they are caught in the act they will be punished. Why is there a system of punishing transgressors in the first place? Well, because some (particularly those with the requisite powers and authorities) thought it a good idea to install such a system. This way of explaining social, aggregate behaviour in terms of the intentions and expectations of the people producing the behaviour seems to make perfectly good sense. Nobody would deny that this is roughly how desirable social, aggregate behaviour (in this case, people by and large abiding to the rules) often is secured. But it also has a trivial ring over it. If this type of explanation were the only type available to (and actually practised by) social scientists, then this would make them vulnerable to the oft-cited charge that social science is not able to go beyond mere common sense (or 'folk') understandings.

But, happily, there is a venerable tradition in social theorising in which another, non-trivial type of explanation of social, aggregate behaviour is put forward: invisible-hand explanation. In a first rough approximation, invisible-hand explanations explain social patterns of behaviour as *un*intended consequences of human actions and interactions. The challenge is to show that useful institutions such as money and, more generally, orderly patterns of aggregate behaviour can emerge and persist even though no-one intended to produce these outcomes. Invisible-hand explanations are far from trivial. There is even something deeply puzzling about such explanations. It is easy to see why. The unintended consequences that invisible-hand explanations refer to typically are useful or mutually beneficial. How do mutually beneficial consequences come about if there is no-one intending to bring about the consequences? Moreover, what seems to explain certain patterns of aggregate behaviour in invisible-hand explanations is the fact that the patterns have mutually beneficial consequences. This seems to have things backwards: some situations are explained by their effects, rather than by their causes.

This book does a lot to resolve this and other puzzles surrounding the notions of invisible hand, unintended consequences and invisible-hand explanations. Indeed, the book goes well beyond the scope and depth of earlier work on these notions. N. Emrah Aydinonat covers a wide variety of topics and fields, ranging from the ways in which Adam Smith invoked the notion of the invisible hand over actual examples of invisible-hand explanations in social-scientific practice to recent work in philosophy of science on models. The combination of competences that Emrah brings to this is at once impressive and rare. Here is a scholar who does not rest content with an easy understanding of what he is talking about. Here is a scholar who goes to the bottom of things. This is done in a highly readable, accessible prose and in a lucid style. To be sure, this is not going to be the definitive word on these issues. But subsequent work can ignore Emrah's groundbreaking work only to its own peril.

It might seem that in praising Emrah's achievement, we are indirectly congratulating ourselves. For the two of us supervised Emrah's Ph.D. project, from which the present book is an updated outgrowth. But we are well aware that our own contribution was not very large. Emrah had the basic ideas of the thesis in his mind right from the start of his project. They just needed some more polishing, refining and more precise articulation. Let us assure you (in case you do not know him already) that, like all talented students, Emrah was much too unyielding to be steered in directions that he did not want to go. During the project we saw Emrah grow from an insecure freshman into the confident and mature scholar with strong preferences and firm ideas that he now is. But his initial insecurity did not prevent him from developing the rough and embryonic outlines of the ideas in this book quite quickly. It was a delight to work with Emrah, not only in our many encounters on the work floor, discussing his work and that of others, but also afterwards, in the bar. All those fine hours were definitely well spent.

Maarten Janssen, Professor, Erasmus University Rotterdam
Jack Vromen, Professor, Erasmus University Rotterdam

Acknowledgements

This book is the revised version of my PhD thesis that was written at the Erasmus Institute of Philosophy and Economics (EIPE), Erasmus University Rotterdam, the Netherlands. This book would not have been possible without the help and support of many people. I would like to begin with the following acknowledgements.

I started my academic career in economics with the encouragement of Professor Yahya Sezai Tezel and Professor Aykut Kibritçioğlu. Yahya Tezel opened my eyes to what we may call the world beyond neoclassical economics and helped me in my struggle to find my way in philosophy and methodology of social science and history. His intellectual capacity and creativity influenced me in a way that I cannot express in words. I learned from him, even before reading the wonderful books by Douglas Adams, that the answer to the question of the meaning of life, the universe and everything (which is 42) did not mean anything if you did not ask the appropriate questions. While Yahya Tezel gave me an 'it don't mean a thing if you ain't got that swing' attitude towards research, Aykut Kibritçioğlu taught me how to do good research and to be organised. Under their supervision I was able to write my master thesis, entitled the *Theoretical Roots of the Heckscher-Ohlin Theorem*. Another influential person in my life was Professor Güven Sak. He encouraged me to work on 'weird' issues, such as the knowledge problem in Hayek's works, while every other student was working either on the Turkish banking system or on moral hazard. I have learned a lot from him – not only about finance, but also about life.

I have to thank everyone at EIPE, especially all my professors, who provided the best guidance concerning philosophy and methodology of economics. Although some of them will reappear below, I cannot resist mentioning their names: John Groenewegen, Maarten Janssen, Albert Jolink, Uskali Mäki, Arjo Klamer and Jack Vromen. Thank you all. Another important person at EIPE was Loes van Dijk. Without her help, it would have been difficult to live in Rotterdam. Loes, thank you for everything.

EIPE staff provided an environment where I had to present and discuss my papers in several places. I had invaluable comments from many people at these presentations. Uskali Mäki provided detailed comments on my papers on invisible-hand explanations and J. S. Mill. Albert Jolink provided many comments and encouragement. Petri Ylikoski had very useful remarks on my Menger paper, James Wible and John Davis gave detailed comments on my Schelling paper. I have also benefited from Mark Blaug's comments and encouragement. I owe them much, for they were unlucky to read the first drafts of this book, and I was lucky to give them the first drafts.

Thanks to internet technology, my papers reached several mailboxes worldwide. Warren Samuels, Greg Ransom and Robert Sugden took the trouble to read my attachments and provided helpful comments for my research. Thanks again.

In the beginning of 2001 I spent two months in South Bend, Indiana, USA, at the University of Notre Dame. I thank Philip Mirowski and Esther-Mirjam Sent for their comments, criticism and encouragement. During my visit I also had a chance to meet Thomas Schelling, Robert Axtell and Peyton Young. They provided invaluable guidance and encouragement. Again, thank you.

The most important thing in a Ph.D. student's life is friendship. Without friends one would not be able to finish a research project without suffering from a mental trauma. Luckily, I had many friends. I thank all of them for their friendship. I need to mention one of them for she contributed much to this book: Caterina Marchionni saved me in my most desperate moments with her friendship and careful reading. She encouraged and challenged me, corrected my English, provided good friendship and she made me laugh. Cate, thank you very much.

Of course, the most important people for my research were my supervisors, Maarten Janssen and Jack Vromen. Without their precise and interesting questions and comments I would not have been able to finish this book. Knowing that their life was hectic with a lot of teaching and research, I can only hope that this book demonstrates that they did not waste their time with me. I also thank the members of my doctoral committee: Mark Blaug, Uskali Mäki and Robert Sugden.

Finally, Yonca, thank you for your love and patience. Without you, I would not have been able to finish the revision of this book.

Some parts of Chapter 5 were published as 'Is the invisible hand un-Smithian? A Comment on Rothschild', in *Economics Bulletin* 2 (1). Chapter 8 is a revised version of the following article: 'Game Theoretic Models as a Framework for Analysis' in C. C. Aktan (ed.) *Advances in Economics: Theory and Applications*, 2nd ICBME Conference Proceedings, vol. 2, 2006, İzmir: Anıl Matbaacılık. I thank the editors of *Economics Bulletin* and *ICBME Conference Proceedings* for not limiting authors' rights to reuse and reproduce their own work. Parts of Chapters 4 and 7 were published as 'Models, conjectures and exploration: an analysis of Schelling's chequerboard model of residential segregation', in *Journal of Economic Methodology* 14 (4): 429–454, by N. Emrah Aydinonat (2007, published by Taylor & Francis Ltd, www.informaworld.com), reprinted by permission of the publisher. I thank the editors of *JEM* for allowing me to use this article.

I hereby gratefully acknowledge the financial support of The Council of Higher Education, Turkey (YÖK), Nederlandse Organisatie voor Wetenschappelijk Onderzoek (NWO), Trustfonds (Vereniging Trustfonds, Erasmus Universiteit Rotterdam) and Erasmus Institute for Philosophy and Economics (EIPE).

1 Introduction

Everyone is familiar with the (aesthetically) unpleasant walking paths on public green fields. Usually, around these fields there are constructed paths for the service of the beloved citizens. However, citizens seem to deviate from these designed paths and take shortcuts passing through the green fields, through the zones they should not pass. The outcome of this behaviour is the death of the plants and the emergence of paths on these fields. Is there a good and acceptable explanation of why these walking paths emerge? What kind of an explanation would be acceptable?

One scenario about the emergence of these paths may be the following. The first person to take a shortcut through the aforementioned field believes that his behaviour will do no harm to the green field and proceeds to take the shortcut. Obviously, he will in fact have a negligible impact on the field. Subsequently, a second person, without being aware of the fact that he is the second person, takes the same shortcut,[1] with similar thoughts in mind. However, although the second person's impact is still negligible, in time the damage on the plants accumulates. After some time, if other people take that very same shortcut the damage will soon become visible. As a consequence, even if some earlier citizens did not intend the outcome of their behaviour, the damage to the plants will become visible; and hence the emergence of the path.

The plants may recover if everybody ceases to use the path. However, people generally continue taking shortcuts given the fact that they see the damage previously done by others. What are they thinking? Do they intend to bring about and maintain a visible path with no plants on it? Probably (and hopefully) not. Their reasoning is perhaps the following: 'Even if I do not take the shortcut, some others will, thus the path will stay there anyway. Then, why should not I take the shortcut?' So in this scenario, although individuals do foresee the outcome of their behaviour, they think that the creation of the path would be inevitable, given that everyone takes the shortcut. Moreover, if one particular individual is the only one taking the shortcut, the consequence of his or her action would be negligible. Thus, she takes the shortcut without intending to create such a nasty path. In this scenario, the damage done to the plants has reinforced other people to use that same path, for it already exists. Thus, unless someone intervenes, the path is there to stay.

Would this scenario explain the emergence of these walking paths? It sounds reasonable, but definitely not everyone is likely to accept this explanation. In fact, this is just one possible explanation. By introspection, some of you may think that there may be other motivations behind taking these shortcuts. You may argue that people do not think about the consequences of their actions at all when taking these shortcuts, or you may think that these paths are intentionally created. For example, you may argue that one of those paths emerged due to a poorly planned city park, which resulted in people intentionally taking a particular shortcut to show their dissatisfaction with the planning of this particular public space.

The above is yet another possible way to explain the emergence of these paths. More importantly, this explanation contradicts the previous general scenario. Thus, one is tempted to ask, 'unless some evidence is provided, why should we believe in such speculations?' If we would ask different people to explain this fact, we would get many different explanations. Many of these explanations would contain a story that makes the emergence of the path plausible. Hence, we would be left with many different (but not necessarily mutually exclusive) speculations or conjectures instead of 'proper' explanations. The fact that there are walking paths on public green fields is ostensibly simple to explain; however, it seems we are left only with conjectures.

Broadly speaking, conjectures and their explanatory characteristics are the subject matters of this book. It examines one particular type of explaining practice in social sciences, namely explaining the emergence of institutions (e.g. conventions and norms) and macro-social structures as unintended consequences of human action, from a methodological and philosophical perspective.

Explanations of 'unintended consequences' show numerous similarities with the above example of walking paths on public green fields. The basic similarity, however, is that they seem to lack empirical content and as such they can be criticised as being simple conjectures with no explanatory value. This book illustrates the merits and demerits of such explanations by examining some of these attempts to explain institutions and macro-social structures as unintended consequences.

Unintended consequences

This is a book about 'unintended consequences of human action' and the mechanisms that bring about these consequences. It investigates the explanatory role of the models that characterise institutions and macro-social structures as unintended consequences of human action.

Many economists argue that certain institutions and/or social structures, such as money, language, rules of the road, fairness, segregated city patterns and localisation, are unintended consequences of human action. Similar ideas can be found in other disciplines, such as in philosophy of language (e.g. Lewis 1969), political philosophy (e.g. Nozick 1974), linguistics (e.g. Keller 1994), philosophy of science (e.g. Hull 1988) and sociology (e.g. Merton 1936, Boudon 1982). The above list, which can definitely be extended, illustrates the importance of the problem of explaining unintended consequences in social sciences and in (political) philosophy.

Popper (1962: 342) acknowledged the significance of unintended consequences for social sciences by arguing that the explanation of unintended consequences of human action is 'the main task of theoretical social sciences'. Yet, unintended consequence is a vague concept and as such it may denote many different things. In this book we will be concerned with a subset of the set of possible unintended consequences; one which is of paramount importance to economics in explaining institutions and macro-social structures. The rough description of this subset is as follows (see Mäki 1991 and Chapters 2 and 5):

(a) individuals do not intend to bring about a social phenomenon (e.g. a social institution, or a macro-social structure);
(b) the consequence of their action is a social phenomenon (i.e. an institution or a social structure); and
(c) one individual alone is not enough to bring about the 'social' consequence – that is, independent actions of similar (in the sense that they do not intend to bring about the consequence) agents are needed.

An explanation of unintentionally hurting your hand, or an explanation of the unintended consequences of a government tax plan, does not fall under the above definition and hence are not examined in this book. The former is not examined because the consequence is not a social phenomenon (violates condition (b) – and (c) depending on the case); the latter is not examined because the intention is about a social institution (condition (a) is violated). Given our definition, 'unintended consequence' is a crucial component of the 'theory of spontaneous order', of Adam Smith's 'invisible hand', of Carl Menger's notion of 'organic phenomena' and of 'invisible-hand explanations'. It lies at the core of many contemporary models of institutions and macro-social structures.

The theory of *spontaneous order* finds its origins in eighteenth-century Scottish thought and it is defined with its characterisation of the social order as 'unintended consequence of countless individual actions' (Hamowy 1987: 3). Adam Smith, the founder of modern economics, who was a part of this tradition, presented a metaphorical statement of 'spontaneous order' with the 'invisible hand'. Menger, on the other hand, presented a closely related account of spontaneously created social institutions, where he considered them as being similar to 'organic phenomena'. Many social scientists and philosophers followed Smith and Menger by trying to answer versions of their questions about institutions and macro-social structures. Their aim was to show how institutions and social structures *could* emerge (or persist) without any design. Generally, it is believed that neoclassical economists followed Smith's lead and tried to prove his insights. However, the neoclassical economist's approach is only one of the possible ways to interpret Smith's insights. There are at least two different interpretations of the 'invisible hand': the one that stresses *processes* and the other that emphasises *end-states* (see Chapter 5). The neoclassical economist's approach is an end-state interpretation: the 'invisible hand theorem' in economics stresses the consequences (e.g. optimum allocation of resources), rather than the processes that bring about these consequences.

This book is mainly concerned with what may be called the process interpretation of the invisible hand. Under this interpretation, the 'invisible hand' represents causal and structural relationships and processes that may bring about unintended social consequences. Explanations under this particular interpretation can be gathered under the notion of 'invisible-hand explanations'. An invisible-hand explanation aims to show the *process* that brings about the unintended consequence. Rather than merely focusing on the properties of the end-state (e.g. equilibrium), it explicates the way in which the end-state may be reached (Nozick 1974; Ullmann-Margalit 1978). It is possible to find many examples of such process models in the contemporary literature. The most prominent examples are game-theoretic models of institutions that show how institutions may emerge (or persist) as an unintended consequence of human action (e.g. Bicchieri 1993; Sugden 1986; Ullmann-Margalit 1977; Young 1998, etc.). Another area where unintended consequences are important is the emerging field of agent-based computational economics (e.g. Axelrod 1997; Axtell *et al.* 2001; Epstein and Axtell 1996; Marimon, McGrattan *et al.* 1990, etc.). The pioneers of the game-theoretic literature acknowledge David Hume, Adam Smith and Carl Menger as their forefathers (see Lewis 1969; Sugden 1986; Ullmann-Margalit 1977; Young 1998) and, similarly, agent-based computational economists generally acknowledge their intellectual debts to Adam Smith and Thomas Schelling (see, for example, Epstein and Axtell 1996; Tesfatsion 2002).

Carl Menger's (1892a) story of the emergence of a medium of exchange, Thomas Schelling's (1969, 1971a, 1978) models of residential segregation, Peyton H. Young's (1993a, 1996, 1998) model of emergence of the rules of the road and Joshua Epstein's and Robert Axtell's (1996) 'Sugarscape' (where they grow artificial societies from the bottom up) are well-known examples of such models. These examples range from verbal models (or stories) to formal game-theoretic and computational models. Despite the differences in their methods and research tools there is an important similarity between them. Each shows how institutions and macro-social structures *may* emerge (or persist) as an unintended consequence of the (inter)actions of individuals. In order to do this, the authors conjecture (in the latter case with the help of computers) about the conditions under which individual actions may lead to the social phenomena in question. The common feature of these examples is that the observed social phenomenon (i.e. institution or macro-social structure) is 'produced' within the model world by conjecturing about the initial conditions (e.g. environmental conditions, characteristics of the agents, etc.) that *may* bring about the social phenomenon in question as an unintended consequence of the interactions of the agents. Moreover, these models are typically ahistorical in the sense that historical facts about the social phenomenon in question do not seem to play any role in these models. They are general and, thus, they are supposed to be applicable to all instances of the social phenomenon in question in different times and places. These models illustrate the possible ways in which certain mechanisms may interact (or may have interacted) to produce the types of institutions or macro-social structures in question.

More strikingly, some of these models seem to challenge the common sense and the historical knowledge about these social phenomena. For example, while many believe that money is a matter of design and was issued by central authorities in the past, Menger argues that it was brought about by the (inter)actions of individuals who were pursuing their self-interests without the intention to bring about a commonly acceptable medium of exchange. Schelling, on the other hand, shows that if individuals cannot tolerate living as an extreme minority in their neighbourhood, then residential segregation cannot be avoided even if they are happy in a mixed neighbourhood. Of course, this seems to go against our belief that strong discriminatory preferences (e.g. racism) and economic factors (e.g. wealth differences among ethnic groups) are the main causes of residential segregation. A more recent example is Young's model of the rules of the road. He shows that the rules of the road may emerge with the accumulation of the precedent as an unintended consequence of the (inter)actions of the individuals. This model also goes against the belief that the rule that specifies on which side of the road one should drive was designed and imposed by central authorities, like other traffic rules. Finally, Epstein and Axtell show how, under certain conditions, fundamental social structures and group behaviours (e.g. institutions, segregation, cooperation) could emerge from the micro level. In this example, social phenomena are quite literally grown by the authors.

All of the aforementioned examples pose difficult questions for social scientists and philosophers. How could these models explain anything if they are simply speculations about the initial conditions under which social phenomena may be brought about as unintended consequences? In other words, we understand that these authors are able to 'produce' a certain social phenomenon in their model world as an unintended consequence of the interactions of model agents. However, given that these models are so abstract, ahistorical and speculative, how could they be used to explain something about the real world?

The philosophical and methodological challenges posed by these models created many debates in related areas. These debates constitute a significant part of the controversies about the role of abstract modelling in social sciences. The related question here is whether we can learn anything about the real world by studying highly abstract models. This is one of the basic questions of philosophy of science. The usual defence of scientific models is the claim that they isolate the relevant parts of the real world and that such realistic representations of the real world give a close to true account of the phenomenon in question when other things are absent or constant. Some economists also use the argument from realistic representation in defence of their models. For example, Young (1998: 10) argues that the assumptions of his models 'represent a fairly accurate picture of reality'.

However, the main criticism to these models is that they ignore the relevant facts, such as the history of the social phenomenon in question – and, therefore, they do not realistically represent the relevant parts of the real world. For example, Menger's 'the origin of money' does not take into account the way in which money was issued and introduced in history. Moreover, Schelling's model of segregation seems to sidestep two of the most important facts about segregation – the presence of strong discriminatory preferences and the role of economic factors. Thus, the argument concerning realistic representation either has to demonstrate why history is irrelevant, or show the complementarity between these models and history.

As mentioned above, these models start with the problem (e.g. that there is residential segregation) and try to produce the conditions under which this problem may emerge as an unintended consequence of human (inter)action. This methodology invites criticism for the following two reasons. First, it seems to be one-sided, for it tries to construct a model that shows something that the author wishes to see (e.g. residential segregation as an unintended consequence). Second, it may be argued that if one devotes enough time and energy it should be possible to construct a reasonable model that is able to show whatever we wish.

Given the focus of this book, the relevant place to start seeking solutions to these problems is the literature on 'invisible-hand explanations'. It is argued in this literature that invisible-hand explanations are valuable independently from their truth for they *explicate* the process that may have brought about the social phenomenon at hand (Ullmann-Margalit 1978). This suggests that these models are valuable even if they are false, or even if they do not get the facts right. It is argued that explication of a hypothetical process that is sufficient to bring about the social phenomenon in question is valuable for its own sake. However, it is not explained why this explication would be valuable, or in what sense it would help us understand the real world. Simply, this argument does not help us much unless it is explicated. This is merely a statement of the author's intuition about the value of these 'explanations'. Economists use these models because they believe that they are valuable. For example, Robert Sugden (2000) argues that the reason we believe that these models are conveying a true message about the real world is that we find them 'credible' – by way of examining Schelling's segregation model. He argues that these models are credible like a good story or a novel. Sugden's basic argument boils down to the statement that we think these models are valuable because we find them plausible. This argument cannot demonstrate the value of these models unless it explains why they are plausible and in what sense plausibility of a conjectural process sheds light on the real world.

Another type of justification comes from computer simulations. It is argued that artificial environments (models, computer models) are used to gain *insights* about the social phenomenon in question (see, for example, Gilbert and Doran 1994a; Liebrand *et al.* 1998) or that models and computer simulations are like experiments where we test our ideas (see, for example, Drogoul and Ferber 1994) or that they are similar to thought experiments (e.g. Liebrand 1998; Liebrand *et al.* 1998). Briefly, it is argued that models and simulations help us in finding out the necessary conditions under which certain results (e.g. segregation) are brought about (within the computer model) and in easily exploring the properties of these model environments. In this account these models are not for explanation but for *exploration*. However, this does not answer our question about moving from the model world to the real world. Specifically, a satisfactory defence of these models would have to tell us how to translate the results of the model in order to interpret the real world.

The discussions about the interpretation of game theory are also relevant in this respect. As mentioned above, some economists, such as Young, use the argument from realistic representation to justify game-theoretic models. Yet not every game theorist would agree with this. One of the most prominent scholars of this field, Robert Aumann (1985) (also see van Damme 1998), argues that realisticness of the models does not matter that much. According to him, the conclusions are much more important: if the model is applicable to many situations and is productive, then it is a good model. He also argues that game theory (and other sciences, in his opinion) 'is not a quest for truth, but a quest for understanding'. He says, 'science makes sense of what we see, but it is not what is "really" there' (van Damme 1998: 181, 182). Aumann basically argues that game-theoretical models help us in putting together what we observe in a coherent framework, that they help us in fitting things together. He also argues that they lead to prediction and control. However, if we accept Aumann's interpretation of game theory (and science) we are still faced with the following questions: firstly, if the models of institutions and macro-social structures do not represent the reality, how could they lead to prediction and control, and most importantly to understanding? Second, Aumann emphasises the productiveness and applicability of the models. Yet the current state of the modelling of institutions and macro-social structures (as unintended consequences) cannot be considered to have many real-world applications or satisfactory predictive power. Should we then conclude that these models are not valuable?

It is the argument of this book that all of the interpretations expressed above convey justifiable intuitions about these models. That is, 'realisticness', 'explication', 'credibility', 'exploration' and 'fitting things together' are all parts of a framework that would help us in making sense of these models. However, there is no existing framework where these things are presented coherently and satisfactorily. It is the main task of this book to develop such a framework and to use it to gain new insights into the contemporary literature that characterises institutions and macro-social structures as unintended consequences of human action.

Plan and summary of the book

To be able to develop such a framework one has to understand what these models really accomplish. The most obvious way to do this is to carefully examine these models and their methodology. But before doing this, a clarification of the very idea of 'unintended consequences' is needed. Chapter 2 analyses and explicates the concept of unintended consequences to prevent misunderstandings that may be caused by its vagueness. In particular, the subset of the set of possible unintended consequences, which is relevant for understanding the models of institutions and macro-social structures as unintended consequences, is specified.

In Chapters 3 and 4, the most prominent examples of such models, Menger's 'origin of money' and Schelling's 'chequerboard model of segregation', are examined. It appears that they are natural candidates for several reasons. First of all, these models are paradigmatic examples of 'explaining unintended consequences of human action' and of invisible-hand explanations. Contemporary authors consider these models as conveying the key insights about their subject matter and about the way in which related issues should be handled. Their models are the predecessors of contemporary research in modelling institutions and macro-social structures as unintended consequences of human (inter)action. Menger is considered to be one of the founding fathers of the theoretical approach to institutions as opposed to the historical approach (see, for example, Rutherford 1994; Schotter 1981). Schelling's model is one of the main predecessors of agent-based computer models (see, for example, Epstein and Axtell 1996; Blume and Durlauf 2001; Pancs and Vriend 2003; Rosser 1999; Casti 1989) and it is considered to be the paradigmatic example of explaining with mechanisms in social sciences (see, for example, Hedström and Swedberg 1998). Briefly, since their models play an important role in the history of 'explaining unintended consequences of human action' understanding Menger's and Schelling's models should shed light on the related areas of contemporary research.

Another good reason to start our examination with these models is the fact that both Menger and Schelling are explicit about their methodology. In their work they explain why they prefer the type of research they are engaged in. Moreover, in the literature there is a considerable amount of philosophical discussion about their methodology. As previously mentioned, their works are predominantly considered to be paradigmatic examples of invisible-hand type of 'individualistic' explanations (see, for example, Nozick 1974; Pettit 1996; Rosenberg 1995; Rutherford 1994; Ullmann-Margalit 1977). It is also common to examine the recent game-theoretical models of institutions alongside invisible-hand explanations (see, for example, Langlois 1986b,c; Mäki 1993; Rutherford 1994; Vanberg 1994). It has also been stated that the authors of these models (including the authors of the computational models) consider themselves as following the invisible-hand tradition, or providing the mechanisms behind the invisible hand. For these reasons there is a considerable amount of resources that may help us in our quest. Thus, we are more likely to find hints about the nature of similar models by starting our examination from Menger and Schelling's models and their relation to the invisible hand. In addition, this choice makes it easier to see the common misunderstandings about 'explaining unintended consequences of human action', for the literature on invisible-hand type of 'individualistic' models is abound with controversies. Chapter 5 undertakes the task of examining 'invisible-hand explanations' in light of the chapters on unintended consequences, and Menger and Schelling's models. This examination sheds light on the nature of invisible-hand explanations. Particularly, an important misunderstanding about the relation between 'unintended consequences' and the 'invisible hand' is removed. By way of removing this misunderstanding, the chapter prepares the ground for examining contemporary examples of invisible-hand explanations.

Menger and Schelling's models and insights were reconsidered and remodelled by contemporary authors. For this reason, there is an explicit link between these models and the contemporary literature we wish to understand. This gives us a chance to evaluate the progress of this 'research programme'. For example, some of the papers directly related to Menger's account of the medium of exchange can be listed as follows: Duffy and Ochs (1999), Gintis (1997), Kiyotaki and Wright (1989), Marimon, McGrattan *et al.* (1990), Schotter (1981), Selgin and Klein (2000), Townsend (1980) and Young (1998). Some of the follow-ups to Schelling's segregation model are the following: Clark (1991), Epstein and Axtell (1996), Sander *et. al.* (2000a,b), Young (2001) and Zhang (2000, 2004a,b). By examining these reconsiderations we may indeed see whether there is any progress or whether contemporary tools (e.g. game theory and computer modelling) improve the way in which we understand and explain the origin of money and segregation. Accordingly, Chapter 6 examines the more recent models of the emergence of money in detail, while recent reconsiderations of 'residential segregation' are used as examples in Chapter 7.

Particularly in Chapter 6, it is argued that we should not evaluate models that characterise macro-social phenomena as unintended consequences in isolation from other related models of the same phenomenon. In order to substantiate this proposition, recent reconsiderations of Menger's explanation of the origin of money are examined. The chapter shows how Menger's intuitions are further explored in the modern literature in various ways. It is argued that these recent models test the logical soundness of Menger's arguments but do not bring us any closer to the real world. Recent models of the origin of money do not introduce new mechanisms but test the plausibility of the mechanisms that were suggested by Menger. While these models increase the plausibility of the idea that media of exchange may be brought about unintentionally, it is argued here that the idea that fiat money may be considered as an unintended consequence of human action does not appear to have a firm basis. Moreover, the chapter examines and demonstrates the relation among these models. This examination supports the thesis that different models have different functions and different models of the same phenomenon may be considered as forming a loose framework for explaining particular instantiations of it.

Chapter 7 explores the philosophical literature on models and explanation to provide a firmer basis for the arguments of the previous chapters. In particular, first, the concept of partial potential explanations is explicated. Second, it is argued that models help us explain by way of providing a proper way to conceptualise the phenomenon under question. Yet this further implies that the relationship between the model world and the real world is rather complex. Third, it examines this complex relationship by way of discussing the related philosophical literature in light of the previous chapters. Fourth, it is argued that similarity between models that are examined in this book and the real world amounts to the existence of certain (known) tendencies (individual mechanisms) in the model world. For this reason, these models may be interpreted as revealing the possible ways in which these tendencies may interact, even if some of the assumptions of these models do not hold. Fifth, the chapter emphasises the importance of exploration. Particularly, it shows how one may gain confidence about the implications of an existing model by way of further exploring its premises and results. To do this, the chapter discusses reconsiderations of the chequerboard model. Finally, the chapter fortifies the idea that no model of this sort should be evaluated in isolation from other related models.

Chapter 8 examines the modern game-theoretical models of conventions in light of the ideas developed in previous chapters. These models may be considered as attempts to provide a general theory of the emergence of conventions. The chapter reviews some of the existing game-theoretic literature on conventions and shows that existent conventions and norms, particular institutions and history are crucial for explaining the emergence of conventions. Six arguments are put forth in the chapter:

1   Static models of coordination (and convention) are concerned with examining the conditions under which certain outcomes are plausible, rather than explaining why and how such outcomes are brought about. Hence, such models are in line with the end-state interpretation of the invisible hand.
2   Dynamic models of coordination provide partial potential (theoretical) explanations of the emergence of coordination and conventions, hence such models are in line with the process interpretation of the invisible hand.
3   None of these models rule out the possibility that coordination and conventions may be brought about intentionally. Rather, they examine whether successful coordination and conventions may emerge as unintended consequences of human action. The interpretation of these models as providing partial potential explanations is well in line with this remark;
4   Explaining particular cases (e.g. explaining the emergence of a particular convention) necessitates empirical research. Nevertheless, general models of coordination and conventions need not be empirical or historical;
5   The collection of different models of coordination and conventions may be considered as providing a general framework for empirical research and providing singular explanations.
6   Game-theoretic models in general may be interpreted as providing a framework for analysis, rather than providing ultimate explanations concerning social phenomena and individual behaviour.

Chapter 9 concludes the book with questions for further research.

2   Unintended consequences

Introduction

The concepts of 'invisible hand' and 'unintended consequences' are closely related to each other. It is the task of this chapter to identify the type of unintended consequences implied by the 'invisible hand'. The paradigmatic examples of invisible-hand explanations, as well as Adam Smith's 'invisible hands', are concerned with a small subset of the set of all possible types of unintended consequences. A good understanding of this subset is a prerequisite for a good understanding of the invisible hand and of the wide variety of models and explanations that employ the concept. Yet, common interpretations of the invisible hand do not clarify the relation between 'unintended consequences' and the invisible hand. By explicating the exact relationship between these two concepts this chapter prepares the ground for the rest of this book.

Previously, Mäki (1991: 162) characterised the type of unintended consequences that are implied by the invisible hand as *invisible-hand consequences* and argued that they have the following characteristics:

1 a single individual's action is not sufficient to bring about invisible-hand consequences;
2 unintended consequences of collective action do not count as invisible-hand consequences;
3 invisible-hand consequences relate to macro-social phenomena; and
4 invisible-hand consequences are generally beneficial.

Although Mäki's characterisation is appropriate, it does not present the full range of invisible-hand consequences. Moreover, it is not clear why invisible-hand consequences should be characterised in this manner. The present chapter develops and explicates this account and gives a more detailed picture of the type of unintended consequences that count as invisible-hand consequences. Particularly, the chapter presents a classification of unintended consequences of human action which clarifies the exact relation between intentions of the individuals and consequences of individual action. It is shown that an invisible-hand consequence is a specific type of unintended consequence that rests on a special relation between individuals' intentions and consequences brought about by individuals' actions.

The plan of the chapter is as follows. The second section introduces and classifies unintended consequences with respect to the targets of individual intentions relative to the level at which the consequences are materialised. The third section identifies the type of unintended consequences that are implied by the invisible hand. The fourth section concludes the chapter.

Types of unintended consequences

Apparently 'unintended consequences of human action' is not an ambiguous notion. Simply, unintended consequences were not intended to be there. Yet, as one starts using this notion, it becomes confusing. In fact, the set of possible unintended consequences of human action is very large and when social scientists and philosophers talk about unintended consequences they generally indicate a subset of it. This subset, however, is commonly left unexplained. Similarly, the invisible hand implies a specific subset of the set of possible unintended consequences. In the following pages we will try to identify this subset.

We may identify three elements of confusion in 'unintended consequences'. The first element is confusion is the concept of 'intention'. When we talk about an individual's intention, we imply that the individual has a *purpose* or a *plan*. Yet, these two meanings should be distinguished. Although intention necessarily indicates that one has a purpose, it is not necessary that one has a plan. Simply, individuals may fail to plan to do something, although they have a purpose (Keller 1994: 11). To prevent confusion, throughout this book, when we talk about someone having an intention, we mean someone having a purpose, rather than having a plan.

The second element of confusion is the concept of 'consequence'. In order to be clear about what is implied by unintended consequences of intentional individual action, we need to distinguish between 'consequences' and 'results'. As Keller (1994: 64) argues, 'the result of an action "A" is an event which has to happen for the action to be considered as having been executed at all'. So, if I intend to have some fresh air by opening the window, my *action* will be 'opening the window', and this is the *result* of my intention. However, there is something more about my opening the window – that is, 'having some fresh air' – and this may be considered as a 'consequence' of my action. Moreover, since I intend this consequence, it is an intended consequence of my action. Thus, the effects of my action (the result of my intention) are the consequences of my action. However, not all the possible consequences of my action are intended. For example, when I attempt to install new hardware on the motherboard of my computer, my intention is to upgrade my computer's components. However, I might obliterate my computer by (mistakenly) discharging the static electricity from my hands to the components of my computer. Although this damage is a consequence of my action, it is not the one that I intended. Thus, we may consider this as an unintended consequence of my action. In this book, the *result* of one's intention should be considered as his or her *action*, and the *consequences* of an action should be regarded as the things *caused by that action*, either intended or unintended.

The third element of confusion is the concept of 'unintended consequence' itself. It implies a vast variety of relations between intentions and consequences. One can think of many examples to see how broad the notion of unintended consequences is. For example, one day I was trying (intending) to lock my bike: however, I happened to pick up the wrong key and as a consequence both the key and the lock were broken. This was, I can assure you, an unintended consequence of my action. But it is also true that if I were to plan the railway timetables as strictly as possible to make everything work perfectly, and if there were a deviation from this tight schedule by a train, the accident on the railway would possibly be the unintended consequence of my obsessive planning which left no place for deviations. Now, on a broader scale there may be the unintended consequences of a government tax plan that may develop due to the 'unexpected' responses of the citizens to this plan. These simple examples suggest that unintended consequences can be observed in various forms, and behind different kinds of unintended consequences we may find different combinations of causal factors.

For an identification of the different types of unintended consequences, one may wish to start from these different causal factors or one may analyse the relationship between intentions and consequences. Since a causal classification presupposes the knowledge of all causal mechanisms that may bring about unintended consequences, it is impractical. However, it is still useful for a better understanding of the relation between intentions and unintended consequences. Thus, before we start our analysis of this relation, it is beneficial to discuss Merton's attempt to classify unanticipated consequences according to their causes.

Merton on unintended consequences

Robert K. Merton, in his classical piece, *The Unanticipated Consequences of Purposive Social Action*, clearly states what is included in the 'consequences' of an action:

> Rigorously speaking, the consequences of purposive action are limited to those elements in the resulting situation which are exclusively the outcome of the action, i.e., those elements which would not have occurred had the action not taken place.
>
> (Merton 1936: 895)

Thus, unless the action of an individual (or individuals) is, at least partially, causally responsible for the 'consequence', it is not the consequence of that action. If there is an unintended consequence of an action, it is plausible to think of other unseen or neglected (disturbing) causal factors which prevented the action from bringing about the intended end. However, for us to consider this as an unintended consequence of an action it is also necessary that if the action had not taken place, the unintended consequence would not have occurred. That is, the action in question is a necessary condition for the unintended consequence, *ceteris paribus*.

When unseen or neglected factors interfere with one's action, unintended consequences may be brought about. It is because of these unseen or neglected causal factors that Merton uses 'unanticipated consequences' interchangeably with 'unintended consequences'.[1] It seems natural to think that all of the unanticipated consequences are unintended. However, one can think about cases where the consequence was intended but unanticipated. For example, people buy lottery tickets with the intention to win the lottery, but most of them do not anticipate that they will win the lottery, for if they did we could hardly be able to explain the surprise of the winners. Moreover, anticipated consequences may be unintended. I may anticipate that if things go wrong my action may bring about certain consequences; however, it is not my intention to bring about those consequences as a part of my action. Now, for the sake of the argument, let us assume, like Merton, that in most cases an unanticipated consequence is unintended, or vice versa. Later, this distinction will prove to be useful for understanding invisible-hand arguments (especially in Chapter 5).

Merton classifies unanticipated (unintended) consequences according to the factors that *help* their coming about. Merton (1936: 898–901) argues that 'the most obvious limitation to a correct anticipation of consequences of action is provided by the existing state of knowledge'. He then lists some of the different kinds of factors, such as ignorance, error and the 'imperious immediacy of interests', that may cause unintended consequences.

We might think of different cases to conceive of Merton's distinctions. For example, let us assume that agent A intends to achieve X and he believes that by doing Y he can achieve X. Thus, A thinks that Y causes X and acts accordingly. Of course, A can be wrong in supposing that Y causes X. It might be the case that Y causes V. If this is the case, when A does Y to achieve X, V will happen. V is an unintended consequence of A's action and this consequence can be accounted with A's *lack of knowledge* of the causal determinants of X.

Another case might be that A is right about the causal determinants of X but he or she is ignorant of the other factors which may change the course of events that will follow his or her action. Therefore, we can say that A is ignorant of the fact that Y causes X *ceteris paribus*. If this is the case, when A does Y to achieve X, because of other interfering factors some other event, say Z, will happen. Z is an unintended consequence of A's action and it can be accounted with A's *ignorance* about the other possible interfering factors.

Of course, A can be right in assuming that Y causes X but he might be unable to execute Y successfully. So, A might fail to do Y or might do something different than Y (as in the example where I used the wrong key to open the lock), say T, which in turn causes V. If this happens to be the case, we can consider V as an unintended consequence of A's action and account it with A's *error*.

If A is stimulated by his or her 'imperious immediacy of interests', he or she might want to achieve X by doing Y without thinking about the other further consequences of his or her action. If it is the case that Y causes X, V and Z, when A does Y to achieve X, he or she will bring about X, V and Z all together. Because V and Z were no part of A's intentions, they are the unintended consequences of A's action and can be accounted with A's *imperious immediacy of interests*, or his or her short-sightedness.

These four cases lie beneath Merton's account of unanticipated consequences of purposive human action.[2] Merton focuses on the reasons why agents might be unable to anticipate the consequences. Accordingly, it suffices for him to discuss the agents' lack of knowledge, ignorance or 'imperious immediacy of interests'. Yet, even if we consider these as causal factors that may bring about unintended consequences, we have to realise the fact that they are only partially responsible for the generation of unintended consequences. In each and every case other factors help unintended consequences in coming about. For every case we may think of many possible (and different) causal processes that may lead to unintended consequences – in combination with agent's ignorance or lack of knowledge, etc. Thus, we may not explain exactly why the unintended (or unanticipated) event had taken place by merely referring to the factors specified by Merton. Such an explanation should be able to give an account of the course of events leading to the unintended consequence – in addition to the specification of the agent's inability to anticipate them. The type of classification presented by Merton only alerts us to those different relations between the agents and other causal factors that are responsible for the unintended consequence, but it does not tell us anything about the 'other causal factors'. In other words, this is a partial 'causal classification' and a full causal classification would require that all types of causal mechanisms that agents could not foresee, or do not know about, are spelled out. Such a classification would also require a specification of different types of actions (e.g. individual, collective, etc.), different types of consequences (e.g. physical, individual, social, etc.) and different ways in which these consequences affect the individuals.

In this line of inquiry, Merton attempts to classify unanticipated consequences according to the sum-total consequences of action:

> These sum-total or concrete consequences may be differentiated into (a) consequences to the actor(s), (b) consequences to other persons mediated through (1) the social structure, (2) the culture and (3) the civilization.
>
> (Merton 1936: 895)

The actions, he tells us, may be differentiated 'into two types: (a) unorganized and (b) formally organized' (Merton 1936: 896).[3] This distinction implies that there are differences between consequences of individual action and what we might call consequences of social action.[4] Similarly, Raymond Boudon points to different configurations of the consequences of social action:

> The number of possible recombinations of the following criteria does, there-fore, define the number of possible configurations: 1. No participant (1a),

some participants (1b), all participants (1c) attain their individual objectives; 2. Producing, at the same time, benefits (2a), or problems (2b), or else collective benefits and problems (2c); 3. Each of these applying only to some (3a), or to all the participants (3b).

(Boudon 1982: 6)

These configurations, together with Merton's classifications, suggest that there is a vast variety of unintended consequences and that there might be different causal processes behind the unintended consequences of organised action and unorganised action, or individual and social action. Although it may not be possible to determine the different types of causal mechanisms behind different unintended consequences, it is still important to understand the possible range of unintended consequences. One possible way to classify unintended consequences is from the perspective of their places of *materialisation* relative to the target of agent's intentions. The next section discusses this classification by introducing a table of possibilities, which combines Merton and Boudon's tentative classifications to give a broader picture.

The table of possibilities

It is not easy to categorise different types of unintended consequences and any kind of categorisation will have problems in locating some of them, for unintended consequences might come about in a variety of contexts and some of these contexts may be very complex. This section aims to develop a framework that makes the task of identifying different 'unintended consequences' easier and shows the subset to which the examples examined in this book belong. Although this account might have some limitations, it gives a better understanding of the explanations of unintended consequences of human action.

To begin with, we have to consider the possible methodological problems we may face in developing an account of a possible set of unintended consequences. As Merton (1936: 897) suggests when we start talking about unintended consequences, we are confronted with two possible methodological pitfalls. The first one is the 'problem of ascertaining the extent to which "consequences" may justifiably be attributed to certain actions'. Simply to argue that unintended consequence X was caused by A's intention to bring about Y, we need to know whether A's action caused X or not. The simplest way to get a grip of this problem is to ask whether X would have occurred in the absence of A's action. If the absence of A's action prevents the consequence X, we may justifiably argue that X was an unintended consequence of A's action, for A's action was partially or fully responsible for X. Any satisfactory explanation of X as an unintended consequence has to provide a justifiable connection between A's action and X. The second problem is 'that of ascertaining the actual purposes of a given action'. That is, if we want to show that X is an unintended consequence of A's action, we need to know the actual intention of A in doing Y, or we need to show that whatever A's intention might be, it is not that of bringing about X. It is always problematic to talk about another person's intentions, for we have no way of reading that person's mind. Again, this is an important problem for any explanation of 'unintended consequences'. However, these do not cause problems for our attempt to categorise unintended consequences, for in what follows we will be talking about *possible* types of unintended consequences.

We may start our analysis of unintended consequences by distinguishing between different levels, such as the social and the individual level. Then, we may consider the relation between the 'target' of the intention and the level on which the consequence is materialised. That is, we may examine *spaces of materialisation* relative to the *target of intentions*. Let me introduce some examples to clarify what is meant by spaces of materialisation. If I break a vase in an attempt to kill a mosquito with a newspaper, the unintended consequence is materialised on a non-living object. If I were to miss the mosquito and strike my wife with the newspaper, then the unintended consequence is materialised on a human being. On the other hand, unintended consequences of the government tax plan would have larger and broader effects. Suffice it to say, unintended consequences would be materialised at the social level in contrast to the unintended consequences of my mosquito hunt. We may add to this that the nature of my intention in the mosquito hunt is different than government's intentions. We may characterise this difference by saying that the target of the first intention is at the individual level, and the target of the second is at the social level.

Let us start by thinking about the possible things one might be intending to do.[5] Table 2.1 lists some possible intention–action pairs.

Out of the seven actions listed in Table 2.1 (a_1–a_7), (a_1) and (a_2) change the state of a physical object and (a_3–a_7) are about my relations with others. However, while I am concerned with myself when I am doing (a_3), (a_4) and (a_5), I am trying to change something *social* when I am doing (a_6) and (a_7). Given the intentions, it is possible to divide these actions into two groups: ones that are about the state of the physical objects, animals, flowers, myself, any other person, etc., and others that are about the society partly or as a whole. For the first one, we can say that the intention is about the *individual level* (I), and for the second that the intention is about the *social level* (S).

Table 2.1 Possible intention–action pairs

Action	Intention
(a_1) Poison an insect	(i_1) To get rid of it
(a_2) Kick a piece of rock	(i_2) To get it out of my way
(a_3) Buy flowers	(i_3) To please my mother
(a_4) Call a friend	(i_4) To arrange a meeting
(a_5) Write a letter	(i_5) To complain about a product
(a_6) Give a speech	(i_6) To convince to take social action
(a_7) Design new rules for the social security system	(i_7) To improve the conditions of my country

We may also consider mixed intentions as a possibility. Suppose that you are able to change the rules of the social security system and you know that there is no effective monitoring or sanctioning mechanism for the abuse of this power (as might be observed in some of the underdeveloped countries). With this in mind, you may take your chance to change the rules of the social security system in order to improve your own social security, even though you also know that the new rules will have a bad influence on the whole social security system. If you do change these rules, we are confronted with a complex situation that is easy to classify: your intention is definitely about the individual level, but you are intending to change something social as well. Here we have an example of an intention that is both about the individual level and social level $(I + S)$ at the same time. It is not necessary that both intentions have equal weight. In this example, it is clear that the intention is more about the individual level than about the social level. $(I + S)$ captures the possible set of mixed combinations.

We have seen that intentions might be about different levels. Accordingly, consequences might be *materialised* at different levels. Like intentions, consequences may be at the individual or social level. Say that I want to clean my walking path (i_2) and kick a piece of rock which is in my way (a_2). If I am able to 'move it away', this is an intended consequence at the individual level $(I \oplus)$. However, what appeared to be a small rock might be unmovable because it is actually a piece of a larger rock lying under the ground. In this case, my kicking might cause an injury to my foot. This is an unintended consequence at the individual level $(I \otimes)$. Moreover, I might as well move the rock away and injure my foot at the same time. In this case, there are two consequences: the former intended and the latter unintended $(I \oplus, \otimes)$. Now suppose that I am giving a speech (a_6) to a crowd to convince them to vote against a new tax policy (a_6). In this case, if I am able to convince them to vote against the new tax policy, this would be an intended consequence at the social level $(S \oplus)$. Of course, there could also be unintended consequences at the social level $(S \otimes)$.

We could also talk about these consequences with respect to their desirability. We may assume that the intended consequences are desirable, whatever we think about them. So, while moving the rock out of my path would be a desirable (intended) consequence at the individual level $(I+)$, hurting my foot would not be desirable $(I-)$. If both of these things happen, I would have desirable and undesirable consequences at the same time. It is possible to talk about the social consequences in the same manner. Of course, talking about the desirability of the social outcome is not easy, but we are just trying to think about the possibilities.

Now we can gather the aforementioned possibilities to form a table of possibilities (Table 2.2). For simplicity, this table takes into account only one individual's intentions and actions. Later, we will go on to discuss other possibilities for multiple individuals and for the cases where many individuals intend to achieve the same consequence.

Now let's try to locate some of our examples in Table 2.2. Consider (i_2) and (a_2) above: the case where I kick the rock on my way to clear my walking path. Here, the intention is changing something at the individual level (row I). Thus, we are

Table 2.2 Table of possibilities

Consequence

		I ⊕	I ⊗			S ⊕	S ⊗		
		+	+	−	+, −	+	+	−	+, −
Intention	I	1.1	1.2	1.3	1.4	n/a	1.6	1.7	1.8
	S	n/a	2.2	2.3	2.4	2.5	2.6	2.7	2.8
	I+S	3.1	3.2	3.3	3.4	3.5	3.6	3.7	3.8

Notes

I: Intention is about the individual level.
S: Intention is about the social level.
I+S: Intention is about the individual and social level at the same time.
⊕: Actual consequence is the intended outcome.
⊗: Actual consequence is not the intended outcome.
+: Actual consequence is desirable.
−: Actual consequence is not desirable.
+, −: Actual consequences include both the desirable and undesirable outcomes.
n/a: If there is no intention, then the cell showing the consequence's relation with the intention is
    filled with 'n/a'. (If the individual is intending to bring about consequences at the individual level,
    it does not make sense to talk about his intention to bring about social results.)

concerned only with the first row of the table. Now, if the consequence is that 'the rock moved away', it would be an intended consequence at the individual level (column I ⊕, +), which corresponds to cell 1.1 in Table 2.2. However, if I cannot move the rock away and hurt my foot in the process, this would be an undesirable unintended consequence, that is, cell 1.3 in Table 2.2. If I move the rock away and injure my foot at the same time then we would have one intended and one unintended consequence at the same time: a combination of cells 1.1 and 1.3 represents this situation. Similar examples may be given for other rows (i.e. for intentions about the social level and for mixed intentions) but this is not necessary for, by definition, the cells corresponding to the columns with unintended signs (⊗) indicate unintended consequences. Thus, out of twenty-four cells, eighteen denote unintended consequences.[6] Broadly, there are three types of intentions that may bring about two types of unintended consequences. Any explanation or examination of unintended consequences should at least specify the type of intention and consequence in these terms.

Remember that Table 2.2 represents the possible set of unintended consequences for one individual. In the case that the actions of many individuals are involved in the generation of the unintended consequence, the set of possible unintended consequences expands. But, of course, we can represent other situations with this table. For example, if we are interested in unintended consequences of collective behaviour, we can interpret the table as representing the collective intentions of the people involved. These possibilities increase the number of possible types of unintended consequences. However, we need not reproduce the table including these cases. An understanding of the broadness of the notion is sufficient for our purposes. Neither the examples examined in this book nor the other examples in the relevant literature are about cases where collective intentionality exists or with cases where one individual's actions bring about unintended social consequences (see Chapter 1). Roughly, in our examples, unintended consequences emerge out of the actions of multiple individuals whose intentions are targeted to the individual level. Since the above table represents only one individual's intentions, we may interpret the rest of the individuals who are acting at the same time as a part of the environment within which the individual is acting. As in our example about the walking paths (see Chapter 1), in some cases other individuals' actions are necessary for the consequence to come about. We will see more of this in the following section.

We may now tentatively tell the subject matter of this book. The examples in this book are explanations of the emergence of institutions or macro-social structures as the unintended consequences of intentional human action. In neither of these examples individuals intend to bring about a social consequence, or act collectively to bring about a social consequence.[7] Generally, we are concerned with models that 'explain' the consequence at the social level as an unintended product of the intentions that are directed to the individual level. Thus, we are concerned only with the first row and with the right-hand side of Table 2.2. In Table 2.2, cells 1.6, 1.7 and 1.8 show the focus of these explanations relative to other possible unintended consequences. Thus, we are not concerned with unintended consequences of government intervention or with cases similar to the example that I unintendedly hurt my foot.[8] The theory of spontaneous order, invisible-hand explanations and Menger's so-called organic phenomena, are all concerned with cells 1.6, 1.7 and 1.8 in Table 2.2. However, this does not fully explicate what this book is really about, for Table 2.2 is concerned with one person's intentions. The next section further specifies the characteristics of the type of unintended consequences we are concerned with.

Invisible-hand consequences

We have seen that the type of unintended consequences we are interested in form a small subset of the set of possible types of unintended consequences. We are interested in social unintended consequences that were brought about by individuals who were intending to bring about consequences at the individual level. But why? Does the invisible hand really imply this specific set of unintended consequences? This section answers this question with a sneak preview of the following chapters.

As a first step to locate invisible-hand type of unintended consequences (invisible-hand consequences) in our table of possibilities let us consider Adam Smith's most quoted sentences concerning the invisible hand:

> By preferring the support of domestic to that of foreign industry, he intends only his own security; and by directing that industry in such a manner as its produce may be of the greatest value, *he intends only his own gain*, and he is in this, as in many other cases, *led by an invisible hand* to promote an end which was no part of his intention. Nor is it always the worse for the society

that it was no part of it. By pursuing his own interest he frequently promotes that of the society more effectually than when he really intends to promote it.

(Smith 1789: IV.2.9, emphasis added)

Here, individuals pursue their own interests and hence they do not intend to bring about consequences at the social level. That is, individual intentions are directed to the individual level. However, the (unintended) consequences are at the so-cial level: by pursuing their own interests individuals promote the interest of the society. Thus, the consequences generated by the invisible hand correspond to cell 1.6 in Table 2.2 (see Chapter 5). Similarly, Carl Menger's (1892a) explana-tion of the origin of money, which is a paradigmatic example of invisible-hand explanations, portrays individuals as having intentions targeted to the individual level which bring about consequences at the social level: cell 1.6 in Table 2.2 (see Chapter 3).

Generally speaking, the invisible hand implies beneficial unintended social consequences of the actions of individuals who are pursuing interests concerning the individual level. Some of the paradigmatic examples (e.g. Menger 1892a) confirm this idea. Yet the invisible hand is also associated with what we may call disadvantageous social consequences of human action. This type of invisible hand may be called *invisible backhand* (Brennan and Pettit 1993). While the invisible hand produces beneficial consequences, invisible backhand produces undesirable or disadvantageous consequences. Well-known examples of invisible backhands are Prisoner's dilemma type of situations: individuals pursuing self-interests bring about a consequence which is not desired by any of these individuals. Another paradigmatic example of invisible hand explanations, Thomas Schelling's (1978) explanation of racial residential segregation, may also be considered as an ex-ample of the invisible backhand. In Schelling's model, mildly discriminatory individuals who are willing to live in mixed neighbourhoods end up living in segregated neighbourhoods because of their intolerance to living as an extreme minority in their neighbourhood. Again, in this example, intentions are directed to the individual level. Individuals' actions are not based on intentions concerning the formation of neighbourhoods or the state of the city in terms of distribution of different races in distinct neighbourhoods. Yet, an unintended consequence emerges out of their actions: ethnically segregated neighbourhoods (see Chapter 4). The pure invisible backhand consequences correspond to cell 1.7 in Table 2.2.

In fact, it is not exactly true that segregation is undesirable in Schelling's model. In his explanation few individuals who are unhappy (i.e. because they live as an extreme minority) move to other neighbourhoods where they can be content. By moving they trigger a process that slowly changes the states of other individuals. Hence, while individuals who were initially living as an extreme mi-nority may consider an ethnically segregated neighbourhood as a desirable result, other individuals may consider it as being disadvantageous.[9] This alerts us to the fact that different individuals may evaluate the desirability of invisible-hand type of consequences differently. For this reason, from a purely methodological point of view, we may consider both desirable and undesirable social consequences of invisible-hand type of mechanisms altogether under the heading of invisible-hand consequences. That is, in principle the invisible hand and the invisible backhand are similar in all respects apart from the desirability of the consequences they produce. Methodologically speaking, they imply similar explanatory mechanisms and for this reason they may be discussed together. Consequently, cells 1.6, 1.7 and 1.8 in Table 2.2 imply invisible-hand consequences.

It is important to note here that even under this broader conception, invisible-hand consequences form a small subset of the set of possible types of unintended consequences. Failure to see this relation may drastically change our opinion concerning the invisible hand and invisible-hand explanations.

Some peculiarities of invisible-hand consequences

Although it is not our task to define the notion of 'social phenomena', it is obviously necessary to have a rough understanding of what is meant by 'social phenomena' to understand unintended social consequences. Indeed, a common sense understanding of it would do. Institutions, conventions, norms, coded rules of the society, etc. are all social phenomena. In Table 2.2 S-consequence represents such phenomena. But it also represents what we may call macro-social structures, that is, the properties of a collection of individuals or the 'things' that are produced by many individuals, such as the aggregate statistics of the society. At first glance, social institutions and social statistics may seem like two distinct categories that should not be handled together. However, the two are not so distinct. In order to see this let us examine the distinction between the two levels that were used above, specifically, the distinction between the individual level and social level.

At the individual or microlevel we observe individual agents acting independently from, or interdependently with, other individuals. The relevant variables here are the individual characteristics of the agents, their actions, their strategies, etc. in isolation, and not the characteristics of a collection of individuals, or what may come about from their interactions. We consider the latter as belonging to the social level. At the social level we have characteristics of a collection of individuals, the collective or aggregate consequences of their actions, etc. For example, while we consider the shopping decisions of an individual, his actions at the marketplace, etc. as belonging to the individual level, we consider the aggregate level of prices, which is a consequence of many individuals' shopping behaviour, as belonging to the social level. Or, the housing decision of one individual is at the individual level, but the residential distribution of different types of individuals (e.g. according to age, sex, income) in a city is at the social level. Quite simply, properties of social phenomena (social level) cannot be attributed to single individuals (individual level).

Take the classic example of money. Apparently valueless coins and papers are considered as a medium of exchange or as a store of value by the collective belief of individuals, thus 'money' is a social phenomenon (i.e. we consider it to belong to the social level). On the other hand, one single individual who believes that certain coins and papers can be used as a medium of exchange belongs to the individual level. His belief alone cannot make those coins and papers 'money'. If, and only if, there are other individuals who share the same belief, we can talk about money.

Now, consider ethnically segregated city patterns. When we look at a city and how different ethnic groups are distributed in the city we may see whether there is segregation or not. Segregation is a property of a city, a collection of individuals living close to each other and for this reason we consider it as belonging to the social level. We cannot find out whether there is segregation or not by examining single individuals and their properties, we have to look at them from above, we have to consider a collection of individuals to observe the phenomenon of segregation. Thus, one individual's housing decision, and her properties, belong to the individual level. Segregation, on the other hand, is a social phenomenon, by our definition.

Similarly, inflation, level of unemployment in a country, rates of interests, characteristics of a group, behavioural regularities in a society, etc. fall under our definition of 'social phenomena'. This is, of course, not a fully developed definition of the social, yet it is consistent with many accounts of 'social phenomena'. For example, Finn Collin defines 'social' as follows:[10]

> 'Social' here simply means collective: a phenomenon counts as social if it involves a plurality of human agents whose actions or plans are *somehow* mutually related.
>
> (Collin 1997: 5, emphasis added)

Given our definition of the social level, what are the properties of an invisible-hand consequence? First of all, it necessitates a multiplicity of individual agents. One individual may change or bring about social phenomena: for example, a dictator may force the segregated individuals in the city to move and thereby bring about an integrated city. Government intervention into the economy may be considered in a similar way. Yet, although one or a couple of individuals may be enough to change or bring about social phenomena, social consequences are always mediated through a multiplicity of individuals. In the first case, many individuals are moved by force to other houses. In the case of a government policy to decrease inflation the expected consequence (low inflation) can only be achieved if the individual agents in the economy give the 'right' responses to the policy. Simply put, social consequences are mediated through a multiplicity of individuals. It does not matter whether the social consequence was intended or not (or whether it was designed or not); it involves a multiplicity of individuals.

To sum up, an unintended social consequence has the following important characteristics:

1 The consequence is located at the social level (or, it is a social consequence).

2   Consequence was not intended by any individual.
3   It is mediated through a multiplicity of individuals.

Yet, our focus is on unintended social consequences that were brought about by individuals who were not intending to change or bring about social consequences – that is, we are concerned with the first row in our table. Thus, we may add the following condition:

4   Individual intentions are directed to the individual level.

This condition excludes unintended social consequences that were brought about by actions of individuals (or of an individual) who were intending to bring about social consequences. Thus, condition 4 is the first step to differentiate invisible-hand consequences from unintended social consequences.

Some unintended consequences that satisfy conditions 1, 2, 3 and 4 may not be considered as invisible-hand consequences. We may think about the following possibility: could the action(s) of one individual who does not intend to bring about social consequences bring about social consequences? Since human beings live in a society this seems to be a possibility. I may fail to recognise that my actions may bring about social consequences. That is, although I may think that my actions would have no consequences at the social level, this may not be true. An example of this appeared in Radio Netherlands. I will adapt the story as follows: a filmmaker goes to a former colony of a country with the intention to shoot a documentary about the culture of the residents. He learns that before the colonisation, these people were cannibals but they were forced to stop this practice. Willing to include scenes of cannibalism, he asks the locals to reconstruct their headhunting rituals for the cameras. However, the locals, once forbidden to perform such rituals, interpret this request as a permission to return back to their old practice of cannibalism and radically change their behaviour after this incident. In this example, a single individual brings about a social consequence, although he was not intending to do so. Although this may seem to be an extreme case, it alerts us to the possibility that a single individual's intentions directed to the individual level might bring about unintended social consequences. Proposed examples of the invisible-hand mechanism (e.g. paradigmatic examples of invisible hand explanations) do not indicate such cases. To exclude these, we may add the following condition:

5   The action of one individual is not sufficient to produce the unintended (social) consequence.

Conditions above do not exclude cases where many individuals collectively act in a similar manner. Since the invisible hand does not imply collective behaviour, we may add the following condition:

6 Individuals do not pursue the same end collectively (that is, collective intentionality is excluded).

Note that this does not out rule the cases where individuals pursue the same end independently – that is, without a collective decision to do so. We ruled out unintended social consequences of a single individual's actions and unintended social consequences of collective action. Given the above conditions, the only way that an unintended social consequence may be generated is through multiplicity of individuals with intentions directed to the individual level. Invisible-hand consequences are defined as the set of unintended consequences implied by the conditions 1 to 6. This book focuses on the models and explanations concerning the set of unintended consequences as described by these conditions.

Lastly, we have to distinguish between 'unintended' and 'unanticipated' and release our assumption that unintended consequences were equivalent to unanticipated consequences. It seems reasonable to think that if a consequence is unanticipated it should be unintended, and vice versa. But this is not the case. First of all, an unanticipated consequence might be intended. For example, when I buy a lottery ticket I intend to win (or intend to increase my chances of winning) the lottery. However, I do not anticipate that I will win. If I win, this would be an unanticipated intended consequence. Second, an anticipated consequence may be unintended. For example, when I take a shortcut through a public green field, I may anticipate that if others do the same, the plants may be irrecoverably damaged. Yet, I do not intend to bring about this consequence when I take the shortcut – I may be ignorant about other people's behaviour and about the final consequence. Or, when someone drives home, despite the fact that he has consumed three glasses of whisky, he may anticipate that if things go wrong he may end up at the police station. However, this is not his intention to do so. He is simply intending to go home by means of taking the risk of being stopped by the traffic police. Thus, in some cases we may have unintended but anticipated consequences. Invisible-hand consequences may be anticipated or unanticipated.

Concluding remarks

In this chapter we have explored the notion of 'unintended consequences' and specified the subset of possible unintended consequences we are concerned with. We are concerned with invisible-hand consequences that are defined by conditions 1 to 6. These conditions hint at the contrast between designed outcomes and unintended outcomes. It is commonly argued that people tend to associate orderly things with design. Showing that what seems to be designed is indeed unintended is an important achievement because it frees our mind from our presuppositions. Yet, it is sometimes hard to distinguish the designed from the unplanned. Consider the example about the unpleasant paths on green fields (see Chapter 1). There was such a path at Erasmus University Rotterdam Campus. It is on the way to the sports hall. Students used to take a shortcut to the sport-hall through the green field and hence the path emerged. However, after some time university management covered this unpleasant path with stones, and it now looks like the other constructed paths, which were in the original plans of the Erasmus University Campus. Newcomers might think that this shortcut path had been planned, while the truth is that it was brought about by the dispersed actions of the students who were just intending to get to their sports hall as quickly as possible. It was not planned. But after the 'spontaneous' emergence of the path, the university management – probably – thought that it was a good idea to have a constructed path there, either because of the impossibility of recovering the path as a green area, or because they saw the need for such a path. In any case, it looks designed, although it is not, and the correct explanation of its emergence should explicate this fact. This example alerts us to two things. First, human beings tend to associate order with design. Second, sometimes order might have some historical evidence that it was designed but it is still not necessary that it emerged as the consequence of some design. Something that looks like a design, like the rules of the road, might indeed be an unintended consequence of human action. In the next chapter, we start examining the plausibility of explaining what looks like to be the product of design as an unintended consequence of human action.

3 The origin of money

Introduction

The aim of this chapter is to examine Menger's explanation of the origin of money and his methodology. According to Menger, money is an unintended consequence of several individuals' actions pursuing their own economic interests. In his story money evolves 'spontaneously' from an environment where individuals directly exchange their goods at the market. There are at least two alternatives to Menger's explanation. First, many believe money is a matter of design and it was introduced by central authority or by common will. Second, some anthropologists argue that money evolved from gift exchange or, more broadly from social relations other than market exchange. The general objection to Menger's account is that it ignores the institutional and historical factors that affected the development of money. To be able to understand the merits and demerits of Menger's account one needs to examine Menger's explanation in comparison to these views. Accordingly, the first part of this chapter introduces Menger's explanation and discusses it in comparison to its 'rivals'.

The second part of the chapter focuses on Menger's methodological views and his justification of theoretical economics in contradistinction to 'historical' economics. This will help us reach an interpretation of Menger's explanation and put forward some of the constituents of the framework presented in Chapter 7.

There are two important conclusions of this chapter: first, different stories of the genesis of 'money' focus on different aspects of 'money' and these are not necessarily contradictory. Money takes various forms and serves several functions and its distinct characterisations may lead to different explanations of its origin. Second, the objection that Menger ignores some of the crucial institutional factors is correct. Yet, this does not immediately permit us to dismiss Menger's explanation, for his exposition is different from the other accounts in that he is trying to find out some of the general mechanisms that may have brought about a generally acceptable medium of exchange. Menger could be justified in abstracting from some institutional factors if his exposition adds something to our knowledge base. It is argued here that Menger should be considered as uncovering a possible way in which certain factors may have interacted in the process of emergence of money and thereby providing some of the possible mechanisms that may help us explain the emergence of money for particular cases. Menger's story is of value for it points to certain mechanisms and a possible way in which they may have interacted in a particular society at a certain time in history. Yet, we should not overestimate the explanatory power of Menger's story. We may consider it as a conjecture about the way in which money could have emerged. Menger only alerts us to some factors and presents a partial potential explanation. It is partial in the sense that it focuses on some of the explanatory factors and potential in the sense that there is no guarantee or proof that these factors were indeed effective in the process of emergence of money.

Menger's explanation

Menger's explanation of the origin of money is an example of what Menger's calls an *organic explanation*. Menger thinks that natural organisms are analogous to certain social phenomena in a limited way. He argues that 'there exists a certain similarity between natural organisms and a series of structures of social life, both in respect to their function and their origin' (Menger 1883: 129). Yet, although he admits that there is a similarity between the origin and function of these phenomena and of natural organisms, he alerts us to the limitations of this analogy:

> it is not an analogy that covers the entire nature of the phenomena concerned, but only partial aspects of them. In this respect it is again only a partial analogy.
>
> (Menger 1883: 131)

According to Menger, only *some* social phenomena are analogical to social organisms and with respect to the applicability of this analogy, social phenomena can be divided into two categories:

1 Social phenomena that are 'the result[1] of purposeful activity of humans directed toward their establishment and development' (Menger 1883: 131) or 'are the results of common will directed toward their establishment (agreement, positive legislation, etc.)' (Menger 1883: 133).
2 Social phenomena that 'are the unintended results of human efforts aimed at attaining individual goals' (Menger 1883: 133).

In Menger's terminology the first type is called *pragmatic phenomena* and the second type *organic phenomena*. (In Table 2.2, cells 1.6, 1.7 and 1.8 represent organic phenomena, and cell 2.5 represents pragmatic phenomena.) Those phenomena that are similar to natural organisms, with respect to their function and their origin, are in fact the social institutions and structures that are the unintended consequences of human action, that is, organic phenomena.

> Similarly we can observe in numerous social institutions a strikingly apparent functionality with respect to the whole. But with closer examination they still do not prove to be the result of an intention aimed at this purpose, i.e., the

result of an agreement of members of society or of positive legislation. They, too, present themselves to us rather as 'natural' products (in a certain sense), as *unintended results of historical development.*

(Menger 1883: 130)

According to Menger (1883: 130), money, law, language, markets, etc. are examples of such phenomena.

As for the incompleteness of the analogy, Menger stresses the differences between realms of natural organic phenomena and social organic phenomena. For example, Menger (1883: 133) rejects the idea of mutual causation between the whole and its parts, he thinks that this is a vague idea and it is inadequate for our laws of thinking. More importantly, he argues that forces acting in nature are different than the ones in the social realm. In nature purely mechanical forces are working but the social realm is dominated by intentional activities:[2] 'they are, rather, the result of human efforts, the efforts of thinking, feeling, acting human beings' (Menger 1883: 133).

Menger (1883: 135) notes the following consequences of these limitations in the organism analogy. First of all, the limitations of this analogy necessitate an understanding of organic as well as of pragmatic phenomena in the social realm. Second, the application of the organism analogy cannot provide a full understanding of organic phenomena:

The mechanical application of the methods of anatomy and of physiology to the social sciences is therefore not permissible even within the narrow limits indicated above.

(Menger 1883: 136)

Given the limitations, it is not possible to study organic phenomena as biological organisms. Menger argues that we do not need to examine organic phenomena as a whole[3] and that the study of the constituent parts of organic phenomena is more appropriate.

The acknowledgement of a number of social phenomena as 'organisms' is in no way in contradiction to the aspiration for exact (atomistic!) understanding of them.

(Menger 1883: 141)

From the above argument, it should be clear that even though Menger thinks that *origin* and *function* of the unintended consequences of human action resemble that of organisms, he does not argue that they should be examined in a holistic way. The relevant question, generally known as the *Mengerian question*, that should be asked when examining such phenomena is:

How can it be that institutions which serve the common welfare and are extremely significant for its development come into being without a *common will* directed toward establishing them? [. . .] What is the nature of all the above social phenomena [. . .] and how can we arrive at a full understanding of their nature?

(Menger 1883: 146, 147)

Here, although Menger does not rule out disadvantageous unintended consequences, he is clearly interested in explaining the beneficial ones (cell 1.6 in Table 2.2). The answer to this question would constitute an organic explanation, according to Menger. That is, an organic explanation explains either the *origin* or the *function* of the social patterns or institutions that are the unintended consequences of human action.

Money

Before examining Menger's explanation of money as an organic phenomenon, it is useful to discuss the meaning of 'money'. According to the Merriam-Webster's dictionary, 'money' is something that is generally accepted as a medium of exchange, a measure of value, a unit of account or a means of payment. We know that there are officially coined or stamped metal money, paper money, commodity money, and money of account. Yet the definition and identification of this ubiquitous phenomenon might be troublesome if one wants to explain its origin. Shall we trace the history of the coins and banknotes? Or, shall we go further to look for items that served the functions of 'money' in earlier tribes, cultures, and civilizations? What are the essential functions of money? Is it possible that what is an essential function of 'money' today may not be a function of 'money' in its earlier forms?

Dalton (1971b: 172) argues that modern money (e.g. US dollars, euros, Turkish liras, etc.) performs all the functions of money (as defined above) for all types of transactions and payments. Modern money is, so to say, all-in-one money. We do not consider precious metals or jewellery as money because they 'come into existence for reasons other than money-ness. Each is capable of one or two money uses' but they do not serve all the functions (i.e. a medium of exchange, a store of value, a unit of account, etc.) of modern money as the euro does. Dalton argues:

primitive monies or valuables used in reciprocal and redistributive transactions are the counter parts of these limited or special purpose monies [e.g., jewellery, bonds, stocks etc.], and not of dollars as media of (commercial) exchange; they resemble dollars only in non-commercial uses (paying taxes and fines, and in gift giving).

(Dalton 1971b: 172)

Briefly 'primitive' money is limited-purpose money in comparison to modern money. They are similar, but still so different that an explanation of the origin of limited-purpose money (e.g. bride money) should be different from an explanation of the origin of a generally accepted medium of commercial exchange.

The differences between modern money and 'primitive' money should alert us to the fact that the delineation of the explanandum phenomenon, in our case 'money', is essential in understanding its explanation. To be able to understand an explanation of the origin of 'money' one needs to know what is considered as 'money' in that explanation. Explanations of the origin of money may take different forms under different definitions of money. This is particularly important if we want to understand how different explanations of the origin of money stand next to each other.

The origin of money

In his 'On the Origins of Money' Menger (1892a) tells us at the outset that he is interested in explaining the emergence of a generally acceptable medium of exchange:

> There is a phenomenon which has from of old and in a peculiar degree attracted the attention of social philosophers and practical economists, the fact of certain commodities (these being in advanced civilizations coined pieces of gold and silver, together subsequently with documents representing those coins) becoming *universally acceptable media of exchange.*
>
> (Menger 1892a: 239, emphasis added)

He observes that at different times and places different types of commodities served as 'money' to different cultures and nations. Yet, he focuses his attention on one property of those tokens, on their functioning as a generally accepted *medium of exchange.*[4] His explanation basically states that (contrary to the common belief)[5] a medium of exchange could have emerged out of the dispersed actions of the individuals, without a 'common will directed towards establishing' it. This is, in fact, the type of research project that he wants to promote in social sciences, especially in economics, as stated in what we have called the Mengerian question. To answer this question with respect to money, he first observes that different commodities have different degrees of saleableness at different times and places. Some goods are available at some places and cannot be found in others. Given the cultural background and environmental conditions some commodities are needed more than others. Some goods have certain properties that others do not have, such as durability. Briefly, in any place there is a certain set of goods some of which are more saleable than others. Indeed, Menger wants to show that the phenomenon of 'money' can be understood in terms of its high saleableness:

> *The theory of money necessarily presupposes a theory of the saleableness of goods.* If we grasp this, we shall be able to understand how the almost unlimited saleableness of money is only a special case, – presenting only a difference of degree – of a generic phenomenon of economic life – namely, the difference in the saleableness of commodities in general.
>
> (Menger 1892a: 243)

The question is, 'given the differences in the saleableness of goods,[6] how could a generally accepted medium of exchange emerge?'

After registering the fact that saleableness of goods differs, Menger goes on to register another 'fact': the fact that individuals act according to their economic (self-)interests. Yet, he neither argues that this is the only motivation of human beings, nor that this statement is valid under every condition. Menger considers economic interests as *an* important factor in explaining economic phenomena given that individuals are market dependent. Thus, market dependence is another important factor in his explanation.

Among the factors that increase the differences in the saleableness of goods we find factors such as the 'development of the market', 'development of commerce', 'the permanence of the need', 'periodicity of the market', etc. (Menger 1892a: 246–247) Moreover, whenever Menger talks about 'economic interests' he is interested in showing the individuals' interest in overcoming the inconveniencies of direct exchange, which necessarily implies individuals' dependency on their exchanges at the marketplace:

> when any one has brought goods not highly saleable to market, the idea uppermost in his mind is to exchange them, not only for such as he happens to be in need of, but, if this cannot be effected directly, for other goods also, which, while he did not want them himself, were nevertheless more saleable than his own. By so doing he certainly does not attain at once the final object of his trafficking, to wit, the acquisition of goods needful to *himself*. Yet he draws nearer to that object. By the devious way of a mediate exchange, he gains the prospect of accomplishing his purpose more surely and economically than if he had confined himself to direct exchange.
>
> (Menger 1892a: 248)

It is evident that the need for certain goods is essential here. If individuals do not need to exchange goods to meet their needs then it is less likely that they will try to solve the problem of 'double coincidence of wants'.[7] Likewise, Menger explicitly states that an important requirement for the need for a generally acceptable medium of exchange to emerge is *increased market traffic*. That is, if individuals exchange goods infrequently they may not need to find ways out of the existing situation of troublesome direct exchange and they may not be able to learn how to improve their situation:

> With the *extension of traffic* in space and with the expansion over ever longer intervals of time of prevision for satisfying material needs, each individual would *learn*, from *his own economic interests*, to take good heed that he bartered his less saleable goods for those special commodities which displayed, beside the attraction of being highly saleable in the particular locality, a wide range of saleableness both in time and place.
>
> (Menger 1892a: 248, emphasis added)

As the market traffic increases, the possibility that individuals could find out the advantages of using more saleable goods in their exchange increases. Thus, Menger takes into account the increase in needs and the gains and losses of individuals in relation to the market traffic. When there are more goods to exchange:

1 the trouble of bringing them to the market;
2 the loss resulting from no exchange;
3 the loss resulting from exchanging goods with less saleable goods that are not demanded immediately; and
4 the gain from exchanging goods with goods that may be easily exchanged with other goods increases.

Menger assumes that the conditions under which a medium of exchange emerges should have started from an environment where individuals already exchange goods. Moreover, the individuals could not have continued their life solely on the goods that they individually produce. This is an environment where there is no medium of exchange (since he wants to explain its emergence) and individuals exchange their goods directly. In sum, the initial conditions are defined by the existence of direct exchange and by the intentions of the market-dependent individuals to exchange good(s) for good(s) to which they have immediate need. Menger assumes that direct exchange is not convenient for market-dependent individuals and that in such an environment the owner of the most saleable goods should have exchanged her goods much more easily than others – because, by definition, there is more demand for more saleable goods.

Given these assumptions, the most significant force in the process of the emergence of money is the 'individuals who are engaged in economising actions'. In principle, Menger's individuals are capable of observing the state of affairs at the market. They are hypothesised roughly to observe:

• that it is not always possible to exchange one's goods with the goods one needs;
• that it is inconvenient to exchange one's goods with the goods one does not need, if that good is not easily accepted by others at the market;
• that it is easier to exchange some goods than others, and that there is greater demand for those goods; and
• that it would be more convenient to exchange one's goods with these easily saleable goods in the case that one cannot exchange her goods with goods that she immediately needs.

Menger portrays individuals as being more or less 'rational' and focuses on their economising actions concerning direct exchange. He argues that the *tendency* to behave 'rationally' (to economise) increases as the intensity of exchange increases – although economising does not solely depend on market traffic. He also observes that individual capacities of 'realising' facts can differ and that this may take time and the process within which a medium of exchange emerges involves some mechanism like imitation:[8]

> It is only in the first instance a limited number of economic subjects who will recognize the advantage in such procedure, an advantage which, in and by itself, is independent of the general recognition of a commodity as a medium of exchange, inasmuch as such an exchange, always and under all circumstances, brings the economic unit a good deal nearer to his goal, to the acquisition of useful things of which he really stands in need. But it is admitted, *that there is no better method of enlightening any one about his economic interests than that he perceive the economic success of those who use the right means to secure their own.*
>
> (Menger 1892a: 249, emphasis added)

Menger thinks that some individuals may realise the state of affairs earlier than others and adopt the strategy of exchanging their goods with more saleable goods when they cannot find the goods they need. Some others may 'imitate' or 'learn' from these individuals. Thus, the process of money's emergence can also be considered as a discovery process.

Given the high level of market transactions and economic dependency of the individuals on market transactions, individuals try to minimise their 'transaction costs' by using more saleable goods in their exchange. Individuals intentionally prefer some goods to others (e.g. more saleable or more needed goods), but they do not have the intention to bring about something as a generally accepted medium of exchange. In a way, in this process some of the goods are *filtered* as 'nominees' for being a generally accepted medium of exchange. Moreover, there is positive feedback from the market environment that encourages more people to use more saleable goods in their exchange; that is, if more people start using more saleable goods in their exchange the market traffic increases and the increased market traffic necessitates the use of a more saleable good in exchange.[9] This feedback also keeps the process going:

> When the relatively most saleable commodities have become 'money', the great event has in the first place the effect of substantially increasing their originally high saleableness. Every economic subject bringing less saleable wares to market, to acquire goods of another sort, has thenceforth a stronger interest in converting what he has in the first instance into the wares which have become money.
>
> (Menger 1892a: 248)

This implies that the range and relative ranking of the goods (according to their saleableness) are changing in the process of the emergence of money. What is becoming a generally accepted medium of exchange becomes more and more saleable.[10] The end state of this process is the emergence of a generally accepted medium of exchange. This institution is the unintended consequence of the actions of economising (self-interested) individuals:

It is clear, rather, that the origin of money can truly be brought to our full understanding only by our learning to understand the *social* institution discussed here as the unintended result, as the unplanned outcome of a specifically *individual* efforts of members of a society.

(Menger 1883: 155)

Reconstruction of Menger's story

To understand Menger's explanation it is important to comprehend his story and its stages. Menger observes that there is a curious social phenomenon called 'money' and he wants to explain its emergence in order to understand its nature. He wants to find out how it could have emerged. Briefly, his task is to explicate how a no-medium-of-exchange world transforms into a world with a medium of exchange.

To explain the origin of money, Menger needs to find out the conditions under which a generally acceptable medium of exchange could emerge. That is, he needs to tell us about the state of the world at time t, World (t), and the processes that transforms World (t) into a state with a generally acceptable medium of exchange at time $t+n$, World $(t+n)$. In Figure 3.1, the box in between the two states of the world represents the process (or mechanisms) that transforms World (t) to World $(t+n)$. The description of World (t) and of its transformation into World $(t+n)$ is needed to explain the emergence of money.

The problem here is the representation of the states of affairs in World (t) and World $(t+n)$. It is relatively easy to represent World $(t+n)$ for it is possible to observe the characteristics of a medium of exchange and how it is used. However, direct evidence about World (t) is limited. Menger states a couple of facts about different uses of money in history and in different cultures. He informs us that cattle, shells, etc. have been used as money, and that the concept of a medium of exchange is not limited with coins and papers issued by the state. He also mentions other general facts, such as the fact that different goods have different degrees of saleableness. In Menger's explanation World (t) is described as a place where market-dependent individuals directly exchange goods at the market. Individuals of World (t) act according to their economic interests and are capable of improving their situation given the conditions of the market. The 'economising actions of these individuals and their interaction' is the main driving force in the transformation of World (t) into World $(t+n)$.

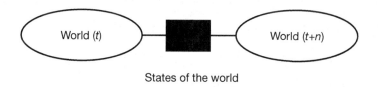

States of the world

Figure 3.1 States of the world.[11]

Objections

There are at least two important objections to Menger's account.[12] The first suggests that 'money was introduced by a communal agreement or political decree or legislative action that is external to the exchange process' (Iwai 1997: 1). The second argues that the origins of modern money should be traced back to gift exchange, or more generally to transaction forms other than commercial exchange (see, for example, Clark 1993: 381–382). These accounts have different characterisations of World *(t)* and its transformation into World *(t + n)*. The common denominator of these alternative accounts is their criticism of Menger's description of the initial state within which money developed, that is, of World *(t)*. It is argued that barter, the direct exchange of goods, is hardly found in earlier and primitive societies and the tendency to exchange goods, which is the main driving force in Menger's explanation, is almost non-existent in those societies.[13]

The design view may have various forms. Its simplest form focuses on coined money and justifiably argues that 'money' was designed. Obviously in this form, design view does not contradict Menger's account of the origin of a generally acceptable medium of exchange.[14] Those who argue that it does are confusing the different delineations of money used in these different views.[15] Indeed, Menger would definitely agree that coined money is a matter of design:

> by state recognition and state regulation, this social institution of money has been perfected and adjusted to the manifold and varying needs of an evolving commerce, just as customary rights have been perfected and adjusted by statute law. Treated originally by weight, like other commodities, the precious metals have by degrees attained as coins a shape by which their intrinsically high saleableness has experienced a material increase.
>
> (Menger 1892a: 255)

A more sophisticated form of the 'design view' argues that money neither developed out of direct exchange, nor that it functioned as a medium of exchange in its early stages. Such an account can be found in Henry (2002). Henry argues that in Egypt money first developed as a 'unit of account' (also see Ingham 1998a,b, 1999). Henry (2002: 5) informs us that 'up to about 4400 BC. Egyptian populations lived in egalitarian, tribal arrangements.' Then he explains how these non-exchange, non-propertied societies with collective production and collective consumption are transformed into a society with hierarchical (unequal) structures. The idea is that after the specialisation of some individuals on agricultural techniques, a classed structure emerges slowly and obligations that were internal to the tribe are transformed into obligations to other classes. 'Under the new social organization, tribal obligations were converted into levies (or taxes, if one views this term broadly enough)' (Henry 2002: 9).

At some early point in the Old Kingdom, the growing complexities of the new economic arrangements required the introduction of a unit of account in which taxes and their payment could be reckoned and the various accounts in the treasury could be kept separate and maintained. This unit was the *deben.*

(Henry 2002: 11)

In Henry's story World *(t)* is a place where there is no market-dependent exchange, but obligatory payments. Moreover, individuals are not equals in their transactions (e.g. they do not exchange goods voluntarily) as they are in Menger's story. For this reason, Henry proposes this historical study as a case that contradicts Menger's account, and supports Innes's (1913) account that sees money as a product of obligations that existed in every society. Yet Henry's account does not really contradict Menger's account. Henry and Menger are *explaining different things*. Henry explains the emergence of money *as a unit of account* at a specific place and time. Menger, on the other hand, presents a general account of money as a medium of exchange. Thus, two differences can be identified: first, they explain different things; second, their explanations are at different levels of generality. As it stands, Henry's design argument only holds for the emergence of money as a unit of account. It leaves the possibility that money as a medium of exchange developed spontaneously in that environment – that is, after the introduction of *deben* – with the development of market-dependent exchange. In any case, although we cannot establish that it contradicts Menger's account in its conclusions, we have to acknowledge the fact that it presents a totally different perspective on the environment within which money as a medium of exchange developed. Thus, his evidence on Egypt arouses doubts about Menger's description of World *(t)*.

The inaccuracy of the traditional description (such as Smith's, Menger's, Jevons's, etc. – as presented in Appendix I) of the stages of society prior to development of money has been emphasised by many anthropologists and historians alike. Such criticism heavily builds on Polanyi's (1944) influential book *The Great Transformation.*[16] The main argument here is that primitive cultures do not have the institutions that modern cultures have and that the traditional view is wrong in imposing the modern definitions of similar institutions to those cultures. The fifth chapter of *The Great Transformation* is full of such examples. Polanyi argues, for example, that money is not necessarily an important institution in primitive cultures, for money serves different functions in such cultures that are of secondary importance to the functioning of those societies. He argues, contrary to the common view in economics that trade most probably developed as long-distance trade and not as internal trade, the exchange of goods was not in the form of barter but in the form of gift exchange. According to Polanyi, it is hard to find a state in history or in primitive cultures where individuals bartered goods and acted in line with their disposition to exchange goods selfishly 'maximising' their 'utility'. Polanyi's argument undermines the description of World *(t)* as presented by Menger.

Similarly, Bohannan and Dalton (1971) argue that 'the familiar dichotomy of barter versus money transactions does not reveal the mode of transaction in goods changing hands' in primitive cultures. In Polanyi's terminology reciprocity, redistribution and market exchange are three different ways in which goods may change hands. Goods changing hands by reciprocal gift exchanges or by redistribution of the goods that were collected at a centre is essentially different from market exchange. Thus, in considering the evolution of money the traditional view (e.g. Menger's) makes the mistake of ignoring these institutional arrangements by assuming the existence of barter.

Departing from similar arguments, some anthropologists argue that money evolved out of the gift system, which can be considered as a type of obligation.[17] In this view money was not intentionally brought about; rather, it developed out of social obligations, that is, gift exchange. For example, Clark argues:

> Menger omits the fact that money, as a social institution, existed thousands of years before the rise of the market economies. Just as the exchange of commodities has its origins in the ceremony of the gift, where the motivation is the exact opposite of that which we assume today, the origins of money are not connected in most instances to trade goods, but to a change in function of an object usually of ceremonial value.
>
> (Clark 1993: 381–382)

Clark argues that different forms of money existed prior to the emergence of a generally acceptable medium of exchange and that Menger should have taken this into account in his explanation of origin of a medium of exchange. The serious challenge here is that Clark argues that Menger does not get the facts right about the state of the world, World *(t)*, within which a medium of exchange was brought about, and that Menger's depiction of the earlier stages of the society ignores the most relevant institutions (e.g. prior forms of money). The objection is that explanation of the origin of a generally accepted medium of exchange or origin of modern all-in-one money cannot be independent from the prior forms of money (i.e. limited purpose money). It is, therefore, not possible to argue that Menger explains the genesis of money in a truly historical sense. If the description of the initial stages, that is, World *(t)*, is not accurate, one is tempted to argue that Menger's explanation is wrong and for that reason it is valueless. Is it possible that Menger's exposition is still valid although its description of the initial stage, World *(t)*, does not correspond to any stage of any society in history?

To answer this question we have to see the differences between the objections and Menger's account. The main difference is that Menger's account is general and that of the objections are not. The objections are pointing to some *particular* facts about the origin of money and their exposition is specific to certain locations and times. Menger's account, on the other hand, is pointing out some *general* mechanisms that may have been working in particular histories of the emergence of a medium of exchange.[18] In Menger's terminology, objections provide a historical understanding, and Menger provides a theoretical understanding. The difference between historical and theoretical understanding is examined in the next section.

Before going further it is useful to emphasise a couple of points. First, because of its general nature, Menger's explanation may be compatible with other scenarios.

For example, in distinct societies different commodities may have served some functions of money prior to the emergence of a generally accepted medium of exchange. Menger's theory of saleability abstracts from the particularities of these societies and tells us that these commodities could be considered as goods with relatively high saleability. The development of market dependence, then, transforms some of these highly saleable goods into generally accepted media of exchange. Thus, Menger's theory is general in the sense that it allows for the existence of prior forms of money. In other words, existence of prior forms of money does not necessarily conflict with Menger's explanation. Second, Menger's story alerts us to the dynamics of market exchange as an important factor in explaining the origin of a generally acceptable medium of exchange. It is important to note here that none of the criticisms above object to (commercial) market exchange as an important factor in the evolution of a medium of exchange. Rather, they object to Menger's description of the initial stage of the society, World *(t)*, within which money emerges. Once we see the fact that Menger is trying to explain the origin of a generally acceptable medium of exchange, and that his emphasis is on market traffic and economic interests, we may see that Menger's exposition studies the emergence of a medium of exchange in isolation from other factors. Third, it is possible to justify his approach by considering him as trying to show *some* of the mechanisms behind the origin of a medium of exchange – not all of them. In this interpretation, Menger's explanation should not be considered as a full-fledged explanation of origin money, but as a partial potential (theoretical) explanation that presents a *possible way* in which *some* of the existent mechanisms may interact (or, may have interacted) in the process of emergence of money. These issues get more attention in the next section.

Exact understanding

Menger (1883: 35–36) makes a distinction between concrete phenomena and empirical forms. Concrete phenomena and their concrete relationships are specific to time and place. For example, the coins and banknotes in your pocket are examples of concrete phenomena. Empirical forms, on the other hand, represent the general aspects of phenomena. What we know as money, supply, demand, price, etc. in economics may be considered as empirical forms in Menger's terminology. They do not denote the properties of phenomena that are specific to time and place; rather, they signify some general aspects of them. For example, while specific characters, pictures and colours printed on a 100 euro banknote, as well as its several uses may be counted as its properties, only the properties that are common to several forms of money can be counted as properties of 'money' (e.g. being a medium of exchange, a means of payment, etc.).[19] Menger (1883: 36) argues that it is the 'investigation of types and typical relationships' that gives us deeper understanding of the real world.

One way to understand a concrete phenomenon is to study its properties and its relationship with other phenomena. According to Menger, such an understanding would constitute individual knowledge of that phenomenon and history and statistics provide such individual (as opposed to general) knowledge about concrete phenomena. General knowledge of phenomena can only be attained theoretically by studying empirical forms and laws. Thus, according to Menger (1883: 43–45), there are two ways of understanding phenomena: the *historical* way and the *theoretical* way. Menger maintains that both of them are necessary for understanding phenomena.[20] Obviously, the evidence presented by the anthropologists (and historians) provides a historical understanding of the origin of particular exemplifications of 'money'. Menger's exposition of the origin of money, on the other hand, is supposed to provide a theoretical understanding of it. Menger says:

> we become aware of the basis of the existence and peculiarity of a concrete phenomenon by learning to recognize in it merely the exemplification of a conformity-to-law of phenomena in general.
>
> (Menger 1883: 45)

That is, we can interpret Menger's explanation as providing some sort of a general law about economic phenomena such as the following: under the conditions of market-dependent direct exchange, self-interested economising individuals would be inclined to use more saleable goods as a medium of their exchange and upon the implementation of this idea by some individuals others will follow – hence the emergence of a medium of exchange.

The problem here is that the historical understanding of the origin of money *seems* to be in conflict with this theoretical 'knowledge'. Anthropologists and historians exhibit many exceptions to this 'law', and since laws are usually conceived as being exceptionless, it may be doubted whether Menger accomplishes what he is supposed to. The answer lies in Menger's conception of laws. Briefly, for the science of economics, Menger is generally talking about *exact laws* that cannot be tested by empirical reality and historical records.

But what are exact laws? According to Menger (1883: 50) there are two types of laws with respect to their strictness: laws of nature and exact laws that hold with no exceptions; and empirical laws that allow exceptions.

There are two different types of laws that have no exceptions: laws of nature and exact laws. According to Menger (1883: 59), 'the laws of theoretical economics are really never laws of nature in the true meaning of the word'. That is, laws of theoretical economics are exact laws.[21] Exact laws cannot be tested and they state strict relationships among phenomena *in isolation* from other factors. Exact laws can only be true for the abstract world created by the researcher. Although the consideration of Menger's explanation as providing exact laws avoids objections that refer to particular facts, a more serious problem arises: if Menger is talking about an abstractly conceived world, and if we cannot test his claims, how can he explain something about the real world? How can he explain the origin of money by presenting an imaginative world and an imaginative scenario about the genesis of money in that imaginative world? To be able to answer these questions we need to see what is really meant by exact laws.

According to Menger (1883: 56–59), there are two orientations of theoretical research, the realist–empirical orientation[22] and the exact orientation. Realist–empirical orientation aims at investigating phenomena 'in their "full empirical reality", *that is, in the totality and the whole complexity of their nature*' (Menger 1883: 56). Menger thinks that this is not feasible. Because 'there are no strict empirical types in "empirical reality", i.e., when the phenomena are under consideration in the totality and the whole complexity of their nature' (Menger 1883: 56–57). What can be accomplished with realist–empirical orientation is limited to the knowledge of 'real types (basic forms of real phenomena)' and to 'empirical laws, theoretical knowledge, which makes us aware of the actual regularities (though they are by no means guaranteed to be without exception) in the succession and coexistence of phenomena' (Menger 1883: 57). Thus, the realist–empirical orientation cannot arrive at 'strict (exact) theoretical knowledge' (Menger 1883: 58). For the realist–empirical orientation starts from concrete phenomena and their relationships, the applicability of the theory achieved by their investigation alone is limited by the spatial and temporal considerations. Note here that Menger is implicitly talking about the problems of induction. With respect to the origin of money the implication is the following: it is not possible to reach a general understanding of the origin of money by studying some of its exemplifications in history or in certain cultures.

While the realist–empirical orientation cannot provide strict laws, the exact orientation of research provides exact laws that strictly specify the relationships among phenomena. The following lengthy quote from Menger reveals the basic constituents of the exact orientation:

> it [the exact orientation] seeks to ascertain the *simplest elements* of everything real, elements which must be thought of as strictly typical just because they are the simplest. It strives for the establishment of these elements by way of an only partially empirical-realistic analysis, i.e., without considering whether these in reality are present as *independent* phenomena; indeed without even considering whether they can at all be presented independently in their full purity. In this manner theoretical research arrives at empirical forms which *qualitatively* are strictly typical. It arrives at results of theoretical research which, to be sure, *must not be tested by full empirical reality* (for empirical forms here under discussion, e.g., absolutely pure oxygen, pure alcohol, pure gold, a person pursuing only economic aims, etc. exist in part only in our ideas). However, these results correspond to the specific task of the exact orientation of theoretical research and are the necessary basis and presupposition for obtaining *exact laws*.
>
> (Menger 1883: 60–61, fourth emphasis added)

Thus, an explanation of the origin of money from the point of view of the exact orientation starts with specifying the simplest elements, such as the differences in the saleableness of goods, inconveniency of direct exchange for market-dependent individuals and the disposition of individuals to act according to their economic interests. It creates, so to say, an abstract world where these 'simplest' elements are isolated from others. For this reason, it is not important whether this abstract world exists in reality exactly as it is defined in the theory or explanation. Menger (1883: 71) says that exact laws cannot be true from an empirical point of view. He (1883: 72) argues that exact laws 'are absolutely true [. . .] as soon as' they are 'considered from the point of view which is adequate for exact research'. Perhaps more radically, he (1883: 73) says '*exact economics by nature has to make us aware of the laws holding for an analytically or abstractly conceived economic* world'.[23]

Consider Menger's argument that exact laws are different from laws of nature.[24] Laws of nature hold strictly for the real world, yet exact laws strictly hold for the model world created by the scientist. Theoretical economics, according to Menger, produces such exact laws. They are strict but cannot be tested directly. The realist–empirical orientation cannot reach such strict laws for it tends to consider phenomena in all their complexity. Suppose that it is possible to see all the stages of the development of money in a certain culture. Even in this case, according to Menger, it would not be possible for the scientist to give a general explanation of the origin of money, for the evidence and observed regularities would be specific to this culture. The exact orientation does not examine the succession of phenomena in this way:

> Exact science, accordingly, does not examine the regularities in the succession, etc., of *real* phenomena either. It examines, rather, how more complicated phenomena develop from simplest, in part even unempirical elements of the real world in their (likewise unempirical) isolation from all other influences, with constant consideration of exact (likewise ideal!) measure. It does this without taking into account whether those simplest elements, or complications thereof, are actually to be observed in reality uninfluenced by human art; indeed, without considering whether these elements could be found at all in their complete purity.
>
> (Menger 1883: 61)

Thus, Menger argues that exact theory works with 'isolation'.[25] The researcher isolates the factors that seem (usually, to the researcher) to be responsible for the phenomenon in question from the other complexities of real life. In Menger's exposition of the genesis of money as a medium of exchange, the simplest elements of this phenomenon is singled out: a medium of exchange is used in market transactions in order to supply economic needs. Of course, Menger is aware of the fact that money is used for other purposes and that the economic side of the transactions does not represent every aspect of the earlier stages of society. Yet, he thinks that studying these factors in isolation would give us a theoretical understanding of the origin of money. This knowledge may be about a special side of this phenomenon (e.g. economic side), but it is useful in understanding its origin. That is, the exact orientation gives us an

> understanding of a *special side of phenomena* of human activity (abstracted from the empirical reality). [. . .] only the *totality of such theories* [which are

produced by exact orientation], when they are once pursued, will reveal to us *in combination with* the results of the *realistic orientation* of theoretical research the deepest theoretical understanding attainable by the human mind of social phenomena in their full empirical reality.

(Menger 1883: 62–63, emphasis added)

Since exact orientation works with 'isolation', a full understanding of the phenomenon of money and its origin cannot be provided by a single exact theory of its genesis. First of all, 'only the exact sciences in their *totality* are able to offer us such things, since each of them opens up only the understanding of a specific side of the real world' (Menger 1883: 77). Second, both the exact and realist–empirical orientations of research are necessary in understanding phenomena, and 'each of them contributes in its own way' to this understanding (Menger 1883: 64). Thus, we have to consider his explanation of the origin of money from this perspective. It specifically focuses on the effects of market-dependent direct exchange and economising individuals. Its explanatory value, therefore, comes from its explication of this special side of the genesis of money. Remember from the objections that each and every one of them focused on Menger's description of World *(t)*, yet none of them objected to the argument that individuals acting according to their economic self-interests would bring about a medium of exchange under the conditions of World *(t)*, as described by Menger. The focus of Menger's story is on the economic factors behind the origin of a medium of exchange.[26]

The theoretical understanding of money as an organic phenomenon is no different from other phenomena. According to Menger (1883: 140), 'all theoretical understanding of phenomena can be the result of a double orientation of research, the *empirical-realistic* and the *exact*'. Similarly, understanding the origin of money cannot solely be dependent on exact research. Menger is by no means opposed to the historical understanding of phenomena or to the realist–empirical orientation of research. What he opposes is the conception of money as a pragmatic phenomenon: he argues that although the exact and 'realist–empirical' orientations of research may go hand in hand, the pragmatic interpretation of unintended consequences of human action is inadmissible (Menger 1883: 145). He shows that money could be considered as an unintended consequence of human action. He challenges those who consider money as a matter of design by showing theoretically that it indeed follows from the basic elements of market-dependent direct exchange that money as a medium of exchange may develop as an unintended consequence of individual action. He also states that the argument that some social phenomena are the unintended consequences of human action should not be considered as a mystic argument, for he thinks that it is the task of the social scientist to *explicate* how it developed as an unintended consequence. He basically shows how a social phenomenon develops from the individual factors of the society. His omission of the institutional factors presented by anthropologists and historians is for the sake of generality. He solely focuses on economic factors and their effects on the development of a medium of exchange. Briefly, he explicates a possible way in which certain mechanisms (i.e. economising actions dispersed individuals, imitation and learning) may interact and bring about money. In fact, economising actions of separate individuals can be considered as separate (individual) mechanisms and their interaction as an aggregate mechanism (or a process). In the next chapter it will become clear why this interpretation of mechanisms is useful. It is sufficient here to note that these individual mechanisms (i.e. economising individuals) and their interaction brings about a commonly accepted medium of exchange under the conditions of market-dependent direct exchange.

Explanatory value

We have seen that Menger's story is an abstract one. It isolates market mechanisms and economising individuals to explain the emergence of money as a medium of exchange. We have also seen that he is abstracting from some institutions that seems to be important in the genesis of money, such as gift giving, obligatory payments, etc. But, although Menger is ignoring some relevant factors in his story, he may be interpreted as providing a new look at the process of emergence of money by showing how interaction of economising individuals may bring about money. There is nothing in Menger's explanation that would prevent us from integrating particular institutional factors to our explanation of particular cases of emergence of money. The apparent conflict between Menger's account and its rivals disappears if we consider Menger as providing a partial potential *theoretical* explanation of the genesis of money.

In contrast to Figure 3.1, Figure 3.2 illustrates two different worlds: the real world (*R-World*) and the model world (*M-World*). M-World is an abstract world (i.e. a model) that is supposed to represent the real world, R-World. R-World *(t)* shows a particular state of the real world where a medium of exchange is non-existent. R-World *(t + n)* illustrates a later state of the real world where individuals are using a medium of exchange.

In order to explain the emergence of money one has to tell us how R-World *(t)* was transformed to R-World *(t + n)*. There may be two different ways of doing this. First, one may give a historical explanation of this transformation by narrating the time- and place-specific details of the process of emergence of

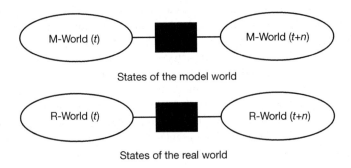

Figure 3.2 Real world vs. model world.

money. Although such an explanation would use some abstractions and isolations, we would expect it to be accurate about the time- and place-specific details. In Merger's terms such an explanation could be considered as conforming to the realist–empirical orientation. A second strategy that could be followed in order to explain the transformation of R-World *(t)* into R-World *(t + n)* is to abstract from the complexities of the real world and try to find out some of the general mechanisms that may have brought about a medium of exchange. This would involve creation of a model world (M-World) which is isolated from the particular details of the real world. Such an explanation would tell us how the model world is transformed from a state with no medium of exchange (M-World *(t)*) to a state with a medium exchange (M-World *(t + n)*). In principle, we would not expect this explanation to give us a fully accurate representation of states of the real world, as long as it unearths some of the causal mechanisms that drive the process of emergence of money. In Merger's terms, such an explanation could be considered as conforming to the exact orientation.

Note that Menger's arguments about exact understanding indicate that a *theoretical explanation* is an explanation in an abstractly conceived world.[27] That is, M-World *(t)* and the mechanisms therein (e.g. economising actions of individuals and their interaction) explain its transformation into M-World *(t + n)*. Then, a theoretical explanation may be defined as an explanation of the state of the affairs in a model world. A historical explanation, on the other hand, is an explanation of the state of the affairs in the real world. The states of affairs in the real world are specific to time and place, thus their explanation would be an explanation of particular facts. To be consistent with the philosophical literature let us call these *singular explanations* (see Ruben 1992: 4). Roughly, singular explanations explain the transformation of R-World *(t)* into R-World *(t + n)* by pointing out the particular facts concerning the particular states of the real world.[28] A theoretical explanation on the other hand needs to uncover the general (causal) mechanisms or 'laws' that drive the process of emergence of money. Certainly, we would expect theoretical explanations to help us explain particular cases.

Since a singular explanation (historical explanation in Menger's terminology) is an explanation that is specific to a certain space and time, the important question is whether there is any sensible relation between states of the model world and the states of the real world that will allow us to use our theoretical explanation to explain particular cases. Or more generally, whether there is a relation (e.g. a certain amount of similarity) between the real world and the model world that will allow us to carry our explanation from the model world to the real world. A theoretical explanation may help us explain particular cases if it satisfies the following conditions:

1 the explanation has to be successful in explaining the states of affairs in the model world; and
2 there must be some similarity between the model world and the real world.

As for the first condition, Menger's story could be considered as a logically plausible story of the transformation of M-World *(t)* into M-World *(t + n)*. To see this we may imagine that there is a possible world that satisfies the conditions defined by Menger for M-World *(t)* and the individuals therein. There is nothing in Menger's story that would make us think that such a possible world would not be transformed into a world with a medium of exchange by the mechanisms defined by Menger. This is why we may consider Menger's story as being logically plausible. Yet we may not argue that it is a logical necessity that M-World *(t)* would be transformed into M-World *(t + n)*. The deficiency of Menger's theoretical explanation is that the workings of the mechanisms of this transformation are not entirely clear.[29] This leaves the possibility that more constraints may apply to his explanation. For example, it may be that under some conditions market exchange does not lead to money as a medium of exchange. Hence, Menger's explanation does *not* establish the logical necessity of the transformation of M-World *(t)* into M-World *(t + n)*. The emergence of money in the model world is not explained in a satisfactory manner because the causal mechanisms and the conditions under which they will bring about a medium of exchange are not fully explicated. Thus, Menger's explanation partly fails to satisfy the first condition. His explanation is just a logically plausible story of the emergence of money.

As for the second condition, we have seen that M-World *(t)* does not really correspond to any particular stage of the real world. Yet we may try to see whether there are some similarities that would let us carry the partial success of the story in the model world to the real world. One of the reasons why Menger's logically plausible story may present some real world mechanisms is that his story is constrained by the relevant facts about the real world. For example, although M-World *(t)* did not exist exactly as it is in history, we know that whatever their institutional structure may have been, cultures that were using a medium of exchange passed from a stage where most of the individuals were dependent on market exchange (e.g. because of specialisation).[30] We know that if individuals have to exchange goods at the market to acquire the necessities of life, it would be inconvenient to exchange goods directly. We also know that human beings have the ability to discover, learn and imitate. Thus, it seems highly plausible that under the conditions of market-dependent direct exchange some individuals would start using more saleable goods to mediate their exchange and others would be following them. Moreover, Menger's theory of saleableness is broad enough to encompass some pre-existing forms of limited-purpose money. That is, if there are certain goods that serve some functions of money (e.g. means of payment for particular institutional obligations) in R-World *(t)*, we may consider these commodities as highly saleable compared to others and start our analysis from there. More importantly, the explanatory mechanisms depicted in Menger's logically plausible story are familiar to us. Individuals tend to economise (e.g. minimise transaction costs) under conditions of market-dependent exchange. Individuals have a tendency to discover new and more economical ways of doing things, and they imitate other successful individuals. Actions individuals who economise, discover and imitate could be considered as the main causal mechanisms in Menger's story. The interaction of these individual mechanisms (i.e. economising individuals) brings about a commonly accepted medium of exchange. The similarity between the abstract world depicted by Menger and the real world boils down to these mechanisms. We find Menger's explanation plausible because of our familiarity with these mechanisms. We feel that what happens in Menger's model world may have taken place in the real world.

We will have more to say about the distinction between individual mechanisms and their interaction in the next chapter. It is sufficient here to state that our familiarity with the individual mechanisms described by Menger turns his logically plausible story into a 'humanly possible' scenario. Nevertheless, 'humanly possible' is still possible. Menger is pointing out to a *possible* way in which certain existent mechanisms (e.g. economising action of separate individuals, imitation and learning) may interact and bring about money. Or more correctly, he is pointing out to *some* of the mechanisms that *may partly* explain the emergence of a medium of exchange albeit the workings of these mechanisms are not well defined.

Until now we have seen that Menger does not fully satisfy the conditions that will help us carry his theoretical explanation to the real world to explain particular facts. However, we should not forget the difficulty of Menger's task. Let us consider Menger's explanation from a different perspective in order to appreciate his partial success. We need to grant the fact that explaining the origin of money is a challenging task. Societies can be considered as a complex web of many institutions, and in every society this web is composed in a different way. The origin of money in a particular society cannot be thought of independently from its web of institutions. Yet, we neither have the complete and detailed historical record of the process of the emergence of money, nor do we have the chance to observe its emergence again. We may have evidence about the types of money used in certain societies and some of their institutions, but we may not start from here and deduce the causal mechanisms behind the emergence of money. In fact, we may consider explaining the origin of money as being somewhat similar to explaining the origin of life. We observe that there are living organisms with certain properties and we have some information about the earlier stages of the world prior to the development of living organisms, but we can do nothing but *conjecture* about the possible processes through which living organisms have developed. We constrain these conjectures with the known facts about living beings, and try to produce a plausible story of the emergence of life on Earth. It may be that our depiction of the earlier stages of the earth is inaccurate, but still these conjectures provide us with a framework within which we may reconsider the way in which certain factors and mechanisms have interacted in bringing about life on Earth. That is, if we are able to produce a plausible story of the emergence of life that would be a story of the transformation of M-World *(t)* into M-World *(t + n)*. This seems to be the only way to proceed unless we are able to observe the state of the real world prior to the existence of life, R-World *(t)*. Once we have a plausible story of the emergence of life, we may proceed to test our conjecture. We may test our conjecture logically by way of examining our model world under various conditions. Or, we may empirically test it by conducting experiments. The aim of this investigation would, of course, be to understand whether the explanatory mechanisms in our plausible conjecture are the real mechanisms behind the emergence of life.

Similarly, we may conceive Menger's story as alerting us to a possible way in which mechanisms of economising, imitation and learning may have interacted in the process of the emergence of a medium of exchange. Menger's conjecture alerts us to certain explanatory factors (mechanisms) that may have been important in the development of a medium of exchange.[31] At the time of its proposal, it presented an alternative way of thinking about the emergence of a medium of exchange. At least, he informed the audience that pointing out to the historical cases of the emergence of coined money cannot be the whole story of the emergence of money. He showed that although coined money may simply be a matter of design, the real problem was in explicating the stages prior to the decision of introducing coining money and the mechanisms that have transformed a no-money world in to a world with a medium of exchange. We may argue that he expanded the mental horizon of his audience by introducing an alternative process of the emergence of money. It was an important proposal to be tested logically and empirically. It was a step – a *partial* and *incomplete* step – forward in explaining the origin of money, an attempt to discover the real story behind the emergence of money.

Despite its deficiencies, the value of Menger's theoretical explanation comes from his exposition of the importance of market-dependent exchange, economising actions and learning.[32] Menger's story is not a full-fledged explanation of the origin of money. It is a partial explanation for it alerts us to *some* of the explanatory factors. It is a potential explanation for it alerts us to a set of *possible* explanatory factors. We do not know whether the proposed individual mechanisms were really responsible for the emergence of money. In this sense, the weakness of the objections to Menger's explanation is that they do not consider the possible importance of proposed explanatory mechanisms in explaining the emergence of money as a medium of exchange. If Menger is right in saying that the exact and empirical orientations of research should go hand in hand, we may argue that historical explanations of the emergence of a medium of exchange should at least consider investigating whether economising actions of individuals, discovery and imitation played any role in its emergence.

Concluding remarks

Before going further to examine Schelling's chequerboard model of segregation it is useful to emphasise some of the important arguments of this chapter. First, contrary to the common belief, Menger takes into account institutional factors in his explanation. Menger presupposes the existence of market-dependent individuals and, therefore, the existence of the market. Market traffic plays an important role in his explanation: increase in market traffic increases the tendency of individuals to act in an economising way, and this in turn increases market traffic. Moreover, the saleableness of goods is also affected by institutional factors. Thus, although Menger focuses on individual factors and provides an 'atomistic' explanation, he does not ignore the importance of institutions. Second, the conception that Menger undervalues historical research is wrong. In fact, one important lesson we should extract from Menger's account of theoretical and historical analysis is that both historical and theoretical research is necessary for a complete understanding of phenomena. Yet, these research methods contribute in different ways to our understanding of phenomena. In this spirit, third, it has been argued in this chapter that different explanations of the origin of money focus on different aspects of money and that they are not necessarily contradictory. It has been acknowledged that the general criticism to Menger's explanation is in fact true: Menger does not take into account known historical facts concerning the genesis of money. Yet this does not imply that his account contradicts these facts, rather Menger focuses on a limited number of factors that may have played important roles in the process of the emergence of a medium of exchange. Menger argues that market-dependent exchange is important in explaining the origin of a medium of exchange. In fact, no-one seems to disagree. Rather, the disagreement is that money in different forms (e.g. as a means of payment) existed prior to market-dependent exchange. Although this is true, it does not contradict Menger's account. We have seen that objections to Menger's account of the origin of money either get his *explanandum* wrong, or do not pay attention to the distinction between abstract reasoning and historical reasoning. When we see that Menger is trying to show some of the mechanisms behind the emergence of money as a medium of exchange we may also see that different explanations of the emergence of money may be complementary. For example, coined money, or money as a unit of account, may be intentionally created, but money as a medium of exchange may have developed as an unintended consequence of human action. We have also seen that Menger's story leaves us some room for plugging in the particular institutions of a certain culture when explaining the origin of a medium of exchange in that culture.

In sum, Menger's explanation is general, yet it focuses on a limited amount of factors. It does not have an explicit characterisation of the workings of the suggested mechanisms (e.g. economising action, learning, imitation) and it does not rest on factual evidence. Menger conjectures about the process of the emergence of a medium of exchange and provides an incomplete partial potential theoretical explanation.

The lack of direct evidence about the process of the emergence of a medium of exchange is a double-sided sword. On one hand, it necessitates conjectures such as Menger's; on the other hand, it does not permit us to historically confirm these conjectures. We are not likely to find direct historical evidence about Menger's explanation, for we cannot go back in time and observe the individuals, or ask them about their intentions and motives. To be able to gain confidence in his explanation we have to find other ways to test his conjecture. Chapter 6 discusses some of the contemporary models of the origin of money and the kinds of progress achieved by these models. But before that, in Chapter 4, the curious concepts of 'mechanism' and 'process' get more attention in our examination of Schelling's models of residential segregation. Chapter 5 explicates the concepts of the invisible hand and invisible-hand mechanisms. And in Chapter 7 we will focus on 'partial potential explanations', where our interpretation of Menger finds its place in a general framework.

4 Segregation

Introduction

All around the world, different ethnic groups live in different parts of a city. It is commonly believed that this type of residential segregation is caused by strong discriminatory preferences (e.g. racism) or by economic factors, such as welfare differences among different ethnic groups. In a series of papers (1969, 1971a,b, 1972) and in *Micromotives and Macrobehavior* (1978), Schelling argues that residential segregation may be compatible with different micromotives; and that *even* mild discriminatory preferences (e.g. trying to avoid a minority status) may bring about residential segregation. Like Menger, he characterises segregation as an unintended consequence of human action and gives a theoretical explanation of its emergence.

Schelling uses a couple of models to show that mild discriminatory preferences may bring about segregation. Among these models, the chequerboard model received the most attention.[1] Schelling's chequerboard model of residential segregation is one of the paradigmatic examples of invisible-hand explanations (e.g. Nozick 1974; Ullmann-Margalit 1978; Karlson 1993; Keller 1994) and one of the predecessors of *agent-based computer models* (e.g. Epstein and Axtell 1996; Rosser 1999; Casti 1989). It is also important to note that Schelling's model is regarded to be one of the examples of *good explanation* in social sciences (e.g. Sugden 2000) and the classical account of explaining with social mechanisms (e.g. Cowen 1998).

This chapter examines Schelling's chequerboard model, focusing on the 'mechanisms' that make the explanation and on its conjectural character that brings about the scepticism concerning its explanatory value. It is argued that Schelling's model suggests only some of the mechanisms behind residential segregation and that the chequerboard model is a conjecture about a possible way in which these mechanisms may interact. In addition, the present chapter suggests a new interpretation for these mechanisms, which makes it easier to understand the nature of chequerboard model.[2]

This chapter argues that Schelling's explanation is a partial potential (theoretical) explanation in the sense that it suggests only *some* of the mechanisms that *may* bring about residential segregation and a *possible* way in which they may interact. As in Menger's case, it is argued that Schelling's models expand our mental horizon. Before Schelling's model, it was believed that residential segregation was caused either by organised, intentional action or by welfare differences among different groups (or, economic processes). The chequerboard model shows the possibility that mild discriminatory preferences might lead to the same result even if the agents are happy to live in a mixed neighbourhood. It is suggested here that these forces might act together in different combinations. Thus, if we would like to investigate residential segregation empirically, including Schelling's results in our model of segregation would increase the explanatory potential (or, the potential explanatory power) of our model. To sum up, although Schelling's model is somewhat incomplete, it seems to enhance our ability to explain an instance of segregation.

The plan of the chapter is as follows. The first section examines the chequerboard model in detail. The second section focuses on its explanatory characteristics, and the third section concludes the chapter.

Residential segregation

The chequerboard model suggests that residential segregation may be an unintended consequence of the dispersed actions of the individuals. The focus of the analysis is on the 'segregation that can result from discriminatory behavior' (Schelling 1978: 138).

That is, in the chequerboard model, segregation is caused by *discriminatory behaviour*. In the context of the model, 'discriminatory' implies that individuals are aware (consciously or unconsciously) of the differences between two (or more) groups of individuals. This awareness may or may not affect individual behaviour. If it does, it may cause a range of discriminatory behaviour from strong discrimination (i.e. very intolerant) to weak discrimination (i.e. very tolerant). Let us suppose that there are two groups of individuals, As and Bs. In the context of the chequerboard model, strong discriminatory preferences imply that As (or Bs) want to be the majority in the neighbourhood, mild discriminatory preferences imply that As (or Bs) try to avoid a certain minority status.

As a first step to get a grip of the chequerboard model, let us start thinking about strong discriminatory preferences. It is quite simple to understand why strong discriminatory preferences (e.g. racism) cause residential segregation. As Schelling nicely argues,

> The simplest constraint on dichotomous mixing is that, within a given set of boundaries, not both groups can enjoy numerical superiority. For the whole population the numerical ratio is determined at any given time; but locally, in a city or a neighbourhood, a church or a school or a restaurant, either [As] or [Bs] can be majority. But if each insists on being local majority, there is only one mixture that will satisfy them – complete segregation.[3]

(Schelling 1978: 141)

If *A*s and *B*s are not content with any minority status, a process of moving to other places where they can be content will start. This process will not stop until every individual is content. This logical constraint implies that the only one mixture that can satisfy every individual is complete segregation – where every neighbourhood is composed of either *A*s or *B*s. Let us consider this as our first model (Model I) of segregation.

The specified mechanism is very simple. For descriptive purposes, we can think of every individual as conforming to a simple rule: IF the ratio of the other group to the neighbourhood population is larger than or equal to $^1/_2$ THEN move, IF NOT stay. More precisely, given that the sum of the ratios of *A*s (α) and *B*s (β) to the neighbourhood population is equal to one ($\alpha + \beta = 1$); the following rules for *A*s and *B*s will bring about complete segregation. *A*s: IF $\beta \geq ^1/_2$ THEN move, IF NOT stay; *B*s: IF $\alpha \geq ^1/_2$ THEN move, IF NOT stay.

These rules can be interpreted as the mechanisms that may bring about complete segregation within this simple model. Although the model is simple, it makes us aware of the fact that strong discriminatory preferences bring about complete residential segregation. If we knew that every individual in ethnically segregated neighbourhoods hold strongly discriminatory preferences, we could easily use this simple model to explain all instances of segregation. This is not the case, however. There are people who have milder discriminatory preferences. Moreover, as far as we can observe, there may be other causes of residential segregation, such as welfare differences among different ethnic groups. It is not possible to explain segregation by simply arguing that racism causes segregation. We need to inquire other possibilities.

Our simple model does not tell us whether different groups of individuals may be segregated even if they are happy with mixed neighbourhoods. Now, let us examine whether such preferences may cause segregation. Assume that agents (individual *A*s and *B*s) are not concerned – and not informed – about the mixture of their neighbourhood, but they care about their *immediate neighbours*. Moreover, assume that both *A*s and *B*s are happy living in a mixed neighbourhood, they can also accept a minority status, but they do not want to live as an extreme minority. These assumptions bring us to our second model of segregation (Model II), which is the widely discussed chequerboard model of segregation.

In the chequerboard world, *A*s and *B*s live in a chequerboard city. This city, like every other city, has boundaries for a given time period and it has a shape. There are sixty-four places to live (e.g. sixty-four houses) and some of these places are not occupied, thus agents can move to these unoccupied places if they wish. The chequerboard city is shown in Figure 4.1, in where every '_' represents a possible place to live in.

The idea is to place two types of agents in this 'city' and see what happens if individuals have mildly discriminatory preferences (i.e. if they could tolerate a minority status, but would not want to live as an extreme minority). Let us suppose that (a) there are twenty-two *A*s and twenty-three *B*s: forty-five people in total; (b) agents are randomly distributed; and (c) 'each [agent] wants that more than one-third of his neighbours are like himself' (Schelling 1978: 148).

Figure 4.1 The chequerboard city.

In Figure 4.2 As and Bs respectively show the places occupied by *A*s and *B*s. Every agent will have at most eight neighbours (the cells surrounding them). Now consider what happens if every agent wants that more than one-third of his neighbours are like himself (Schelling 1978: 148). Given this specification, circled letters (As and Bs) in Figure 4.2 show the nine agents who are not satisfied with their place. Since there are some unoccupied places, residents who are discontent can move to places where they could be content. However, as they move they will change the state of their old neighbours. When an A moves to another place he reduces the number of *A*s for his previous neighbours, so the *A*s in those places are more likely to become discontent after he moves. If they are, they will try to move as well. As dissatisfied residents start moving, the residential distribution of *A*s and *B*s changes.

Figure 4.3 shows one of the possible results. It could be seen in Figure 4.3 that the process that was triggered by dissatisfied residents leads to residential segregation. Note that it is possible for each individual to move to a couple of places, however, though different moves will lead to different outcomes (i.e. different distributions of residents), the overall result – segregation – does not change. In Figure 4.3 all the individuals are content and they are more segregated. 'This is more than just visual impression', says Schelling. If you compare the number of neighbours that are the same type for each group and the average number of unlike neighbours – compare Figure 4.2 with Figure 4.3 – you will see that a lot has changed. Also, the number of individuals who have no neighbours of the other type has increased.

The surprising thing about this demonstration is the following: in Figure 4.2

_ A _ A B A _ B
A A A B _ B A B
_ A B _ _ A B A
_ B A B A B A B
B B B A B B B _
A _ A A A _ _ B
_ A B A B A B _
_ B _ B _ _ A _

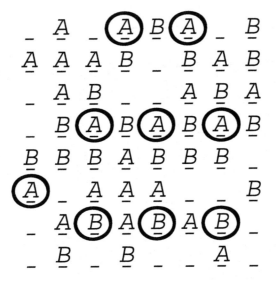

Figure 4.2 Randomly distributed residents.

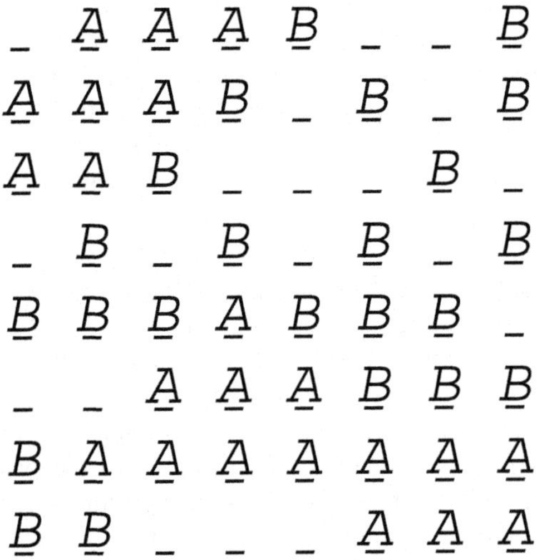

_ A A A B _ _ B
A A A B _ B _ B
A A B _ _ _ B _
_ B _ B _ B _ B
B B B A B B B _
_ _ A A A B B B
B A A A A A A A
B B _ _ _ A A A

Figure 4.3 Segregated city.

there were only nine out of forty-five people who were dissatisfied. Their decision to move to another place triggered a process that brought about residential segregation. The model shows that *even if* people in this representative city have mild preferences about the type of people living around them (e.g. every agent wants more than one-third of his neighbours to be like himself) and *even if* the number of dissatisfied agents is not more than the 20 per cent of the whole population, they might end up living in separate places. If the readers try the model on an actual chequerboard with different preferences of agents and with different initial distributions they can see that various combinations of different individual preferences and initial distributions will bring about the same result: more segregation.[4] But if they coincidentally start with a mixed neighbourhood where everybody is content, they will see that segregation will not occur. However, this is not a stable distribution. If you move a couple of individuals and make some of the residents unhappy, the process will take you to segregation instead of restoring the mixed neighbourhood.

Schelling argues that the 'instructive' thing 'about the *experiment* is the unrevealing process' (1978: 150, emphasis added). We can get a better understanding of this process by examining the constituent mechanisms of the model. As for the first simple model, we can define the mechanisms at the individual level as simple behavioural rules for each agent. Movements of the agents might be defined in terms of use of simple IF . . . THEN rules. We only have to assume that agents are capable of computing the fraction of the neighbours who are their own type (see Epstein and Axtell 1996: 165). If we call this fraction x, agents move according to the following rule: IF $1/3 \geq x$ THEN move. IF $1/3 < x$ THEN stay.

There are no other specifications about where to move. Agents may randomly move until they find a suitable place, or they may know where to move in advance, but this does not change the results of the model. *Interactions* of the agents are also defined according to these basic IF . . . THEN rules. These rules implicitly define the responses of the individuals to the existence of others. As some agents execute these rules, the states of some other agents (e.g. their previous and subsequent neighbours) change from 'content' to 'discontent'. This, in turn, causes further execution of IF . . . THEN rules – until every agent is content.

If we interpret the IF . . . THEN rules as the mechanisms at the individual level, the connection of these individual mechanisms can be interpreted as an aggregate mechanism (or a process). To define the mechanisms at the individual level, we have to define the states of the agents. Every individual could be content or discontent.[5] If the agent is content the mechanism is not triggered and the state of the individual does not change. But if the individual agent is discontent, the mechanism is triggered and the individual moves to another place – thus, the state of the agent changes to 'content'. The inputs are, then, the mixture of immediate neighbours and the preferences of the agent.[6] Because the mixture of the immediate neighbours is one of the inputs, every individual (and, thus, every mechanism) is connected to his immediate neighbours (to its neighbour mechanisms). If the agent moves, not only his own state changes, but also the *states of his previous and subsequent neighbours may change*. This network of mechanisms can be considered as an aggregate mechanism that is responsible for the transition between the different states of the city. The states of the city can be defined by the 'number of neighbours that are the same type for each group', by the 'average number of unlike neighbours' or by the 'number of people who have no neighbours of the other type'. Note that under this interpretation reinforcement is not a mechanism, rather, it is a property of the aggregate mechanism:[7] as individuals execute the rules and move, execution of the rules by other individuals is reinforced (also see Holland 1995). It should be noted that in this model there is no feedback from the state of the city, for the state of the city is not one of the inputs of the individual mechanisms.

Table 4.1 summarises our depiction of the mechanisms in the chequerboard model. The model is mainly based on the individual mechanisms that define the actions of the dispersed individuals and on the interactions between them. The actions of the agents are defined in terms of IF . . . THEN rules. IF . . . THEN rules also implicitly define the response of the individuals to the existence of others. Since there are as many mechanisms as the number of agents and they all react to the distribution of the other agents near them, we can think of the chequerboard city as a network with nodes connecting one agent to at most eight agents. Finally, we can think about the relation between the lower level mechanisms (rules for agents) and higher-level mechanisms, that is, the relation between the changes in individual states and changes in the state of the system as a whole. Under this interpretation basic individual mechanisms can be considered as the building blocks for the higher-level mechanisms (see Holland 1995). The interaction of the individual mechanisms (behavioural rules of the agents) constitutes a social mechanism that may as well be defined as a process, which transforms an integrated city into a segregated one. Thus, what we have here is an 'aggregate mechanism that which takes as "input" the dispersed actions of the participating individuals and produces as "output" the overall social pattern' (Ullmann-Margalit 1978: 270).

We may continue conjecturing by way of changing some of the assumptions of the chequerboard model. For example, we may ask whether the results change

Table 4.1 Mechanisms

		Mechanism	Input	Output	State
Level	I	IF . . . THEN rules	Preferences and the mixture of the immediate neighbours	Change in the states of the agent and agent's previous and subsequent immediate neighbours	Content or discontent
	S	Interaction of the individual mechanisms	The states of the individual mechanisms	Change in the state of the city	Degree of segregation

Notes
I: Individual level.
S: City level.

if individuals are concerned about the composition of their neighbourhood. Or, we may use heterogeneous agents who have different tolerance levels. These conjectures may help us analyse the conditions under which mild discriminatory preferences bring about segregation. In Chapter 7, we will consider some of the recent follow-ups to the chequerboard model and discuss the strength of Schelling's conjecture. Yet, at this step it is important to understand the nature of this conjecture.

Basically, Schelling conjectures about some of the mechanisms that may work at the individual level, investigates how they may interact (the aggregate mechanism), and whether or not they produce segregation. There is one chief reason for the selection of this methodology: while it is possible to observe the effects of an aggregate mechanism for a particular real city, it is not possible to find out what individual mechanisms might be at work from these observations, because the states of the city and effects of the aggregate mechanism give very little information about the underlying individual mechanisms. That is, the observation of the fact that a particular city is residentially segregated does not give us any information about the motivations and preferences of the agents. Moreover, several different types of micromotives may be responsible for residential segregation. For these reasons, Schelling uses the chequerboard model to get an idea about the types of individual mechanisms that may produce segregation. The next section examines these issues in detail.

Explanatory value

In order to examine the explanatory value of the chequerboard model we need to ask the following questions. First, is it necessary to conjecture about the individual mechanisms behind segregation and is this a good starting point for understanding particular cases of residential segregation? Second, are the results of the chequerboard model applicable to the real world? Does the similarity between the chequerboard model and the real world (if any) allow us to carry the results of the model to the real world? Third, does the chequerboard model contribute in any way to our understanding of ethnically segregated neighbourhoods in the real world? Let us start seeking answers to these questions.

Postulating mechanisms

In Figure 4.4 M-World *(t)* depicts the chequerboard model's initial state where there is no segregation. M-World *(t + n)* corresponds to the state of the chequerboard world with segregation. R-World *(t)* and R-World *(t + n)* illustrates the states of the real world (i.e. states of real world cities without and with segregation). Schelling demonstrates how M-World *(t)* can be transformed into M-World *(t + n)* through the interaction of individual mechanisms. The chequerboard model seems to be successful in showing this transformation for it contains a precise definition of the involved mechanisms. Remember that we have questioned the eligibility of Menger's explanation for the model world for it did not contain such explicit

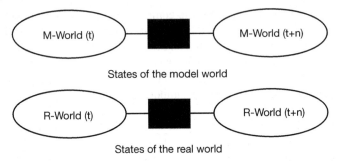

Figure 4.4 Model world and the real world.

mechanisms, only suggestions. Yet, although Schelling successfully demonstrates that even mild discriminatory preferences may lead to complete segregation in the chequerboard city, he does not explore all the possibilities in the model world. He does not argue, for example, that all initial distributions of agents (with different housing patterns and different number of agents) will lead to segregation in the chequerboard city, nor does he examine what would happen if the agents' conception of immediate neighbours is different. That is, it is not a logical necessity that mild discriminatory preferences bring about segregation under every condition in M-World *(t)*. Thus, he only shows the *possibility* that mild discriminatory preferences may cause segregation in the chequerboard city.

The reader may have immediately observed that it is not likely to find a R-World *(t)* where different ethnic groups are distributed randomly as it is in M-World *(t)*, because it is very unlikely that a real city becomes segregated starting from a point where different ethnic groups are randomly distributed. Rather, it is usually the case that a new ethnic group moves into the city at a point in time and the process of segregation gets started like this. As seen in Menger, M-World *(t)* rather represents an ideal starting point to investigate some of the mechanisms of segregation. The question is whether it is necessary to start from such an abstract world, or not.

The chequerboard model is an example of more general *critical-mass* models (Schelling 1978: 99). Critical-mass phenomena have the general property that people's behaviour depends on how many others are behaving in a specific way: 'What all the critical-mass models involve is some activity that is self-sustaining once the measure of that activity passes a certain minimum level' (Schelling 1978: 95). In fact, the chequerboard model is a *tipping model*, a subclass of these critical-mass models. 'Tipping is said to occur when some recognizable minority group in a neighbourhood reaches a size that motivates the other residents to begin leaving' (Schelling 1972: 157). When some *A*s start moving because they are unhappy, leaving some of the old residents unhappy, we can say that they are tipping out. Similarly, when some *B*s move in and make some of the *A*s unhappy enough to make them leave, they are tipping in. The tipping phenomenon was first described and illustrated by Morton Grodzins (1957) and Schelling examines the following question in his 1972 article: 'how do we recognize tipping when we see it?' As he argues, there are two options: one can either make a direct inquiry into the motives and expectations or look at the quantitative data on who moves in and who moves out. However, the first method gives us data – if it is reliable – about specific areas and it would be hard to find the mechanisms that can bring about segregation in this way.[8] The latter data is about aggregate behaviour, so if there are different mechanisms that can produce the same aggregate behaviour, it would not enable us to say something about these mechanisms. Moreover, because different people might have different tipping points (different preferences about living with the other group) it would be hard to find out a tipping point by looking at the aggregate data – because there might be none for a city (or a school, etc.) as a whole. If we are unable to see them from the aggregate data, how can we find these mechanisms? Schelling suggests that we should try different mechanisms and see if they work:[9]

> Rather than trying to infer from empirical data what mechanisms may be at work, we can postulate a mechanism and examine what results it would generate. If we can then verify the mechanism by the empirical identification of its components, we can use the model to explain and predict. Less ambitiously, we can compare the phenomena generated by the mechanism with what we observe, to see whether we can rule the mechanism out or establish its eligibility. Most likely of all, there may be some aspect of the mechanism that alerts us to certain phenomena, or helps to explain bits of what we observe, and sharpens the concepts that guide further research.
>
> (Schelling 1972: 159–160)

Finding out the causal mechanisms and structural relationships that produce the *explanandum* phenomenon is the key to a good explanation. The problem we sometimes face in our search for causal mechanisms is that of not being able to see it through the *explanandum* phenomenon. As early as 1843 Mill directed our attention to this problem with his distinction between 'chemically composed' causes and 'mechanically composed' causes. According to Mill, when causes combine 'chemically' the joint effect is not equal to the sum of the separate effects of the active causes.

> The chemical composition of two substances produces, as it is well known, a third substance with *properties entirely different* from those of either of the two substances separately, or of both of them taken together. [. . .] We are not, at least in the present state of our knowledge, able to foresee what result will follow from any new combination, until we tried it by specific experiment.
>
> (Mill 1843: 211, emphasis added)

In the case of chemical reaction, 'the separate effects cease entirely and are succeeded by phenomena altogether different, and governed with different laws' (Mill 1843: 254). Moreover, in this case it is not easy to find out the components (e.g. hydrogen and oxygen) of the resulting substance (e.g. water) by merely observing it. Thus, Mill thinks that an inquiry into the chemically composed causes should be experimental[10] for 'every new case, stands in need of a new set of observations and experiments' (Mill 1843: 144).

Some of the properties of residential segregation $(s_1, s_2, \ldots s_n)$ have no counterpart at the individual level, and they are not reducible to the properties of the dispersed individuals $(p_1, p_2, \ldots p_r)$. Moreover, aggregation of these individual properties does not give us the properties of residential segregation. The relation between $(s_1, s_2, \ldots s_n)$ and $(p_1, p_2, \ldots p_r)$ is somewhat similar to what Mill calls 'chemical composition' – although it would be totally misleading to think that segregation is similar to chemical substances, for it is governed by different causal mechanisms. If we look at the *properties* of the individuals, we cannot derive the *properties* of the aggregate phenomenon – and vice versa. But it does not follow from this that if we experiment with different combinations of individual agents (e.g. with different preferences), we cannot derive the properties of the resulting aggregate phenomenon. Thus, by experimenting with the individual mechanisms we may discover how the properties at the microlevel are connected to the properties at the macrolevel.

Mill's argument captures the very idea that we should try different combinations (experimental method, in Mill's terminology) to understand the phenomenon in hand. After all, if the system is complex and its properties do not provide enough information about the properties of its constituent parts, it is necessary to conduct experiments. Yet for the case of segregation, it is not easy to conduct an experiment in real cities. For this reason, Schelling states that 'rather than trying to infer from empirical data what mechanisms may be at work, we can postulate a mechanism and examine what results it would generate'. Schelling is proposing a different experiment from what Mill could have thought. He is suggesting a *thought experiment*. Instead of experimenting with real cities and individuals, Schelling invites us to experiment within the model world to see how different micromotives can lead to the same social pattern of residential segregation. If material experiments are not feasible, it is a good idea to try different set-ups (initial conditions, distribution of agents, etc.) and different mechanisms (different transition rules, preferences, etc.) to see which combinations are able to 'produce' the phenomenon in the model world. This is exactly what Schelling is doing. He shows us how properties of the dispersed individuals $(p_1, p_2, \ldots p_r)$ are linked to the properties of residential segregation $(s_1, s_2, \ldots s_n)$ by a social mechanism (by the network of individual IF . . . THEN rules), in the model world. Briefly, instead of examining the transformation of R-World *(t)* to R-World *(t + n)*, Schelling conjectures about the possible mechanisms that may transform M-World *(t)* into M-World *(t + n)*. Remember that Schelling asserts that 'most likely of all, there may be some aspect of the mechanism that alerts us to certain phenomena, or helps to explain bits of what we observe, and sharpens the concepts that guide further research'. He may be interpreted as conceiving these conjectures as a starting point for explaining real phenomena and gaining insights about the possible ways in which R-World *(t)* may be transformed into R-World *(t + n)*.

Sometimes, experimenting with different mechanisms (conjectures) may be a better starting point to understanding the nature of the phenomenon. Because of the complexity of the interactions among agents, conjecturing within the model world might be necessary to gain some knowledge, or at least insights about the real world. As with Menger's case, we may use the analogy of explaining the origin of life to see the nature of Schelling's contribution (see Chapter 3). The chequerboard model alerts us to a possible way in which certain individual mechanisms may interact and produce residential segregation. Stewart, some 200 years ago, defended a similar approach:

> In examining the history of mankind, as well as in examining the phenomena of material world, when we can not trace the process by which an event has been produced, it is often of importance to be able to show how it *may have been* produced by natural causes.
>
> (Stewart 1793 [1858]: 34, emphasis in original)

One of the most respected complexity theorists, John Holland, argues:

> To build a competent theory one needs deep insights, not only to select a productive, rigorous framework (a set of mechanisms and constraints on their interactions), but also to *conjecture* about theorems that might be true (conjectures, say, about lever points that allow large, controlled changes in aggregate behaviour through limited local action).
>
> (Holland 1998: 240, original italics deleted, emphasis added)

Yet how can we trust that these conjectures have any relevance for the real world? How can we jump from the mechanisms of segregation in the model world to the real causes of segregation?

Mechanisms and isolation

We have seen that Schelling thinks that it is not possible to discover some of the underlying causes of residential segregation by examining actual collective segregation – that is, by observing aggregate data concerning R-World $(t+n)$. The argument is that it is not possible to draw inferences about the preferences of individuals by means of examining actual collective segregation. For this reason, Schelling suggests conjecturing about the individual motives that may lead to segregation. Evidently, residential segregation may emerge because of a variety of reasons. First, a group can *organise* itself in a way that every member acts consciously to prevent mixed neighbourhoods, and/or to move into places where no member of the other group exists, and/or to prevent others from entering into the housing market of their neighbourhood. Second, different groups might have different welfare and different living standards, and this may cause segregation.
Schelling writes:

> Lines dividing the individually motivated, the collectively enforced, eco-
> nomically induced segregation are not clear lines at all. [. . .] They are further
> more not the only mechanisms of segregation. Separate or specialized com-
> munication systems – especially languages – can have a strong segregating
> influence.
>
> (Schelling 1978: 139)

Note that he acknowledges that there might be many mechanisms acting together to bring about residential segregation. However, Schelling *isolates* his model
from these factors and *focuses on unorganised discriminatory behaviour*. In other
words, the aim of his models is to find out the kind of unorganised discriminatory
behaviour that may lead to segregation.

Schelling (1972: 161) lists the following observations and insights concerning
segregation:

1 People live in cities which might have complex housing patterns and vaguely
   defined neighbourhoods.
2 People have *preferences* about their neighbours and sometimes about their
   neighbourhood.
3 People have *expectations* about their neighbours and sometimes about their
   neighbourhood. They also have expectations about the dynamics of their
   neighbours and sometimes about dynamics of their neighbourhood. This
   might affect the overall outcome.
4 People might move in and out at different *speeds* and this might affect the
   overall outcome.
5 There are *potential other entrants* to the city – e.g. new people from another
   place – and the population of the residents might change over time, which
   might affect the overall outcome.

However, the chequerboard model is isolated from most of these complexities
of real life. For example, the chequerboard model employs the following isolations:[11]

1′ People live in simple cities with no defined neighbourhoods.
2′ People have similar preferences about their neighbours, and they are not
    concerned about mixture of the neighbourhood, but the mixture of their
    immediate neighbours.
3′ People have no expectations.
4′ Speed is neglected.
5′ Potential other entrants are neglected.

Schelling focuses on one of the properties of the real individuals: that they may
have a range of discriminatory preferences – that is from none to strong discrimination – and he examines the type of discriminatory preferences that may lead
to residential segregation. If we assume that the agents in the model do not care

about the type of their neighbours, segregation does not occur unless the city is not already segregated. If they have strong discriminatory preferences – that is, if they want to be the majority in their neighbourhood – they get segregated. These results represent a possible state of affairs in the real world in a fair way. It seems to be true that strong discriminatory preferences would bring about segregation. Or, if individuals do not care about the type of their neighbours they would not be segregated given that there is no other reason (e.g. economic) that would separate them. Schelling's chequerboard model points out another possibility; that agents with mild discriminatory preferences may cause segregation.

We may start assessing the chequerboard model by asking whether mild discriminatory preferences exist in the real world. We know that some individuals tend to avoid a minority status and need to belong to a certain group. It may even be argued that this tendency has some evolutionary roots. For example, it is commonly argued that the need to belong and tendency to live among a group increases the chances of survival (Baumeister and Leary 1995; Alexander 1974; Barchas 1986). Moreover, it has been argued that many individuals prefer to associate themselves with what they consider to be their own kind (i.e. homophily[12]) (Bowles and Sethi 2006; McPherson, Smith-Lovin and Cook 2001; Tajfel *et al.* 1971). In brief, we may be pretty confident that mild discriminatory preferences exist in the real world. But do they cause segregation?

Sugden (2000: 23) argues that if we are to make inferences from the model world to the real world we must recognise some significant similarity between the model world and the real world. Schelling's M-World *(t)* does not represent any real city in a faithful manner (more on this in Chapter 7). However, the individual mechanisms depicted in the model seem to represent real world tendencies. We do not know if this is a significant similarity but individual mechanisms that embody mild discriminatory preferences are similar to real world mechanisms. Maybe the only aspect of the chequerboard model that is familiar to us (i.e. represented in the model) is these individual mechanisms. In fact, it is due to the familiarity of the individual mechanisms presented in the model that we tend to think what happens in the chequerboard city may happen in the real world. This seems to be the only reason why we think that the possible ways in which the individual mechanisms interact in the model may be considered as possibilities for the real world.

The similarity between the chequerboard model and the real world is limited to some familiar individual mechanisms in isolation from others. We know that these mechanisms exist. We know that individuals have a tendency to avoid an extreme minority status. The chequerboard model alerts us to a possible aggregate mechanism: a possible way in which those individual mechanisms may interact in bringing about residential segregation. It is because of our knowledge of, and our familiarity with, these individual mechanisms that we consider the chequerboard model similar to the real world. The novelty of the chequerboard model comes from showing how these individual mechanisms may interact. Yet we cannot accept the model, merely because it represents certain tendencies that we know about. The chequerboard model may be plausible and interesting, yet it does not tell us whether mild discriminatory preferences bring about segregation in the real world. It is just a thought experiment to illustrate the plausibility of this hypothesis.

Sugden (2000: 25) has one more suggestion about credibility: 'Credibility in models is, I think, rather like credibility in "realistic" novels.' Sugden is right here, at least to the extent that the model gives the account of the *successive stages* to explain the generation of the phenomenon in hand. Especially if the model is trying to explain the emergence or the origin of a phenomenon, credibility of a realistic novel might be required from an explanation. Gallie gives a very nice account and an example of this:

> To follow a story – or a conversation, or a game, or the development and execution of a policy – involves for one thing some vague appreciation of its drift or direction, a vague sense of its alternative possible outcomes: but much more important for our purpose, it involves a relatively clear appreciation of certain relations of dependence of the sort that characteristically historical explanations serve to articulate [. . .] Consider, e.g., what we do when a child complains that he cannot follow the story we read aloud to him [. . .]. We re-read to the child the earlier stages of the story, or re-tell them in simpler language so as to emphasize those incidents which give sense or context to the present, puzzling episode. But in doing this we do not try to show that the present episode was a predictable consequence of earlier events, else the story would have been not un-followable, but unbearably dull as a story.
>
> (Gallie 1955: 395)

If we consider Schelling's account of the emergence of segregation as a story, we should appreciate its full credibility because simple but familiar behavioural rules bring about a surprising result: residential segregation emerges because of people who are trying to avoid a minority status. Moreover, successive stages of the story are clear and comprehensible, but not dull. Moreover, it encourages us to conjecture about other scenarios in order to see other possibilities. That is, we may grant Schelling's model the status of a good story. This, of course, adds to its credibility but the main reason why we feel that the story may have some relevance for the real world is our familiarity with the suggested mechanisms. We conceive the states of the chequerboard model as possible states of the real world because of this familiarity. It is, in this sense, that Schelling's explanation is credible like a realistic novel. Moreover, because it examines the interaction of some known mechanisms, it is more than a 'conceptual exploration' (Sugden 2000: 11, cf. Hausman 1992). Despite its deficiencies, it is a theoretical explanation that alerts us to certain possibilities in the real world. More properly, it is a partial potential theoretical explanation.

Explanatory breath

Note that some authors suggest that 'Schelling is presenting a critique of a commonly held view that segregation must be the product either of deliberate public policy or of strongly segregationist [i.e., discriminatory] preferences' (Sugden 2000: 9). Of course, Schelling's model might be able to convince us about the weakness of the explanations that explain segregation either by organised action or by strong discriminatory preferences. However, Schelling thinks that there might be several different causes of residential segregation. But he does not focus on these aspects of segregation, rather he shows another possibility. In fact, the chequerboard model does not seem to contradict other theories about residential segregation; it is consistent and coherent with the existing body of knowledge about residential segregation. To see this, let us assume that we have a meta-model or theory (i.e. a collection of models) of residential segregation that combines Schelling's model and the other models (or explanations) of residential segregation. According to this meta-model, near to complete segregation will emerge if the following conditions exist separately or in different combinations:

1 If agents have strong (or milder) discriminatory preferences and if they collectively or separately intend to prevent a mixed neighbourhood.
2 If agents have strong (or milder) discriminatory preferences (about the neighbourhood or about their immediate neighbours) and if they intend to live in a place where they can be content, but have no intention to change the mixture of the neighbourhood.
3 If there are other forces (e.g. economic) preventing the two different groups to live in (move to, etc.) close places.

It is generally accepted that other things being equal, we should prefer a model that explains more than the alternative hypotheses (e.g. Thagard 1992: 74). Consider the meta-model before Schelling. It asserts that strong discriminatory preferences, organised action and economic processes are the main causes of residential segregation. Schelling's contributions change the existing meta-model by adding one more explanatory factor to it. Thus, Schelling's contribution improves the explanatory breadth of the meta-model. Or to put it differently, the new meta-model has more applicability. Schelling's model extends our understanding of residential segregation and gives us extra tools to explain particular instances of residential segregation. Thus, we may interpret Schelling as stating that 'if you want to explain residential segregation in Rotterdam, you should search for organised action, economic factors (such as welfare differences among different ethnic groups) and mild segregationist preferences. Then you should use the appropriate tools to see whether any of these causes exist in Rotterdam.' The good thing about Schelling's model is that it makes us aware of the fact that any of these causes (or any combination of these causes) may lead to residential segregation. The chequerboard model does not readily improve our understanding of particular cases of segregation, yet opens our eyes to a new explanatory factor which may explain segregation. Moreover, even in cases where economic factors and/ or organised action are the main causes of segregation, the mechanisms proposed by Schelling may have some relevance. For example, suppose that individuals with strong discriminatory preferences are organised in a way to prevent mixed neighbourhoods. Some *A*s are intentionally forming isolated neighbourhoods. Yet not everyone in the city would be likely to join this organised action. Some of them would have weaker discriminatory preferences. Yet when the number of *A*s in their neighbourhood decreases to a level they would not tolerate, they may consider moving out. In this example, the mechanisms proposed by Schelling are not the main explanatory factors in explaining the resulting residential segregation, yet if we can confirm the existence of these mechanisms they would provide a deeper understanding of this case.

To understand how Schelling's models changed our understanding, we can also think in the following way: previous theories of segregation assumed a linear relation between segregation and the strength of the discriminatory preferences. Schelling argued that the relation is not linear. For example, up to a point within the range of possible preferences – from no discriminatory preferences to mild discriminatory preferences (e.g. tolerant to 25 per cent minority status) – we do not observe segregation. However, after that point – from mild discriminatory preferences to very strong ones – segregation emerges, that is, there is a transition.[13] Thus, Schelling's models, without any contradictory statements about the causes of segregation, improves upon previous models by incorporating a wider range of possible types of preferences. [14]

Concluding remarks

Many authors have stressed the importance of specifying the social mechanisms in an explanation. For example, Hedström and Swedberg (1998: 1) argues that 'the advancement of social theory calls for an analytical approach that systematically seeks to explicate the social mechanisms that generate and explain observed associations between events'. In *Social Mechanisms, An Analytical Approach to Social Theory* (edited by Hedström and Swedberg), many authors (including Jon Elster, Gudmund Hernes, Diego Gambeta and Tyler Cowen) emphasise the explanatory value of specifying social mechanisms in a model. It is possible to find different but compatible descriptions of social mechanisms in the literature. Yet, although these definitions provide a good starting point for social scientists, it is still unclear what these mechanisms look like. In this chapter we have explicated the individual mechanisms in Schelling's explanation by interpreting them as IF . . . THEN rules. The major advantage of this interpretation is that it lets us see clearly how individual mechanisms interact and constitute a more complex (higher-level) mechanism. Under this interpretation a process can be considered as an aggregate mechanism that embodies the interactions of its constituent mechanisms. It is important to note here that interpretation of mechanisms as IF . . . THEN rules does not imply that real mechanisms are IF . . . THEN rules. Rather, such an interpretation is introduced because it helps us see how individual mechanisms are related to each other in bringing about an aggregate consequence.

By way of showing how mechanisms at the individual level interact with each other, Schelling's model expands our mental horizon: it alerts us to new possibilities that may be relevant in explaining particular cases of residential segregation.

Yet the model only focuses on some of the factors that may bring about residential segregation. It leaves out other factors such as welfare differences and intentional organised action. In addition to this, Schelling does not try to explain any particular case of segregation. The suggested explanation is partial for it focuses on some of the explanatory factors and potential for we do not know whether the suggested mechanisms are effective in particular cases of segregation. By providing a partial potential explanation of segregation, he tries to improve upon the existing models and explanations of segregation. Since it does not contradict the existing models, he may be interpreted as improving the meta-model (or theory) of residential segregation by way of showing us new possibilities. Yet, on the negative side, the effectiveness of the suggested mechanisms has to be established to explain particular cases of segregation.

Both Schelling's chequerboard model and Menger's story of the origin of money are paradigmatic examples of invisible-hand explanations. The next chapter examines the notions of 'invisible hand' and 'invisible-hand explanations'. Then, Chapter 6 demonstrates how Menger's explanation of the emergence of a medium of exchange is further explored in the contemporary literature. In Chapter 7 we will come back to the issue of abstract models and how they may expand our mental horizon. The case of residential segregation and recent explorations of Schelling's models are also discussed in Chapter 7.

5 The invisible hand

Introduction

Thus far, we have examined the concept of 'unintended consequences' and two paradigmatic examples of invisible-hand explanations. We have seen that both Menger's explanation of the origin of money and Schelling's explanation of residential segregation are partial potential theoretical explanations that illustrate some of the possible ways in which certain mechanisms may interact and bring about the unintended social phenomenon under consideration. Our analysis of these paradigmatic examples of invisible-hand explanations did not focus explicitly on the concept of invisible hand. The aim of this chapter is to provide a better understanding of the concept of invisible hand and its relation to 'unintended consequences' and invisible-hand explanations.

The notions of 'invisible hand' and 'invisible-hand explanations' are closely related with 'unintended consequences', yet their relation is not clear at all. In fact, there is a general misunderstanding about their relation, which takes it for granted that these concepts presume that individuals are 'blind' in that they cannot see the consequences of their action. Moreover, some argue that providing an invisible-hand explanation presupposes a scientist who knows better and who sees more than any individual can. Similar ideas are entertained in E. Rothschild's (2001) *Economic Sentiments*. Based on these ideas, she argues that the concept of 'invisible hand' is not consistent with Smith's thoughts. This chapter critically discusses Rothschild's argument to provide a better understanding of the relation between 'unintended consequences' and the 'invisible hand', which is important for an appropriate understanding of the contemporary models and explanations of the origin (and persistence) of unintended social phenomena.

The chapter also discusses Smith's thoughts about philosophy (and philosophy of science) in order to clarify the intended meaning of the concept of invisible hand: the invisible hand is a metaphorical statement of the way in which natural and social phenomena should be explained. Smith uses the concept to imply the connecting principles of nature and society that should be explicated to explain natural and social phenomena. Accordingly, it is argued that invisible-hand explanations should be considered as explanations that explicate the mechanisms that may bring about unintended consequences.

Furthermore, the chapter identifies two interpretations of the invisible hand: (a) end-state interpretation, which is generally associated with the general and partial equilibrium models; and (b) process interpretation, which is associated with Austrian economists such as Hayek, and contemporary (evolutionary) models of institutions that utilise the resources of game theory. It is sometimes argued that the end-state interpretation does not make justice to the original metaphor of Smith, for Smith has a process conception in mind. Although this argument is correct, it may prevent us from seeing the connection between these two interpretations. In this chapter, we introduce ideas about how they may be related, and in the next chapter it is shown that the models that subscribe to the end-state interpretation can be considered as particular ways in which the conjectures of invisible-hand explanations may be tested. Another important aspect in this discussion is the emphasis on the conjectural character of invisible-hand explanations. Economists such as Tobin consider modern equilibrium models of economics as tests of Smith's conjectures. On the other hand, Austrian economists seem to follow the tradition of conjectural history in explaining unintended consequences.

The plan of the chapter is as follows: first, the chapter clarifies the notion of 'invisible hand' by means of going back to the original statements of Adam Smith and removes the previously cited misunderstanding about Smith's invisible hand. Second, the relation between modern conceptions of the invisible hand and Smith's conception is examined. Finally, the notion of invisible-hand explanations is explicated.

Smith's invisible hand

Adam Smith's 'invisible hand' is widely used and discussed in economics and other social sciences, as well as in language theories, philosophy of science, ethics, political theory and active politics.[1] Although Smith's 'invisible hand' is considered to be an influential metaphor, he uses the phrase only three times and in different contexts. In 'The principles which lead and direct philosophical enquiries: illustrated by the history of astronomy' (henceforth HA) he refers to those individuals who ascribe the 'irregular events of nature to the agency and power of their gods' (Smith 1795: 49).

> Fire burns, and water refreshes; heavy bodies descent, and lighter substances fly upwards, by the necessity of their own nature; nor was the *invisible hand of Jupiter* ever apprehended to be employed in those matters. But the thunder and lightening, storms and sunshine, those more irregular events, were ascribed to his favour, or his anger.
>
> (Smith 1795: 49, emphasis added)

In *The Theory of Moral Sentiments* (henceforth TMS) he invokes the 'invisible hand' when he tries to show how the selfish behaviour of the rich (in combination with natural forces) 'advance[s] the interest of the society, and afford[s] means to the multiplication of the species' (Smith 1790: IV.I.10):

The rich only select from the heap what is most precious and agreeable. They consume little more than the poor, and in spite of their natural selfishness and rapacity, though they mean only their own conveniency, though the sole end which they propose from the labours of all the thousands whom they employ, be the gratification of their own vain and insatiable desires, they divide with the poor the produce of all their improvements. They are *led by an invisible hand* to make nearly the same distribution of the necessaries of life, which would have been made, had the earth been divided into equal portions among all its inhabitants, and thus without intending it, without knowing it, advance the interest of the society, and afford means to the multiplication of the species.

(Smith 1790: IV.I.10, emphasis added)

In *The Wealth of Nations* (henceforth *WN*) he uses it when he tries to show how merchants support the public interest when they intend to increase their security 'by preferring the support of domestic to that of foreign industry' (Smith 1789: IV.2.9):

By preferring the support of domestic to that of foreign industry, he intends only his own security; and by directing that industry in such a manner as its produce may be of the greatest value, he intends only his own gain, and he is in this, as in many other cases, *led by an invisible hand* to promote an end which was no part of his intention. Nor is it always the worse for the society that it was no part of it. By pursuing his own interest he frequently promotes that of the society more effectually than when he really intends to promote it.

(Smith 1789: IV.2.9, emphasis added)

The fact that Smith uses the invisible hand in three different contexts makes it hard to understand the implied meaning of the phrase. Some argue that there is no conflict in Smith's uses of the invisible hand (e.g. Thornton).[2] Some argue that although they have been used differently there is no inconsistency, but the role of the invisible hand was reversed from HA to *TMS* and *WN* (e.g. Macfie 1971: 596). Some argue that the invisible hand refers to the hand of God (e.g. Denis 1999) or that it is about the wisdom of nature (e.g. Khalil 2000b). Some others argue that the invisible hand is not necessarily providential (e.g. Flew 1987). Some say that we should only be concerned with the context within which the phrase is used and if we do this, the invisible hand in *WN* is simply about import duties (e.g. Persky 1989).[3] Rothschild (1994, 2001), on the other hand, argues that Smith was sardonic in his use of 'invisible hand' and that the concept of the invisible hand is not consistent with Smith's system of thought. Briefly, on the side of the historians of thought, there is no consensus about the interpretation of Smith's invisible hands.[4]

In what follows we will examine the relation between Smith's invisible hands. We will see that the use of the invisible hand in HA provides the methodological background on which we may examine its other uses in *WN* and *TMS*. We will also see that Smith's use of the invisible hand appears to be somewhat ironic, as Rothschild suggests. Yet it does not follow from this that the concept of invisible hand is not consistent with Smith's system of thought. Rothschild's interpretation embodies a misunderstanding concerning the relation between the concepts of invisible hand and unintended consequences. Since it is important to eliminate this misunderstanding we will use Rothschild's (1994, 2001) interpretation as a case in point.[5]

Rothschild (2001) solves the apparent inconsistency in Smith's use of the invisible hand by arguing that the invisible hand was indeed an ironical joke, and she concludes that it is *un-Smithian*. To demonstrate this, Rothschild puts forward several pieces of evidence to show that the idea of invisible hand does not fit Smith's general framework, and that Smith would not have favoured such an argument. Yet she (2001: 122) starts her discussion with an interpretation of the invisible hand that Smith would have favoured. It is argued that the invisible hand provides such an understanding about the economic or social order that it beauti-fully connects the parts of the system to the orderly state of the socio-economic world without the need of invoking a designer who is responsible for this order. Although Rothschild argues that such an interpretation would be supported by Smith, she goes on to argue that the implications of the invisible hand suggest that Smith would not have favoured the invisible hand.

Throughout her argument she takes it as given that the invisible hand is related to three main arguments (also see Vaughn 1987):

1 'that the actions of individuals have unintended consequences;
2 that there is order or coherence in events; and
3 that the unintended consequences of individual action sometimes promote the interests of societies' (Rothschild 2001: 121).

First, Rothschild suggests that the idea of individuals who are not able to see the overall picture and who are acting blindly conflicts with Smith's overall thought. Rothschild (2001: 124) formulates this in the following way: 'to be contemptuous of individual intentions, to see them as futile and blind, is to take a distinctively un-Smithian view of human life.' Second, Rothschild (2001: 124) proposes that because the invisible hand 'is founded on a notion of privileged universal knowl-edge' and 'it presupposes the existence of a theorist [. . .] who sees more that any ordinary individual can', it is un-Smithian. Third, Rothschild (2001: 126–128) presents Smith's proposal in *WN* that merchants should *not* seek their individual interests by political means (particularly by supporting restrictions on imports) as being conflicting with the idea that they would promote the public good by pursuing their self-interests. Fourth, Rothschild (2001: 129–130) suggests that religious connotations of the invisible hand do not go well with Smith's some-what irreligious views. Finally, she (2001: 131–132) argues that the Stoic idea of a *providential* order, which is implied by the invisible hand, conflicts with Smith's general views. It is argued here that Rothschild's first three propositions are wrong because her account contains a misunderstanding about the invisible hand and its relation to unintended consequences.[6] The truths of fourth and fifth propositions heavily depend on Smith's ideas concerning religion on which there is no consensus, yet they are debatable. Our main focus will be on the first three propositions. Whether Smith was ironic in his use of the invisible hand is another matter and Rothschild seems to be right in suggesting that he was. But we will see that the irony in the invisible hand directs us to the philosophy of science behind the invisible hand, it does not suggest that the invisible hand is un-Smithian. Let us go back to the writings of Smith in order to discuss Rothschild's propositions and to develop a better understanding of the invisible hand.

History of astronomy

Consider the paragraph where the invisible hand appears in HA:

> Fire burns, and water refreshes; heavy bodies descent, and lighter substances fly upwards, by the necessity of their own nature; nor was the *invisible hand of Jupiter* ever apprehended to be employed in those matters. But the thunder and lightening, storms and sunshine, those more irregular events, were ascribed to his favour, or his anger.
>
> (Smith 1795: 49, emphasis added)

Smith's use of the invisible hand in HA seems to be radically different from his uses in *WN* and *TMS*. For this reason, we need to examine the context in which he uses it in order to understand how it relates to the invisible hands in *WN* and *TMS*. In HA Smith uses the phrase 'invisible hand of Jupiter' to argue that in the very early stages of the society people used to explain irregular events as the acts of invisible beings such as gods. He states that in those days people had 'little curiosity to find out the *hidden chains of events* which bind together the seemingly disjoined appearances of nature' (Smith 1795: 48, emphasis added). He argues that in the first ages of society individuals would consider the regular and usual acts of nature as given and in need of no explanation, but they would explain the irregular events with reference to the acts of gods.

> With him, therefore, every object of nature, which by its beauty or greatness, its utility or hurtfulness, is considerable enough to attract his attention, and whose operations are not perfectly regular, is supposed to act by the direction of some *invisible and designing power*.
>
> (Smith 1795: 48, emphasis added)

Smith thinks that this behaviour is 'the origin of Polytheism and vulgar superstition which ascribes all the irregular events of nature to the favour and displeasure of intelligent, though invisible beings, to gods, daemons, witches, genii, fairies' (Smith 1795: 48). It is in this context that Smith uses the phrase 'invisible hand of Jupiter'.[7] So according to Smith, savage man would not think about the acts of

Jupiter when he observes the regular events of nature, rather he would explain the apparently irregular events with the invisible hand of Jupiter.

Since some authors interpret the invisible hand in *WN* and *TMS* as the hand of God (e.g. Denis 1999) or associate the concept with some sort of deity (e.g. Davis 1989: 65) it is important here to note that Smith does not approve the explanatory strategy used by the savage man. These individuals failed to see the connecting chains of nature and tried to explain some natural phenomena as the consequences of the actions of invisible and powerful beings. Smith suggests that in order to understand nature one has to search for these apparently invisible chains of connecting events. Thus, it is highly improbable that he would adopt a strategy which is similar to that of savage man in his other works. In fact, Smith is very clear about what he considers to be the proper explanatory strategy. He argues that it is the task of philosophy to explicate the apparently invisible chains of nature.

Smith suggests that with the development of society and specialisation some of the individuals in the society had the security and time to investigate these causes. These individuals became 'less disposed to employ, for this connecting chain, those invisible beings whom the fear and ignorance of their rude forefathers had engendered' (Smith 1795: 50). Strikingly, a similar argument appears in *WN*:

> The great phenomena of nature, the revolutions of the heavenly bodies, eclipses, comets; thunder, lightning, and other extraordinary meteors; the generation, the life, growth, and dissolution of plants and animals; are objects which, as they necessarily excite the wonder, so they naturally call forth the curiosity, of mankind to inquire into their causes. *Superstition* first attempted to satisfy this curiosity, by referring all those wonderful appearances to the immediate agency of the gods. *Philosophy afterwards endeavoured to account for them from more familiar causes*, or from such as mankind were better acquainted with, than the agency of the gods.
>
> (Smith 1789: V.1.152, emphasis added)

Philosophy, according to Smith, 'is the science of the connecting principles of nature' (Smith 1795: 45):

> Philosophy, by representing the *invisible chains* which bind together all these disjoined objects, endeavours to introduce order into this chaos of jarring and discordant appearances, to allay this tumult of the imagination, and to restore it.
>
> (Smith 1795: 45–46, emphasis added)

The proper explanatory strategy for the philosopher (i.e. scientist) is to uncover the apparently invisible chains that connect phenomena, or to replace the 'invisible hand of Jupiter' with connecting principles of nature.[8] Thus, Smith could not have used the invisible hand in *WN* and *TMS* to imply the work of some invisible power or some sort of god. Rather, the invisible hand in *WN* and *TMS* is better understood as a place holder for the connecting principles of nature which have been already explicated by Smith. In HA Smith criticises the use of invisible powers in science but he uses the phrase 'invisible hand' in *WN* and *TMS*. The invisible hand in *WN* and *TMS* is ironic. It is used to express what Smith seems to consider himself to have accomplished; making explicit the apparently invisible chains that connect social phenomena.

In effect, the discussion of the invisible hand of Jupiter in HA provides the methodological background for the use of the invisible hand in *WN* and *TMS*. For this reason it is useful to have a better understanding of Smith's philosophy of science. Smith approaches the questions about understanding nature from a cognitive perspective. He argues that when we see two distant phenomena that seem to be somehow related, our imagination feels uncomfortable and tries to fill in the gap between these phenomena. As the savage man used to fill in the gap by imagining the acts of invisible beings, philosophers fill in the gap by explaining them with more familiar causes, and by trying to find out the chain of events that connects these phenomena, which were invisible to us at first sight.

> [Imagination] endeavours to find out something which may fill in the gap, which, like a bridge, may so far at least unite those seemingly distant objects, as to render the passage of the thought betwixt them smooth, and natural, and easy. The supposition of a chain of intermediate, though invisible, events, which succeed each other in a train similar to that in which the imagination has been accustomed to move, and which link together those two disjointed appearances, is the only means by which, if one may say so, can smooth its passage from the one object to the other.
>
> (Smith 1795: 41–42)

Smith discusses the history of astronomy to demonstrate these points, and to show the several ways in which philosophers tried to discover the connecting principles of celestial appearances. HA is an essay where Smith tries to demonstrate the validity of his arguments about imagination and of his basic argument that wonder, surprise and admiration are the main sentiments behind scientific discovery. In the essay, he tries to abstract from the relation between the several models, which he calls systems, of astronomy and reality. He merely wants to show how these models were created to 'sooth the imagination':

> without regarding their absurdity or probability, their agreement or inconsistency with truth and reality, let us consider them only in that particular point of view which belongs to our subject; and content ourselves with inquiring how far each of them was fitted to sooth the imagination, and render the theatre of nature more coherent.
>
> (Smith 1795: 46)

At the end of his essay (1795: 104–105), he argues that Newton's system is the most successful system in the history of astronomy with respect to soothing our imagination. Yet he cannot resist adding a couple of more comments about the relation of Newton's system to the real world. First of all, he argues that in addition to its coherence, its power of unification, and explanatory breath, Newton's system explains the most distant objects with the 'most familiar' and known property: the gravity of matter. He states that 'we never act upon it without having occasion to observe this property' (Smith 1795: 104). Thus, according to Smith, Newton's system connects with what is real at least at a basic level. Moreover, he also appreciates the explanatory and predictive power of Newton's system which implies that Newton's system may be true:[9]

> They [Newton's principles] not only connect together most perfectly all the phaenomena of the Heavens, which had been observed before his time, but those also which the preserving industry and more perfect instruments of later Astronomers have made known to us; have been either easily and immediately explained by the application of his principles, or have been explained in consequence of more laborious and accurate calculations from these principles, than had been instituted before.
>
> (Smith 1795: 105)

Smith then asks the reader (and himself) whether Newton's theory may be considered as being true about the real world:

> And even we, while we have been endeavouring to represent all philosophical systems as mere inventions of the imagination [. . .] have been drawn in, to make use of language expressing the connecting principles of this one [i.e., of Newton's system] as if they were the real chains which Nature makes use of to bind together her several operations. Can we wonder then, [. . .] that it should now be considered, not as an attempt to connect in the imagination the phaenomena of the Heavens, but as the greatest discovery that ever was made by been and sublime truths, all closely connected together, by one capital fact, of the reality of which we have daily experience.
>
> (Smith 1795: 105)

Since Smith is not conclusive, we cannot have a decisive account of Smith's 'philosophy of science', but given these comments we can speculate about two possibilities. The first possibility is that he has an account of scientific theories that considers them as 'mere inventions of imagination', or as systems that helps us to 'save the observed phenomena', which do not have to be true or false.[10] Thus, they are simply conjectures. The second possibility is that Smith indeed thinks that scientific systems (models, theories) are quests for understanding real relations in nature, but also that we can never be exactly sure about the truth of our theories (see Thomson 1965). Thus, since there is no guarantee of truth, they are conjectures about what may be real. Indeed, Smith's comments about Newton's system suggest the second minimal realist reading. Of course, he may have entertained both of these views, in a sense that the first applies to natural philosophy and the later applies to moral philosophy:[11]

> A system of natural philosophy may appear very plausible, and be for a long time very generally received in the world, and yet have no foundation in nature, nor any sort of resemblance to the truth.[12] But it is otherwise with systems of moral philosophy, and an author who pretends to account for the origin of our moral sentiments, cannot deceive us so grossly, nor depart so very far from all resemblance to the truth.
>
> (Smith 1790: VII.II.106)

Whatever the type of realism he may have entertained, Smith considers philosophical systems (i.e. models, theories) as being somewhat similar to thought experiments:

> Systems in many respects resemble to machines. A machine is a little system, created to perform, as well as to connect together, in reality those different movement and effects which the artist has occasion for. A system is an imaginary machine invented to connect together in the fancy those different movements and effects which are already in reality performed.
>
> (Smith 1795: 66)

A philosophical system is similar to a machine in that the machines, as man-made systems, connect the acting forces of nature; theories and models, on the other hand, connect the forces of nature in our fancy, or in our thoughts. Smith believed that philosophy tries to find out the connecting principles of nature; that it is a quest for a more coherent view of nature; and that instead of powerful and intelligent beings (such as Jupiter) philosophy attempts to explicate the connecting principles of nature. Moreover, this attempt involves conjectures concerning these principles. This is the context within which we should understand his use of the invisible hand in *TMS* and *WN*.

Smith is a philosopher, and considers himself as a philosopher, whose task is to conjecture about the connecting principles of nature and society, to create a coherent body of thought that would render it more easy to our imagination how the nature and causes of the wealth of nations, as well as the basic sentiments and dispositions of man, are related to each other. Smith, both in *TMS* and in *WN*, is at pains to show how things are connected to each other. In *TMS* he tries to explain how the self-regarding actions of the rich may work for society as a whole, despite the fact that the land is unevenly distributed. In *WN* he tries to show why and how, without import restrictions, society may be better off by virtue of the interaction between the self-regarding actions of individuals. In these texts he indeed tries to show how the actions of the individuals (and additionally in *TMS,* that of nature) work for the good of society, although they are acting self-regardingly. He tries to show how two apparently distinct things, self-interested action and beneficial social consequences, are connected to each other.[13] He tries to provide those connecting principles of the society that at first glance were invisible. Smith is at pains to show how people who are following their own interests (intentions targeted at the individual level) bring about unintended social consequences. It is the interaction of these familiar individual mechanisms (i.e. individuals pursuing their own interests) that bring about unintended social consequences. In *WN* and *TMS* there is nothing invisible in the invisible hand. Thus, the phrase that 'individuals are led by an invisible hand to promote an end which was no part of their intention' may indeed be considered as an ironical statement, but not in a sense that conflicts with Smith's own views. Or rather, it can be read as a metaphorical statement that implies the explication of some of the connecting principles of the society (also see Evensky 2004[14]). In HA the invisible hand is the invisible hand of Jupiter, which is called upon by the superstitious savage man. In *TMS* and *WN* it indicates the explication of some of the apparently invisible forces in society by a philosopher: Adam Smith. Briefly, from the point of view of Smith's ideas about philosophy, there seems to be nothing about the invisible hand that is un-Smithian. But this does not yet answer Rothschild's concerns. We should now inquire into the relation between the invisible hand and unintended consequences.

The invisible hand and unintended consequences

Does the fact that Smith refers to individuals who are not aware of the future consequences of their action, and who fail to see the invisible hand, make the invisible hand an un-Smithian idea? Rothschild thinks so. She (2001: 123) argues that the word 'invisible' implies blindness[15] and points out that Smith 'sees the people as the best judges of their interest [. . .] But the subjects of invisible-hand explanation are blind, in that they cannot see the hand by which they are led.' Thus, she concludes: the invisible hand cannot be a truly Smithian idea.

A certain type of 'blindness' may be identified in the argument against import regulations in *WN*[16] in two different forms. First, it is argued that those who try to implement the import regulations cannot judge the interests of the individuals. They cannot observe their interests and the peculiarities of their individual situation. These are invisible to the regulators.

> What is the species of domestic industry which his capital can employ, and of which the produce is likely to be of the greatest value, every individual, it is evident, can, in his local situation, judge much better than any statesman or lawgiver can do for him.
>
> (Smith 1789: IV.2.10)

In fact, a similar argument appears in *TMS*, where Smith talks about a legislator who wishes to rule a society:

> He seems to imagine that he can arrange the different members of a great society with as much ease as the hand arranges the different pieces upon a chess-board. He does not consider that the pieces upon the chess-board have no other principle of motion besides that which the hand impresses upon them; but that, in the great chess-board of human society, every single piece has a principle of motion of its own, altogether different from that which the legislature might choose to impress upon it.
>
> (Smith 1790: VI.II.42)

That is, no individual can know what is good for all the others, and since one is 'blind' to the principles of the motion of other individuals, it is better to let individuals judge on their own what is good for them. We may add to this that since the exact response of the individuals to a regulation cannot be known in advance, the legislator would also be 'blind' to the future consequences of his regulation. The second form of 'blindness' is the 'blindness' of the individuals who do not intend to bring about social consequences. As the legislator, any individual is 'blind' to the decisions taken by the rest of the individuals that may influence the consequences of his action. They may also be 'blind' to some of the other factors that may influence the consequence of their action. These two forms of 'blindness' are essentially similar.[17] 'Blindness' is attributed to all the individuals in the society; to merchants as well as to legislators, tailors, shoemakers, etc. The legislator cannot judge for the individuals, and any individual judges better for himself as long as he is not intending to bring about social consequences. Individuals are 'blind' to the social consequences of their action, but concerning their own interests and their local environment[18] they know better than others.[19] In the terminology employed in Chapter 2, Smith argues that it is good for the society when each and every individual intends to bring about consequences at the individual level (at least for the cases in which he employs the invisible hand). He assumes that when every individual acts in such a way, beneficial social consequences will be brought about.

Remember that Rothschild thinks that the 'blindness' implied by the invisible hand is un-Smithian for this view conflicts with the view that individuals are the best judges of their interest. She argues:

> [T]his independence and idiosyncrasy of individuals is what Smith seems to be denying in his account of the invisible hand; it is in this sense a thoroughly un-Smithian idea.
>
> (Rothschild 1994: 320)

But when we distinguish between interests directed to the individual level and to the social level we may see that Smith's argument is the following: Individuals are the best judges of their interest, but they cannot judge the interests of the rest of the society (i.e. they are 'blind' with respect to the interests of others); therefore they should not try to bring about social consequences. When seen like this, the 'invisible hand' seems to be a truly Smithian idea.[20] It represents the connecting principles of the society, the network of interacting shoemakers, tailors, merchants and all others who, by definition, are pursuing their self-interests, acting somewhat myopically, and who are nonetheless the best judges of their interests. There is nothing in Smith's account of the invisible hand that would deny the 'independence and idiosyncrasy of individuals'.

Rothschild (2001: 126–128) also suggests that Smith's proposal in *WN* that merchants should *not* seek their individual interests by political means (particularly by supporting restrictions on imports) is conflicting with the idea that they would promote the public good by pursuing their self-interests. Yet from the above argument it is obvious that pursuing self-interests by political means (intentions targeted at the social level) is an entirely different matter from pursuing self-interests at the individual level, and, thus, there is no such conflict.[21]

It is also important here to note that Smith does not argue that self-interest promotes the interest of society under every condition (also see Schlefer 1998[22]). According to Smith, if individual intentions are about the social level, that is, if self-interested individuals are intending to change social phenomena, then self-interest would conflict with society's interest. The reason for this is clear. Individuals could not know what is good for others. In *WN* Smith explicitly mentions that interests of merchants who are trying to impose trade restrictions conflict with that of society.

But if no individual knows better than the other what is good for the society, how can Smith know better? How can he be against import regulations?

> To give the monopoly of the home-market to the produce of domestic industry, in any particular art or manufacture, is in some measure to direct private people in what manner they ought to employ their capitals, and must, in almost all cases, be *either a useless or a hurtful regulation*. If the produce of domestic can be brought there as cheap as that of foreign industry, the regulation is evidently useless. If it cannot, it must generally be hurtful.
>
> (Smith 1789: IV.2.11, emphasis added)

How can Smith suggest that import regulations are either useless or hurtful? Rothschild (2001: 24) suggests that because the invisible hand 'is founded on a notion of privileged universal knowledge' and because 'it presupposes the existence of a theorist [. . .] who sees more than any ordinary individual can', it is un-Smithian.

Two important points should be noted. First of all, Smith sees philosophers as products of division of labour (also see Peart and Levy 2005). They are *not* naturally better acquainted than others for inquiring into the connecting principles of nature and society: 'by nature a philosopher is not in genius and disposition half so different from a street porter' (Smith 1789: I.2.5). But, by way of specialisation, they can do better:

> Many improvements have been made by the ingenuity of the makers of the machines, when to make them became the business of a peculiar trade; and some by that of those who are called philosophers or men of speculation, whose trade it is not to do any thing, but to observe every thing; and who, upon that account, are often capable of combining together the powers of the most distant and dissimilar objects. [. . .] [S]ubdivision of employment in philosophy, as well as in every other business, improves dexterity, and saves time. Each individual becomes more expert in his own peculiar branch, more work is done upon the whole, and the quantity of science is considerably increased by it.
>
> (Smith 1789: I.1.9)

Thus, it is quite natural that he thinks that he observes better than the porter, and that he is less 'blind' to the connecting principles of nature and society than others who are specialised in other industries. Yet this does not necessarily imply 'privileged universal knowledge'. Smith, a man of speculation, is conjecturing about those connecting principles. It is also true that Smith thinks that the shoemaker, the tailor, as well as the merchants, are able to understand his argument that it is not to the advantage of a society to produce the goods that are produced less costly in other countries, and that every nation will be better off if they produce the good in which they have advantage. But more importantly, Smith does not presume that he has knowledge of the local situations and interests of particular individuals. Rather, from the argument that this is not possible, he suggests it is better to leave every individual to their own principles of motion.

Moreover, Rothschild implicitly assumes that 'unintended' means 'unanticipated'.[23] It has been argued in Chapter 2 that it is possible to have anticipated but unintended consequences. The absence of foresight and awareness of the social consequence is not a necessary condition for invisible-hand type of explanations. It is entirely possible that one or some of the individuals foresee the unintended consequence that is ahead but fail to act accordingly to change this consequence. There may be many reasons for this, but the most important seems to be that since there are many individuals who are involved in the process that brings about the unintended consequence, it may be costly to deviate from the original intention/action unless others do the same. In some cases, collective action may be costly and/or risky, thus individuals may bring about an unintended but anticipated social consequence. Smith, as well as any other individual, may foresee or recognise unintended social consequences. For this reason, Smith's recognition of the beneficial unintended consequences does not imply that he has 'privileged universal knowledge'.

Rothschild has two other points that may be discussed together. First, she (2001: 129–130) suggests that religious connotations of the invisible hand do not go well with Smith's somewhat irreligious views. As Rothschild nicely argues, the religious connotations come from its previous uses. Moreover, Smith use of invisible hand in HA carries religious connotations. In fact, we have seen that he criticises the practice of associating the apparent irregularities of nature with the 'invisible hand of Jupiter'. This supports Rothschild's argument that Smith used the phrase ironically, in *TMS* and in *WN*. However, if Smith uses it ironically, this means that the latter uses do not necessarily have any religious connotation.[24] We can read the invisible hand as a metaphor conveying a message about the responses of our imagination to the surprising aspects of nature. In *TMS* and *WN*, it may be understood as saying that 'what savage man may have associated with "the invisible hand of Jupiter" is hereby explicated'. Smith used the phrase to indicate that behind the order of things (which we may associate with design) there is some 'invisible' chain of events that brought them about. Yet he suggests that this invisible chain of events needs to be explicated in order to explain the phenomenon. Second, Rothschild (2001: 131–132) argues that the Stoic idea of a providential order, which is ostensibly implied by the invisible hand, conflicts with Smith's views. While it is true that Smith would not agree with the idea of an order that is not caused by the individuals who take part in it (the idea of providential order), we have seen that Smith's invisible hand does not necessarily imply such an idea. In fact, in *WN* and *TMS* social facts are explicated with reference to the interaction of individuals who are pursuing their own interests.

The invisible hand is an important concept in economics and our understanding of it should rest on a good understanding of the subset of unintended consequences implied by it. As we have seen, the invisible hand is neither a mysterious concept, nor does it imply complete blindness on the part of individuals or universal privileged knowledge on the part of the scientist. In fact, on the contrary, the concept of the invisible hand emphasises the will to remove mysteries concerning nature and society; it acknowledges the ability of men to act intentionally and calculate the consequences of their action; and alerts us to the incompleteness of our knowledge concerning other individuals and nature. Unintended consequences are brought about by men who are pursuing their own ends and it is the task of the social scientist to explicate how different individuals are connected to each other in producing those consequences. In the language of the previous chapters, the concept of the invisible hand suggests that we should study how certain individual mechanisms (e.g. the principles of motion of different individuals) are connected to each other.

Modern conceptions of the invisible hand

We have seen that the concept of the invisible hand is about how individuals who are pursuing their own interests bring about unintended social consequences and that there is a certain explanatory strategy attached to it. It is useful now to see how modern conceptions of the invisible hand relate to the original concept. In what follows, two different modern interpretations of the invisible hand are identified: the end-state interpretation and the process interpretation. Mainstream economists use the end-state interpretation that does not pay much attention to the explanatory strategy attached to the original invisible hand. Austrian economists use the process interpretation, which is closer to the original concept. Yet they seem to have amplified the meaning and effect of the invisible hand. Let us start with the mainstream interpretation.

The end-state interpretation

Generally, in standard economics, the invisible hand is associated with the fundamental theorems of welfare economics, which basically state that 'under certain conditions, every competitive equilibrium is a Pareto optimum, and conversely, every Pareto optimum is a competitive equilibrium' (Chipman 2002: 1). A fundamental theorem of welfare economics is sometimes characterised as the invisible-hand theorem. It states that when the distribution of income is given, a long-run perfectly competitive equilibrium will yield an optimum allocation of resources, and that every optimum allocation of resources is a long-run perfectly competitive equilibrium (see Blaug 1997: 577–579). Many mainstream economists conceive the invisible hand as an argument about efficiency and equilibrium (see, for example, Stiglitz 1991; Durlauf 1991; Hahn 1970, 1981; Marris and Mueller 1980, among others.) Indeed, many economists would consider the progress of economics on a line that connects Smith's invisible hand, the concept of equilibrium, and Arrow and Debreu's (1954) proof of the existence and optimality of competitive equilibrium. In fact, general equilibrium models in general are considered either as the proof of Smith's 'invisible hand', or as showing the conditions under which this conjecture holds. For example, Tobin (1991: 6) argues that Adam Smith's invisible hand is a *conjecture* about the workings of the market system which provides simple intuitions (1991: 12–13), and that this conjecture was proven by Arrow and Debreu. Likewise, Stiglitz (1991) considers the modern theorems of welfare economics as a modern representation of Smith's invisible hand. The proof of the existence and optimality of equilibrium indeed connects with Smith's argument that self-regarding actions of individuals would lead to a beneficial outcome for the society. Existence of a Pareto-optimal equilibrium proves the possibility of such a result.[25] However, the conditions under which the existence of equilibrium is proven may also be considered as a proof of the limitations of such an argument. Stiglitz (1991) expresses his worries about the empirical adequacy of these proofs. Both he and Tobin (1991) mention that the proofs of the existence of optimal market equilibrium do not take into account issues such as increasing returns, externalities, imperfect competition, time, uncertainty, incomplete markets, instabilities, etc. (also see Coase 1992; Maskin 1994). These omissions, according to Tobin, show the limitations of the invisible hand argument.[26] Hahn (1981) similarly argues that although equilibrium proofs indicate the *logical possibility* that an optimal equilibrium exists, because of these limitations (i.e. failure to take into account increasing returns, externalities, etc.) we have to consider the invisible hand argument as a remote reference point in comparison to the real world:

> [I]t is logically possible to describe an economy in which millions of agents looking no further than their own interests and responding to the sparse information system of prices only can nonetheless attain a coherent economic disposition of resources. Having made that clear let me nonetheless emphasise the phrase 'logically possible'. Nothing whatever has been said of whether it is possible to describe any actual economy in these terms.
>
> (Hahn 1981: 5)

Veblen (1899) raises a similar point against Smith's conception of economic affairs. He particularly criticises the practice of 'normalising' the state of affairs in the real world, that is, the abstract nature of Smith's argumentation. But, he goes on to argue that modern interpretations of the invisible hand should also be criticised since they work with isolations and abstractions to show that the dispersed self-regarding actions of individual may bring about beneficial consequences. Having said that, there is an important difference between the two. In the previous section we have seen that Smith used the invisible hand to represent the connecting principles of society that are supposed to bring about an orderly phenomenon and that the notion is best understood in terms of its relation to unintended consequences at the social level, which are brought about by the actions of individuals who are intending to bring about results at the individual level. His argument against the import restrictions as well as his explanation of the origin of money (see Appendix I) connects the actions of the individuals to the aggregate outcome. He indeed tries to explicate the process through which those unintended social consequences may be generated. The modern conception of the invisible hand, however, does not explicate how individual actions are related to the aggregate outcome.[27] Moreover, existence proofs do not mention the process through which the equilibrium is reached. Modern economists' conception of the invisible hand is, thus, basically an end-state interpretation. Or, as Hahn (1973: 324) argues, 'general equilibrium is strong on equilibrium and very weak on how it comes about'. It stresses the consequence (i.e. existence of equilibrium), rather than the process that brings about the consequence. Blaug (1997: 60) argues that this end-state interpretation makes no justice to the original metaphor:[28]

> But Smith's faith in the benefits of 'the invisible hand' had absolutely nothing whatever to do with allocative efficiency in circumstances where competition is perfect á la Walras and Pareto; the effort in modern text books to enlist Adam Smith in support of what is now known as the 'fundamental theorems of welfare economics' is a historical travesty of major proportions. For one thing, Smith's conception of competition is [. . .] *a process conception*, not an end-state conception.
>
> (Blaug 1997: 60, emphasis added)

Similarly, Barry argues that the invisible hand is concerned with the process rather than the end-state:

> The notion of the Invisible Hand must be seen as a metaphor that illuminates a continuing process of exchange and competition between individuals which brings about a coordination of plans and purposes. It must not be seen as a picture of an end-state of perfect equilibrium in which all plans have already meshed, since that implies the cessation of human action. The Invisible Hand image refers to an unending process of change and adjustment and not to a perfectly harmonious end-state in which incentives to change have been removed.
>
> (Barry 1985: 138)

Many would agree with Barry and Blaug that the end-state interpretation does not make justice to the original metaphor (e.g. Holcombe 1999, Knudsen 1993: 149–150).[29] Although their propositions are correct, this does not mean that the end-state metaphor is not related with Smith's invisible hand. As we have mentioned above, end-state (equilibrium) models can be considered as tests of the logical soundness of the invisible hand argument. That is, they show/test the conditions under which self-interested actions of individuals would lead to beneficial aggregate outcomes. The next chapter demonstrates this point by way of showing the relation between the end-state models of the emergence of a medium of exchange and Menger's account of the origin of money.

The process interpretation

Another popular interpretation of the invisible hand follows Smith by emphasising the need to explicate the processes through which social outcomes are brought about. Under this interpretation, the invisible hand is a metaphor that represents a process that brings about a harmonious social order as an unintended consequence of the dispersed actions of individuals who are pursuing their own interests. This conception of the invisible hand is mainly related to the idea of spontaneous order,[30] and is generally entertained by Austrian economists.[31] Because the concept of 'spontaneously generated orders' is generally regarded as the most significant sociological contribution made by the Scottish philosophers (such as Hume, Smith and Ferguson), this interpretation is argued to be closer to Smith's own thought (see Hamowy 1987; Barry 1982). Roughly, the 'theory of spontaneous order' asserts that social institutions and patterns – which are highly complex – are the unintended consequences of numerous individual actions. The most prominent advocate of the theory of spontaneous order, Hayek (1967a: 72) argues that the task of social science is to explain the unintended consequences of the dispersed actions of individuals.[32] Yet he criticises the way in which mainstream economists approach unintended social phenomena.

Hayek complains about the equilibrium approach to economics for it is not about the processes through which equilibrium comes about:[33]

> In the usual presentations of equilibrium analysis it is generally made to appear as if the questions of how the equilibrium comes about were solved. But, if we look closer, it soon becomes evident that these apparent demonstrations amount to no more than the apparent proof of what is already assumed. The device generally adopted for this purpose is the assumption of a perfect market where every agent event becomes known instantaneously to every member.
>
> (Hayek 1937 [1949]: 45)

This criticism implies another similarity between Smith and Hayek; that Hayek's interpretation of individual knowledge resembles that of Smith's in certain aspects. He (1937, 1945) argues that the knowledge of society is never available to the individual in its totality because of the peculiarities of individual knowledge.[34]

> This is the constitutional limitation of man's knowledge and interests, the fact that he *cannot* know more than a tiny part of the whole of society and

therefore all that can enter into his motives are the immediate effects which his actions will have in a sphere he knows.

(Hayek 1946a [1949]: 14)

According to Hayek (1946a [1949]: 45) the end-state interpretation of the invisible hand (that of neo-classical economists) is 'an assertion of a tendency toward equilibrium' yet it does not tell us how this equilibrium is reached. To show this, one has to show how the knowledge and expectations of individuals are changed and how these individuals are able to coordinate given their partial information about the market.[35] He states that the process through which market order (or equilibrium[36]) is reached cannot be understood by assuming that every individual knows everything. To be able to understand the process 'we must look at the price system as such a mechanism for communication information' (Hayek 1945 [1949]: 87).[37] According to Hayek, when agents act spontaneously by pursuing their own interests (i.e. at the individual level) more information is utilised in the market and coordination among individuals is rendered more easy in comparison to market where individuals intent to bring about social consequences.

Competition is essentially a process of the formation of opinion: by spreading information, it creates that unity and coherence of the economic system. [. . .] It is thus a process which involves a continuous change in the data and whose significance must therefore be completely missed by any theory which treats these data as constant.

(Hayek 1946b [1949]: 106)

What Hayek means by 'data' here is the information upon which agents act. According to him, by assuming that individuals are hyper-rational, traditional equilibrium analysis in economics misses the point that every individual has limited knowledge about the economy as a whole and about the motives of other individuals. The 'data' changes as the agents act, and the analyses of the process of economic activity has to take this into account. He (1937 [1949]: 38) argues that neoclassical economics abstracts from the fact that the consequences of one individual's actions are dependent on other individuals' actions, and that the overall consequence depends on the interactions of individuals who have limited knowledge and rationality. The aforementioned similarities (between Smith and Hayek) indeed point to the argument that if those connecting principles of society that link the dispersed actions of the individuals to the social consequences are to be explicated, one has to take two things into account: the interaction of individuals who have limited knowledge and the process through which they coordinate their activities. These issues are not dealt with in traditional static equilibrium models of economics. In Chapter 8 we will see that some of the contemporary economists try to deal with these issues by using the tools of game theory.

Another important idea that is common to Austrian economists and Smith is the explanatory strategy by which unintended social consequences are explained.

Hayek emphasises the element of conjecture and conceptual construction in explaining spontaneously generated orders (see, for example, Hayek 1967a: 72). Hayek echoes Smith with the idea that explanation can only be made starting from the regularities of human action.[38] He also echoes Smith in that his explanatory strategy is similar to that of 'conjectural history', the explanatory strategy that is attributed to Smith by Dugald Stewart. In his *Biographical Memoir of Adam Smith*, Stewart argues that historians sometimes find themselves in situations where they cannot find or use direct evidence, and proposed that

> [I]f we can shew, from the known principles of human nature, how all its various parts might gradually have arisen, the mind is not only to a certain degree satisfied, but a check is given to that indolent philosophy, which refers to a miracle, whatever appearances, both in the nature and moral worlds, it is unable to explain.
>
> (Stewart 1793 [1858]: 34, emphasis added)

To this type of investigation Stewart gave the title of 'conjectural history'. Stewart particularly refers to Smith's (1762 [1985]) *Considerations concerning the first formation of languages* as an example of conjectural history. In *Considerations*, Smith conjectures about the way in which language may have developed, and provides an invisible-hand type of explanation of its emergence. He indeed tries to show the 'conditions under which language might evolve naturally (i.e., without human or divine contrivance)' (Land 1977: 677).[39] By way of departing from what he sees to be the facts about human beings and natural languages he constructs a story where language develops naturally. Smith's argument against import restrictions, and his explanation of the origin of money (see Appendix I) contain such conjectural elements.

Yet it is also true that the Scottish philosophers were against 'boundless conjectures'. For example, Ferguson (1767: 7–12) criticises the 'state of nature' theorists, such as Jean-Jacques Rousseau and Thomas Hobbes, by pointing out that their approach is conjectural (see Berry 1997: 23). He criticises the practice of explaining the 'progress of society from a supposed state of animal sensibility' and argues that we should not work with such an analogy.[40] For him (1767: 12) the explanation of the progress of society should start from the 'laws of human nature'. Thus, the 'instinctive propensities' and 'set of dispositions' of man should be used in developing such a theory (Ferguson 1767: 16). Yet Ferguson is also aware that historical records about earlier stages of the society are rare. He only suggests that our theories should be consistent with the 'facts' we know about human nature. Whether we agree with his argument about human nature, or not, his suggestion is clear: our theories have to be constrained by and be coherent with the known facts about our object of inquiry. This is, in fact, what Stewart means by conjectural history:

> In examining the history of mankind, as well as in examining the phenomena of material world, when we can not trace the process by which an event has

been produced, it is often of importance to be able to show how it may have been produced *by natural causes.*

(Stewart 1793 [1858]: 34, emphasis in original)

Conjectural history, then, emphasises the element of rational reconstruction that has to be consistent with the facts we know about the subject matter. It shows a social phenomenon 'may have been produced' by 'natural causes'. We have seen that Smith suggests explanation of unfamiliar phenomena with reference to familiar facts. In fact, in *WN* and *TMS*, Smith provides us conjectural constructions of real-world phenomena which are based on what Smith considers to be familiar facts concerning individuals. Also, Smith's accounts of the origin of money and language are conjectures which are based on familiar individual mechanisms. Hayek, as well as Stewart, sees such activity as an essential component of conjectural history.

Conjectural history in this sense is the reconstruction of a hypothetical kind of process which may never have been observed but which, if it had taken place, would have produced phenomena of the kind we observe.

(Hayek, 1967a: 75)

What is common to Hayek, Menger and Smith is their interest in the processes through which unintended consequences may emerge, but the proposed processes do not have to be the actual ones in every occasion, for we usually have limited information about the actual process (e.g. about particular details of the phenomenon, or about the actual intentions and motives of the individuals). By means of starting from what is thought to be the basic principles or facts about the object of inquiry, rational reconstruction, conjectural history (or a theoretical explanation) may inform us about the connecting principles of society or of the social phenomenon we wish to examine. The element of conjecture is a necessary part of theoretical analysis for they argue that we can neither know the exact interests and intentions of the individuals, nor the peculiarities of their situation.[41]

Austrian interpretation of the invisible hand seems to be a faithful interpretation of the original metaphor. Yet it should be noted here that Smith's use of the invisible hand does not emphasise the explanation of an overall social order as an unintended consequence of human action. It is, rather, concerned with explanation of certain type of phenomena such as the consequences of the actions of merchants, or the rich. Smith explains the origin of money and language as unintended consequences of human action but he does not lay much emphasis on the explanation of overall social order as an unintended consequence of human action. Explaining the overall social order as an unintended consequence of human action would require an explanation with multiple layers of unintended phenomena and intentions therein. Such an explanation would run into difficulties because of the complexity of the social realm, which embodies both intentionally designed and unintentionally brought about institutions.[42] It seems to be fair to say that Austrian interpretation of the invisible hand magnifies the original concept by considering the overall social order as a beneficial unintended social consequence of the dispersed actions of individuals. In Chapter 8 we will consider game-theoretical models of coordination which follow the process interpretation of the invisible hand without endorsing the Austrian emphasis on overall harmonious social order.

Thus far, we have discussed two different interpretations of Smith's 'invisible hand' – that is, the end-state interpretation and the process interpretation – by contrasting the two. But, we should not forget the fact that they are coming from the same source, and that they are alternative ways of justifying Smith's invisible hand. The static equilibrium interpretation tries to show that it is indeed *possible* (under certain conditions) to reach an equilibrium, an orderly state, where every individual acts self-regardingly, and that they may be better off by doing so. Or, the end-state interpretations of the invisible hand focus on the conditions under which unintended social outcomes are compatible with self-interested individual action. The process interpretation, on the other hand, tries to show through which mechanisms this opportunity may be rendered more probable. The relation between these two different justifications[43] of the invisible hand is discussed further in Chapter 6 by means of discussing the contemporary contributions to Menger's explanation of the origin of money. But before doing this we should examine the notion of invisible-hand explanations.

Invisible-hand explanations

The invisible hand is about unintended social consequences that have been brought about by the interaction of individuals who are pursuing their own interests. Moreover, the original concept emphasises the need to explicate the process which brought about the social phenomenon in hand. Explanations that follow this process interpretation of the invisible hand are generally known as *invisible-hand explanations* (see Nozick 1974; Ullmann-Margalit 1978):[44]

> They [some explanations] show how some overall pattern or design, which one would have thought had to be produced by an individual's or group's successful attempt to realize the pattern, instead was produced and maintained by a process that in no way had the overall pattern or design 'in mind.' After Adam Smith, we shall call such explanations invisible hand explanations.
> (Nozick 1974: 18)

Invisible-hand explanations explain how some well-structured social pattern could have emerged, or persists, as an unintended consequence of individual actions (Ullmann-Margalit 1978). An invisible-hand explanation claims to show the pattern in question as the 'output' of a process that aggregates the dispersed actions of numerous individuals who did not intend to bring it about. This process may be called an 'invisible-hand process'.

By the 'invisible hand process' is meant the *aggregate mechanism* which takes as 'input' the dispersed actions of the participating individuals and produces as 'output' the overall social pattern.

(Ullmann-Margalit 1978: 270, emphasis added)

And 'the onus of the explanation lies on the process, or mechanism, that aggregates the dispersed actions into the patterned outcome' (Ullmann-Margalit 1978: 267). An invisible-hand explanation is, thus, an explanation of the unintended social consequences by means of a process that connects the dispersed actions of the individuals to the social phenomenon. It is an explanation that explicates the 'connecting principles' that may have brought about the social phenomenon.

Using our definition of unintended social consequences (see Chapter 2), we can characterise an invisible-hand explanation in the following way:

An invisible-hand explanation explains the emergence (and/or maintenance) of an unintended social phenomenon by explicating a process that may bring it about (and/or maintain it). Its *explanandum* must be an unintended consequence in the sense that:

1 it is located at the social level;
2 it was not intended by any individual;
3 it is mediated through a multiplicity of individuals;
4 individual intentions are directed to the *individual level*;
5 the action of one individual is not sufficient to produce the unintended (social) consequence;
6 individuals do not pursue the same end collectively (that is, collective intentionality is excluded); and
7 the actions resulting from one single individual's intention cannot affect the social level directly, or in isolation.

Our examination of the paradigmatic examples of invisible-hand explanations and their roots implies that invisible-hand explanations are partly based on the assumption that interactions of individuals are essentially complex, and it is usually not possible to have enough information about the motives of the individuals and about the peculiarities of their individual situation. We have also seen that it is usually not possible to gain knowledge about the characteristics of the individuals by simply observing social phenomena – in the sense implied by invisible-hand explanations. For these reasons, invisible-hand explanations usually take the form of conjectures about individual mechanisms that may have brought about the observed social phenomenon. They are somewhat speculative reconstructions of the 'connecting principles' that bring about the social phenomenon at stake. We have seen in the previous chapters that paradigmatic examples of invisible-hand explanations explicate some of the possible ways in which certain 'familiar' mechanisms may interact. They are conjectures, yet good invisible-hand explanations should be different from 'boundless conjectures' in that the depicted mechanisms should be plausible (see Chapter 7 for further discussion of this issue).

Our discussion of Menger and Schelling's explanations indicates that a good invisible-hand explanation should have the following properties:

1   Its explanation of the transformation of the model world (from the state where the explanandum phenomenon is absent to a state where it exists) should be successful.
2   It should explicitly state the proposed individual mechanisms and how they are connected to each other.
3   The description of the initial state of the model may be inaccurate in terms of representing the corresponding particular initial state of the real world (it is better if it is not), but it should at least depict some 'familiar' individual mechanisms. The description of the individual mechanisms (that is, the description of the individuals and how they behave) should be plausible in the sense that under the conditions specified in the model we should expect real individuals to act in a similar fashion.
4   The description of the explanandum phenomenon should represent its relevant characteristics given the interests of the explanation.
5   It should either suggest that a new / previously unexamined sort of (aggregate) mechanism may explain the phenomenon at stake, or provide a firmer basis for the previously suggested invisible-hand process.

We have seen in Chapter 4 that Schelling's explanation satisfies these conditions. But we have also noted that it is a partial potential explanation (i.e. not a full-fledged explanation of the emergence of residential segregation). Obviously, Menger's explanation cannot do as well for it is not explicit enough. As suggested in Chapters 3 and 4, the above characteristics make it more likely that the proposed mechanisms in an invisible-hand explanation have the potential of explaining particular exemplifications of the explanandum phenomenon. Nevertheless, the proof of the pudding is in the eating. That is, we can only know whether an invisible-hand explanation of the singular sort will be successful by inquiring whether the proposed aggregate mechanism is responsible for the phenomenon at stake.

Many invisible-hand explanations aim to explain the origin of phenomena such as money, language and segregation, and it is usually not easy to ascertain whether the proposed mechanisms really explain the phenomenon or not. For this reason, they may live as conjectures for a long time. Yet the quality of alerting the researchers to new mechanisms is an important contribution in itself. It expands our mental horizon, and indicates new paths for further empirical and theoretical research. As these first steps are usually incomplete, like that of Menger and Schelling's, the intuitions provided by such explanations are further explored by other researchers, by testing both the logical soundness and empirical validity of the proposed invisible-hand arguments. Briefly, such conjectures are valuable basically because they point out some of the ways in which certain mechanisms may interact in the real world and therefore they may be considered as attempts to discover the way in which the world works. We will see in the next chapter how these conjectures may be explored, and in Chapter 8 we will see how they may open the way to new research agendas and new results.

Concluding remarks

The invisible hand has many faces and it is open to misunderstandings. It was the task of this chapter to foreclose actual and possible misunderstandings about it and to provide a better understanding of what it amounts to. First, we have seen that a proper understanding of what is implied by the notion of unintended consequences is necessary for a proper understanding of the invisible hand. Second, with the help of textual evidence from Smith, we have established that the invisible hand is not a mysterious concept, on the contrary, it is a metaphor that stresses the need for explicating the mechanisms (and their interaction) that are responsible from the social phenomenon under investigation. Third, two different modern conceptions of the invisible hand were examined. It has been argued that although the end-state interpretation makes little justice to the original metaphor, it is related to it in that end-state models try to test the basic ideas implied by the invisible hand. Fourth, we have seen how the modern process interpretation of the invisible hand is built on the original ideas presented by Smith; such as individuals with limited knowledge, conjectural history and the will to explicate the connecting principles of society. Finally, we have examined the way in which invisible-hand explanations are related to unintended consequences and how they may be evaluated.

The reader may have realised, though, that one of the important issues has been hanging in the air and no serious attention has been paid to it. Simply put, we have not examined whether modern models and explanations of social phenomena as unintended consequences of human action explicate real or imaginary mechanisms, or connecting principles. It can be argued that the paradigmatic examples of invisible-hand explanations, (i.e. that of Menger and Schelling) aim at understanding the real mechanism behind social phenomena. In fact, Mäki (1990a,b, 1991) demonstrates that many Austrian economists, especially Menger, can be considered as realists and that the invisible-hand process should be considered as a causal process. Our examination of Menger and Schelling also suggests that they are searching for the real 'connecting principles' of the society. But we have also argued that invisible-hand explanations are conjectures about some of the possible ways in which certain mechanisms may interact. At first sight, this may be considered as a contradiction: realists providing conjectures and possible aggregate-mechanisms? The solution to this contradiction lies in the differences between aims and accomplishments of a scientist. It is possible that a search for the real connecting principles provides nothing but conjectures. But good conjectures are valuable for they may help us in discovering real mechanisms. In Chapter 7 we will see how these models may be interpreted as conjectures about the way in which certain real mechanisms may interact. But before that we have to make the point that we should not examine these partial potential explanations (or models) in isolation from other explanations and that we should rather see how these explanations are further explored in order to understand how things stand next to each other in the real world. Accordingly, the next chapter examines further considerations and explorations of Menger's conjecture. After this, in Chapter 7, we will see how models and explanations relate to the real world, and to other models and explanations of the same phenomenon. Further explorations of Schelling's conjecture are also discussed in Chapter 7.

6 The origin of money reconsidered

Introduction

We have seen that the paradigmatic cases of invisible-hand explanations, Menger's 'the origin of money' and Schelling's chequerboard model, are partial potential explanations. The present chapter examines contemporary reconsiderations of Menger's account in order to see whether there is any progress in explaining the emergence of a medium of exchange as an unintended consequence of human action and in terms of justifying and/or confirming Menger's intuitions. The question is the following: do we know any better than Menger about the origin of money? This chapter may also be considered as an illustration of a particular way in which research advances in economics, particularly in explaining unintended social consequences of human action.

Menger's account of the origin of money depends on a description of a model world that is rich in terms of the proposed explanatory factors, but vague in its description of how these factors are related to each other. Contemporary models, simulations and experiments[1] about the 'the origin of money' consider model worlds similar to that of Menger's, but they focus on subsets of the factors that were suggested by Menger. They are often restricted versions of Menger's model world. Menger provides the intuition that a medium of exchange may be considered as an unintended consequence of human action. Examination of the economics literature on the emergence of money reveals that economists formulate this intuition with the following hypotheses:

H(1): Commodity money may be an unintended consequence of human action.

H(2): Fiat money[2] may be an unintended consequence of human action.

One may examine H(1) and H(2) by asking the following questions:

1 Is the state where fiat/commodity money exists a possible state of 'the model world' (i.e. Menger's model world), where individuals pursue their own interests (at the individual level)?

2   If money could exist in 'the model world', can we show that it is a consequence of the actions of self-regarding agents? Or, how is it possible to reach a state where a medium of exchange exists in the model world? Or, what are the mechanisms that transform 'the model world' into a state where money exists?

3   Do real individuals behave in a similar manner to the agents in 'the model world'? Or, if real individuals are confronted with a situation similar to the one described by 'the model world', would their dispersed actions bring about money?

4   Do the conditions in 'the model world' hold in the real world? Or, what would happen if the conditions described in 'the model world' do not hold? Could we consider money as an unintended consequence of human action in the presence of other (real world) factors that are omitted in 'the model world'?

This chapter examines some of the attempts to provide answers to these questions. Most of the literature on the 'the emergence of a medium of exchange' focuses on the first two types of questions, and there is little progress in tackling the latter two types of questions. Thus, the above hypotheses, H(1) and H(2), are tested on logical grounds to a considerable extent, but the relation between 'the model world' and the real world has not been examined in a meaningful way, that is, the above hypotheses are not tested against real world data.

The models that examine the first type of question correspond to the end-state interpretation of the invisible hand (see Chapter 5). Generally, they are existence proofs, and they do not examine the way in which money may be brought about. They provide a detailed analysis of the conditions under which money may exist. Yet the proof of the existence of money (i.e. monetary equilibrium) as a possible state of a world of self-regarding individuals does not prove that it may be brought about by these individuals. Models that examine the second type of question explore the ways in which the end-state (i.e. money) may be reached. These models generally have the characteristics of an invisible-hand explanation and they picture the possible aggregate mechanisms that may bring about money as an unintended consequence of human action. Few attempts have been made to examine the third type of question, and the fourth type of question is not examined at all.

It is argued here that the models, simulations and experiments examined in the following pages test the logical soundness of the intuition that 'money may be an unintended consequence of human action'. Yet they fail to bring us any closer to the real world than Menger's account did. While our understanding of the workings of the model world depicted by Menger is improved, we do not know any better than Menger whether money is an unintended consequence of human action or not. This chapter examines the literature on 'the emergence of a medium of exchange' in the following order:

1   end-state models;
2   simulations of the emergence of money;

3 experiments concerning the behaviour of real individuals; and
4 theoretical models of the process of the emergence of money.

The results of this examination are then presented, which is followed by the concluding remarks.

Before going further, two things should be noted: first, most of the models examined in this chapter are quite technical, but to be able to focus on their conclusions they are presented in a simplified manner.[3] Second, the examination is not exhaustive – that is, there are also other models of the emergence of money.[4] Yet this selection of models fairly represents the related literature.

Existence of a medium of exchange

Most of the models of 'the origin of money' are based on, or related to the Kiyotaki–Wright (1989) model that focuses on the existence of an equilibrium where model agents use a certain commodity as a medium of exchange. For this reason it is important to have a fine grasp of the Kiyotaki–Wright environment and its relation to Menger's account. This section starts with a description of the Kiyotaki–Wright model and examines the related models. First, the results concerning the existence of commodity money and then the existence of fiat money equilibrium is examined.

The existence of commodity money equilibrium

Kiyotaki and Wright (1989) present a model economy where there are three commodities (1, 2, 3) and three types of agents (I, II, III). Type I agents consume commodity 1, type II agents consume commodity 2, and type III agents consume commodity 3. No agent produces what he consumes. Thus, to be able to consume (e.g. to eat) they have to exchange their production good with their consumption good – that is, they are market dependent. It is assumed that every agent can only store one commodity at a time, and storing is costly. The storage costs of commodities are different for different types of agents. If we define the cost of storing good j for type i as c_{ij}, then the following condition holds: $c_{i3} > c_{i2} > c_{i1}$.

The agents meet (in pairs) randomly at the marketplace, and when they meet, they have to decide whether or not to exchange their inventories. Exchange entails one-for-one swap of inventories. If an agent is able to acquire his consumption good at the market, he immediately consumes it and produces one unit of his production good. If an agent decides to exchange his inventory with a good he cannot consume, he stores it and waits for the next exchange opportunity to exchange it with his consumption good. Because of these specifications, in every period there are always agents who are facing the double coincidence of wants problem. Every individual is assumed to choose a trading strategy that would maximise his expected discounted utility, which is dependent on the utility of consumption, disutility of production and storage, and a discount factor (see Appendix II). Trading strategies are rules that determine whether one agent is willing to exchange his inventory with another agent. Ideally, a trading strategy should depend on the trading history of the agent. Yet Kiyotaki and Wright isolate their model from the influence of time and of agents' trading history. They only focus on the inventories of the trading agents, on the storing costs and on the economising actions of the agents. The strategies of the agents depend on a comparison of the indirect utility of storing the current inventory and exchanging it with another good. If a type i agent is able to exchange his inventory, good k, with his consumption good i^* ($k \neq i^*$), he always does it. But if he matches with an agent who has in inventory a good j ($j \neq i^*$, $k \neq j$) that he cannot consume, the agent takes a decision, given the storage costs of j and k, and his expectations. That is, if the agent expects it to be easier to exchange j with i^* in the next period than exchanging k with i^*, then he is willing to exchange k with j, given that the storage cost of j is sufficiently low. Trade occurs if both agents are willing to exchange k with j.

Given the assumptions of the Kiyotaki–Wright (1989) model we may examine the way in which it is related to Menger's model world (see Figure 6.1). In the real world there may be many factors that may be effective in the process of the emergence of a medium of exchange. We have seen that Menger focuses on some of them by isolating the influence of others. Particularly, Menger focuses on a set of *explanantia*, which consists of:

1   economising actions;
2   discovery, learning and imitation;
3   absence of double coincidence of wants;
4   storage costs;
5   marketability[5];
6   expectations of individuals;
7   frequency of use; and
8   availability of goods.

Roughly, saleability of a good is influenced by the following set of factors: {4, 5, 6, 7, 8}. Kiyotaki and Wright (1989) make further isolations to examine the

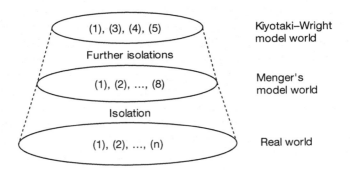

Figure 6.1 Kiyotaki–Wright model vs. Menger's model.

existence of a monetary equilibrium, as shown in Figure 6.1.[6] That is, they focus on a subset of the factors in Menger's model world, particularly on 1, 3, 4 and 5. In this manner, their model is much more specific.[7]

Kiyotaki and Wright analyse the existence of equilibrium for two different economies, Model A and Model B, in this isolated environment. The only difference between the two models is the characterisation of agent types in terms of what they produce, as seen in Table 6.1.

For Model A, Kiyotaki and Wright (1989: 934–935) determine two types of equilibrium for different parameter values:[8]

1   A *fundamental equilibrium* where every agent 'prefers a lower-storage-cost commodity to a higher-storage-cost commodity unless the latter is their own consumption good'.
2   A *speculative equilibrium* where agents 'sometimes trade a lower- for higher-storage-cost commodity [. . .] because they rationally expect that this is the best way to trade for another good that they do want to consume, that is, because it is more marketable'.

Kiyotaki and Wright find that for both economies there is an equilibrium where all agents use fundamental strategies and where commodity money exists. They also find that for certain parameter values some agents speculate and there is an equilibrium where both low-storage-cost and high-storage-cost goods serve as media of exchange. That is, they find that commodity money equilibrium is a possible state of the model world (i.e. H(1) confirmed). However, they do not tell how this model economy is transformed from a state of unmediated exchange to a state with a medium of exchange. Moreover, they do not tell why agents should believe that a certain good is more marketable. They only show that if they expect that a high-storage-cost good is more marketable, then (under some conditions) they will use it as a medium of exchange.

Starr (1999) obtains similar results to that of Kiyotaki and Wright (1989). He presents two different models that examine the importance of 'absence of double coincidence of wants' and 'scale economies in transaction costs' in isolation. The first model establishes that in the absence of double coincidence of wants, goods with low transaction costs are likely to be used as media of exchange. In this model, it is not necessary that the equilibrium gets established with a unique

Table 6.1 Production and consumption in Kiyotaki and Wright (1989)

	Consumes	*Produces*	
Agent	*Model A and B*	*Model A*	*Model B*
Type I	1	2	3
Type II	2	3	1
Type III	3	1	2

Note
1, 2 and 3 denote the type of commodities.

medium of exchange, rather there may be multiple monies. In the second model, the problem of double coincidence of wants is absent. Instead, the model focuses on scale economies. Here, it is assumed that the transaction costs of a commodity (or an instrument) decrease as its trading volume increases. That is, as individuals start using one commodity in their exchanges more and more, the commodity gets more marketable, which in turn leads other individuals to use this commodity in their exchange, and as this commodity gets more marketable its transactions costs decrease. The model shows that when there are increasing returns (i.e. in terms of decreasing transaction costs) to the increase in the trading volume of commodities, a monetary equilibrium with a unique medium of exchange exists (i.e. H(1) confirmed).

Starr, as well as Kiyotaki and Wright, shows that the commodity money equilibrium exists, and that there is a state of 'the model world' where rational individuals who try to maximise their expected utility would use media of exchange. Yet they do not show how exactly 'the model world' is transformed into such a state. To explain the *emergence* of a medium of exchange one needs to explicate the process through which a medium of exchange gets established. These models fail to do this. Later, we will return to this question when we examine the process models of 'the origin of money'.[9]

The existence of fiat money equilibrium

In addition to the existence of commodity-money equilibrium, Kiyotaki and Wright analyse the existence of a fiat-money equilibrium (i.e. H(2)). Fiat money is an object (good 0) which does not have the intrinsic property of providing utility to the agents, but which nevertheless serves as a medium of exchange. Kiyotaki and Wright analyse whether such an object takes on value in their model world. They (1989: 942) find that there exist equilibria in which fiat money does not circulate. The intuition behind this result is the following: if no-one believes that the others will accept a valueless object (i.e. provides no utility), good 0, they will not exchange their goods with good 0. Kiyotaki and Wright show that fiat money can only exist *if everyone believes that every other will accept it*. Aiyagari and Wallace (1991), and Kiyotaki and Wright (1991, 1993) obtain similar results with more general models (see Appendix II). These models try to show that fiat-money equilibrium is another possible state of these model economies, but they *fail* to show that fiat money may emerge in these model economies without the existence of prior institutions or intervention.[10] The reason for this is clear: these models fail to prove the existence of fiat-money equilibrium unless most of the agents (are assumed to) believe in the initial stage that the fiat good will be accepted by others.[11] That is, they show that if everyone believes that everyone else will accept the fiat good in exchange then there is an equilibrium where the fiat good is used as a medium of exchange. Thus, if there is to be an equilibrium with fiat money, the common belief in its existence has to be established in the initial stage in some way. Yet these models do not explain how such a belief may be established or why agents would hold valueless objects and consider using them as media of exchange. In contrast to commodity money, it is not clear whether fiat money can get started if the faith is not imposed from outside (e.g. by a central authority). Thus, although these models assert that H(2) is confirmed, they fail to show this.

Of course, the fact that the proof of the existence of fiat money depends on an assumption of common belief is in line with our intuitions about the nature of money. Money exists only if everyone expects others to accept and use it in exchange. This makes us understand (once again) that money is a stable institution; that is, if everyone believes that every other person will use it, no-one will have an incentive to stop using it.[12] But the formal restatement of this intuition (i.e. about the nature of money) does not tell much about the origin of fiat money, or more precisely, about the way in which such a common belief gets established.

Alternatively, Williamson and Wright (1994) focus on an economy where there is uncertainty about the quality of commodities that are subject to exchange. The aim of the model is to analyse the function of money in reducing these uncertainties, which are caused by private information. The Williamson–Wright model abstracts from the problem of double coincidence of wants and assumes that all commodities of the same quality provide the same utility for all consumers in the model economy. It is also assumed that individuals may not be able to recognise the quality of a commodity in an exchange and thus their problem consists of deciding whether to exchange their commodities with another commodity of an unknown quality.[13] They prove that under these conditions, when there is no private information (i.e. no uncertainty concerning the quality of commodities) only high-quality goods are produced and there is no need for money, but when there are uncertainties concerning the quality of goods, introducing a generally recognised fiat money to the economy improves welfare – as everyone can recognise money and as there are no uncertainties concerning its quality. This model suggests that fiat money cannot be an unintended consequence of human action (H(2) is not supported) when there is uncertainty. Given their analysis, it is probable that while commodity money may emerge in a small economy, because of the uncertainties that may emerge from increased market size and traffic, at a point in history state intervention for refining the existent medium of exchange, or for replacing it by another one (e.g. fiat money) may become necessary (see Chapter 3). This explains the reason why Menger argues that unintended institutions may have to be refined at a point in history.[14]

Concluding remarks

Remember that we have characterised Kiyotaki–Wright (1989) as focusing on certain aspects of Menger's model world by way of leaving out other aspects of it. Other models that follow Kiyotaki and Wright (1989) are essentially doing similar things. They make similar further isolations to examine other aspects of Menger's model world. As it may have been observed, these models also have a more specific description of the trading environment and the way in which the trade takes place. Thus, they provide an understanding of what is possible (i.e. in terms of existence of monetary equilibrium) in such specific environments. Nevertheless, the explication of the process through which agents solve the problem of double coincidence of wants is needed for an explanation of the emergence of commodity money. More broadly, both in fundamental and speculative equilibria, individuals expect others to use a certain good as a medium of exchange, but the above models do not tell much about how such a belief may be established. Thus, the criticism of end-state interpretation of the invisible hand is right on target (see Chapter 5).

Emergence of a medium of exchange

The attempts to examine the process of the emergence of money may be classified under three headings: computer simulations, experiments and theoretical models. First, 'simulations' and 'experiments' are examined, for they are directly related to the Kiyotaki–Wright (1989) model. Next, the theoretical models of the process of the emergence of money are examined, two of which are essentially different from the Kiyotaki–Wright model.

Simulations

Marimon *et al.* (1990) examine how individuals would learn to coordinate, or whether they would be able to coordinate in the Kiyotaki–Wright (1989) environment. They replace the rational agents of the Kiyotaki–Wright environment with artificially intelligent (AI) agents. They examine whether AI agents would play fundamental strategies, or speculative strategies. To do this they consider different exemplifications of the Kiyotaki–Wright model, by considering different model economies with different specifications. It should be emphasised here that these simulations do not directly test hypotheses H(1) and H(2), but indirectly test them by testing the results of Kiyotaki–Wright (1989).

Marimon *et al.*'s AI agents can learn from experience and update their expectations and strategies, given the information about their history. They are endowed with Holland's (1975) classifier systems. A classifier system can be considered as a repertoire of behavioural rules or strategies for an individual.[15] Obviously, for different situations different repertoires of strategies are applicable. However, only two decisions are important in Marimon *et al.* (1990): trading decision and consumption decision. We have seen above how trading decisions are taken. Individuals are matched in pairs randomly, they simultaneously offer their goods, and decide whether or not to accept the trade offer. We know that they always accept their consumption good. But if they are not offered their consumption good, they have to decide whether to accept a good that they cannot consume. Many behavioural rules may be applicable to such a situation, such as 'always accept trade', 'accept if the storage costs are low', 'accept if the storage cost of the offered good is lower than storage cost of the good in inventory', 'accept if other agents accept', and so on. Two types of AI agents are examined in Marimon *et al.* (1990). The first type knows all the available strategies in advance. The second type knows only a portion of the available strategies, which is randomly assigned. Both types of agents attach strengths to every behavioural rule, or strategy, and update these strengths according to what happens when a certain rule is used. We may assume that the first time they trade, agents use one of the available strategies randomly, and if the randomly selected rule is successful, the agents increase its strength; if it is not, they decrease it. That is, every agent keeps track of the success of the available strategies they have executed. The classifier systems of the second type of agents evolve in a similar manner. However, in this case a 'genetic algorithm' is introduced to the system. Roughly, after some time, unsuccessful 'classifiers' die and new ones are introduced into the system, which are similar to the successful classifiers. The process of updating the repertoire of behavioural rules includes four operations: 'creation', 'diversification', 'specialisation' and 'generalisation'. That is, occasionally new rules are added, modified, updated (to make them more specific to certain situations) and generalised. Generalisation is a genetic algorithm that occurs randomly (but the probability of its occurrence decreases in time) in the following way. Occasionally 'rules' are classified into two different groups according to their ranking ('fitness') among classifier systems: 'potential parents' and 'potential exterminants'. Then, two 'potential parents' are randomly matched to create a 'child' rule, which has a strength equal to the average of its parents' strengths. And each time a new child is created, a randomly chosen 'potential exterminant' dies. Given these characteristics, we may call the first type of agents as agents with 'full strategies' (F), and the second type of agents as agents with 'random-partial strategies' (R).

The results of Marimon *et al.*'s simulations are summarised in Table 6.2. The rows of the table summarise the specifications of different models used in the simulations. Production columns show the production of different agents (remember that agent type i consumes commodity i). Storage costs columns show the storage costs of every good for different economies. The utility column shows the utility of consuming commodity i. 'Initial CS' column shows the chosen type of agent in terms of classifier systems: F or R. The last column shows the type of equilibrium reached by the agents after the execution of the simulation. F indicates that the simulation converged to the 'fundamental equilibrium', and S indicates that the model economy converged to a 'speculative equilibrium'. A question mark, '?', denotes that the simulation is not conclusive. The results of the simulations are shown in Table 6.2.[16]

For model economies A.1.1 and A1.2 the simulation converges to the fundamental equilibrium, that is, AI agents learn to play fundamental strategies and commodity money emerges. In B1, the short-run behaviour of the simulation is close to the speculative equilibrium of the Kiyotaki–Wright model, yet in the long run AI agents learn to play fundamental strategies. For all other model specifications the simulations are inconclusive, but Marimon *et al.* report that the results are closer to the fundamental equilibrium than to the speculative equilibrium. Concerning fiat money (economy C), Marimon *et al.* find that if the storage costs of real commodities are sufficiently high, and if a sufficient number of agents are endowed with the fiat good in the initial stage, and if agents do not know all of the available strategies (i.e. economy C2), then we may expect the economy to

Table 6.2 Description of the economies in Marimon *et al.* (1990)

	Production					Storage costs					Utility	Initial CS	Eqn type
	I	II	III	IV	V	1	2	3	4	5			
A1.1	2	3	1	–	–	0.1	1	20	–	–	100	F	F
A1.2	2	3	1	–	–	0.1	1	20	–	–	100	R	F
A2.1	2	3	1	–	–	0.1	1	20	–	–	500	F	F/S?
A2.2	2	3	1	–	–	0.1	1	20	–	–	500	R	F/S?
B1	3	1	2	–	–	1	4	9	–	–	100	F	F/S
B2	3	1	2	–	–	1	4	9	–	–	100	R	F/S?
C1	2	3	1	–	–	0.1	20	70	0	–	100	R	?
C2[a]	2	3	1	–	–	9	14	29	0	–	100	R	F
D	3	4	5	1	2	1	4	9	20	30	200	R	?

Source: Marimon *et al.* (1990: 350).

Note
a   Although Marimon *et al.* (1990: 368) present the results for this model (C2), they do not present it in the table. Moreover, they do not talk about C1. It is assumed here that the specifications for C1 are correct.

converge to the fiat money equilibrium. On the other hand, simulation results are inconclusive for economy C1.

Briefly, Marimon *et al.* (1990) test Kiyotaki and Wright's insights, and conclude that the fundamental equilibrium is more likely to be selected. In addition, Marimon *et al.* show that although not every model economy (e.g. models with different specifications and assumptions) converges to a monetary equilibrium, when the economy converges it is more likely that the good that is less costly to store emerges as a medium of exchange. On the other hand, the simulation fails to show that fiat money may be considered as an unintended consequence of human action, because there is no reason for the agents to hold a fiat good and use it in their exchanges – unless this assumption is made.

An important characteristic of Marimon *et al.*'s work is that it does not use a traditional analytical model to analyse the problem of the emergence of money. Instead the authors simulate the behaviour of artificially intelligent (AI) agents in different model worlds (e.g. economy A, B under different parameter values) and observe the agents' behaviour with the help of a computer. This strategy helps the researcher to *observe* what happens (the process) under particular states of the model.[17] Simulations usually show what is possible given the assumptions and specifications of the particular models used for the simulation. Briefly, Marimon *et al.* (1990) test Kiyotaki and Wright's results with an *artificial experiment*. As with every experiment, these results are specific to certain model specifications (i.e. artificial laboratory conditions). In this way, they suggest that although the speculative equilibrium is a possible state of the model economy, it is less likely to be reached.[18] For Marimon *et al.*'s results are valid for particular specifications of 'the model world', they partly support hypothesis H(1) – that commodity money may be an unintended consequence of human action – and cannot be argued to support H(2).

Gintis (1997: 24) argues that it is indeed true that Marimon *et al.*'s results are sensitive to the choice of parameters for the Kiyotaki–Wright (1989) model and he (1997, 2001) presents an alternative simulation.[19] His simulation is different in that it is based on Darwinian notions, such as natural selection, mutation and adaptation, and it dispels AI agents. Gintis argues that both fundamental and speculative equilibria are possible (i.e. H(1) is supported), and that unless we assume at the outset that a very high percentage of agents accept a fiat good, a fiat good equilibrium will not emerge (i.e. H(2) is not supported). Gintis characterises the individuals as agents who are endowed with genomes that define their strategies, which evolve mimicking a Darwinian evolutionary process – that is, a process that selects the successful genomes (see Appendix II for a description of the model). Therefore, Gintis's model does not give much role to the rational decisions of the agents. This may, of course, suggest that the emergence of commodity money equilibrium does not need much intelligence or rationality. But since the model is characterised by the introduction of new agents who inherit the successful strategies of their parents, we may argue that that there must be some role for individual learning or imitation.

Experiments

A more plausible way to test these results is to conduct a 'real' experiment with real agents, rather than an artificial experiment. Such experiments have been conducted by Brown (1996), and Duffy and Ochs (1999). They examine the behaviour of real individuals in the Kiyotaki–Wright (1989) environment. The question is whether the results of the Kiyotaki–Wright model hold if rational model agents, or Marimon *et al.*'s AI agents, are replaced with real individuals. Here, we will focus on Duffy and Ochs (1999) who have general a set-up where they consider different specifications of the Kiyotaki–Wright (1989) environment.[20] In particular, they inquire into the motivations of the real agents in such an environment.

In contrast to Marimon *et al.*'s artificial experiment, Duffy and Ochs (1999) conduct a 'quasi-material experiment' (see Morgan 2000), that is, they let the real individuals act in the Kiyotaki–Wright environment. The actions of the real individuals are constrained by the Kiyotaki–Wright environment, real individuals are supposed to act in an imaginary environment, and they are supposed to act according to the given specifications of the model world, and their behaviour is isolated from the effects of other factors by the model specifications. Yet the agents are real and Duffy and Ochs (1999) expect that their behaviour may help us in understanding whether the results of the Kiyotaki–Wright model and its variants may be carried to the real world. That is, they inquire what is possible in the real world. The conclusion of the experiment is as follows:

Our subjects showed a pronounced tendency to play fundamental strategies regardless of treatment conditions [. . .] At the individual level, behaviour reflected a response to differences in past payoffs – as assumed in reinforcement models, but did not reflect any response to differences in marketability conditions – as required by full rational Bayesian agents.

(Duffy and Ochs 1999: 873)

Duffy and Ochs's experiment with human subjects supports Marimon *et al.*'s conclusion that agents are more likely to play fundamental strategies. Two conclusions follow from this: commodity money may emerge as an unintended consequence of human action (i.e. H(1) is supported), and real individuals do not play speculative strategies (i.e. Kiyotaki and Wright's results concerning speculative equilibrium are not supported). Considering the first conclusion, we may say that this experiment takes the plausibility of the idea that 'money may be an unintended consequence of human action' one step further, for it shows that real individuals may bring about commodity money in an environment similar to that of the Kiyotaki–Wright model. Yet the results of Duffy and Ochs (1999) (as well as Marimon *et al.*) are valid for particular specifications of the Kiyotaki–Wright (1989) environment and there is nothing in this experiment that suggests that these results may be extended to more complex real-life situations. This also relates to the interpretation of the second conclusion: the specifications of the experiment may be responsible for the absence of speculative equilibrium. That is, although they show that the speculative equilibrium does not emerge in the Kiyotaki–Wright environment, this result does not necessarily apply to other environments. This suggests that other experiments are necessary to see whether 'marketability' or 'individual expectations' are important in the process of the emergence of money.

Following this line of argumentation, Sethi (1999) argues that both fundamental and speculative equilibria may be stable, and that the economy may evolve into these equilibria if agents are able to observe at least one strategy (and its success) other than their own. That is, it is argued that if agents are able to observe others' behaviour and imitate them, they may end up in both kinds of equilibrium. Sethi argues that Duffy and Ochs's experiment fail to test this idea for they only provide information about every individual's own past payoff. He suggests an alternative experimental set-up, where individuals may observe some other individuals' performances:

An alternative experimental design, which could test both for convergence to speculative equilibria and for the propensity of individuals to respond to observed differences in payoffs, would provide each subject with his own average historical payoff as well as the average payoff obtained by the other players in the same sub-population. This could serve as a standard of comparison that may induce players with below-average performance to experiment with alternative patterns of behaviour.

(Sethi 1999: 245)

Simply, Sethi suggests that the role of imitation and learning has to be examined further, both theoretically and experimentally, to get a better understanding of the process of the emergence of money. Yet imitation and learning are only examined theoretically in this literature and there is lack of empirical evidence. We now turn to the theoretical examination of the process of the emergence of money.

Theoretical models

Imitation

An evolutionary model which focuses on imitation is Luo (1999). Luo (1999) presents a model that emphasises the role of imitation in the process of the emergence of money. His model is based on the Kiyotaki–Wright environment, but there are three important differences. First, he introduces imitation to the model. Second, he assumes that all goods are perishable, and that they perish at the end of the day. Third, there are two trading sessions every day. Given these assumptions, agents can exchange their production good with any good in the first trading session, but in the second session it is only rational to acquire their consumption good. Because of the absence of double coincidence of wants, it is rational for every individual to exchange their production good in the first period with a good that they may use to acquire their consumption good in the second period. When the trading sessions are closed every individual *i* meets another individual *j* with positive probability and if the agent *j* is more successful in acquiring his consumption good, *i* imitates *j*'s trading strategy with positive probability, and otherwise continue using the same strategy. Luo demonstrates that this process brings about one or many media of exchange. Different versions of the model (i.e. with different specifications) are set to examine different questions about the emergence of money. The examination of these different versions show that storability (e.g. low storage costs) as well as the initial trading strategies (i.e. initial beliefs of the agents), the proportion of the agents who are specialised in producing different goods (i.e. relative number of different type of agents), and the amount of mutation (i.e. probability of agents playing arbitrary strategies) may determine the medium of exchange and the type of equilibrium (i.e. fundamental or speculative). That is, the initial conditions of the model determine the good that will serve as a medium of exchange. Moreover, there are more equilibria than there are in the Kiyotaki–Wright (1989) model and several goods may serve as media of exchange at the same time, but if the mutation rate is sufficiently high the equilibrium that emerges out of the process of imitation is the fundamental equilibrium.

Obviously, this model is closer to Menger's account of money for it takes into account 'imitation' in an explicit manner. Luo not only demonstrates that money may be considered as an unintended consequence in such a model world (i.e. H(1) confirmed), but he also explicates a possible mechanism that may bring about money. Briefly, he shows that Menger was right in suggesting that 'imitation' is an important mechanism in the process of the emergence of a medium of exchange. Yet by examining a restricted version of Menger's account, he does not take us any closer to the real world.

Each and every model we have examined above is about coordination of individual activities. Yet the above models mainly focus on the problem of selecting the commodities that may serve as media of exchange. An alternative (but related) line of research explicitly focuses on the coordination of individual activities by way of abstracting from the selection of 'candidate' commodities that may serve as media of exchange. In this line of research two (or more) commodities are presupposed as candidate goods and individuals are supposed to face a coordination problem: they have to select a good (or a combination of goods) to use as a medium of exchange. Roughly, individuals know that they have to use a good as a medium of exchange but they do not know which of the two (or many) goods would be used by others and would serve as a medium of exchange. So they have to focus on coordinating their behaviour.[21]

Coordination

Schotter (1981) emphasises coordination and the role of learning from the history of play in the process of the emergence of money. He characterises individuals as agents who learn from experience and presents the problem behind the emergence of a medium of exchange as a coordination problem among agents in an agrarian society. (Schotter 1981: 35–39) It is assumed that there are no transaction costs, and that trade is mediated by a marketing agent. In this model economy there are m plastic chips (e.g. with different colours) that do not provide any utility to the farmers. Nevertheless, these chips are *supposed* to be used as media of exchange. When farmers come to the market they bring a number of chips that equals the equilibrium price of the goods they have supplied[22] to the marketing agent, and they use these chips to buy goods from other farmers. The problem is that there is no uniform way to pay with these chips.[23] For the market exchange to take place, every farmer has to decide in what chip or in what combination of chips he should be paid for, and in what chips he will pay for the goods he wants to buy. Thus, the farmers in the economy are confronted with a *coordination problem*. If two farmers cannot fully coordinate they can only exchange the goods that correspond to the number of chips they agree on. If they completely fail to coordinate they cannot trade. If, in time, all farmers coordinate on one type of chip, this chip emerges as the generally accepted medium of exchange. Of course, it may well be that the farmers coordinate on a mixture of chips, then this mixture of chips will function as the medium of exchange.

Let us present a highly simplified version of Schotter's model. Assume that there are only two types of chips, A and B. The coordination problem facing the farmers every time they need to trade can be represented with the chip game in Table 6.3. If farmer I (seller) wants to be paid in As (i.e. plays A) and if farmer II (buyer) wants to pay in Bs (i.e. plays B), they cannot trade. Thus, we assume they receive zero payoffs. But if they are able to coordinate on either A or B, they will be able to trade and get what they want – hence the positive payoff, a. They have to make their choices simultaneously and they cannot communicate.

We have said that if both players choose the same option they have positive

Table 6.3 Money game

Farmer I
(Player I)

		A	B
Farmer II	A	a,a	0,0
(Player II)	B	0,0	a,a

Note
a>0

payoffs and otherwise they get nothing. That is, if they choose (A, A) or (B, B) they are able to coordinate. In game theory these two coordination points are called Nash equilibria. In a Nash equilibrium, the players' strategies are the best responses to the other players' strategies, that is, players get the highest payoff given others' strategies (see Gintis 2000: 6–14; Bierman and Ferandez 1998: 16). Or more intuitively, 'no player has any incentive to deviate unilaterally from it', so 'players do not regret their strategy choices' (Colman 1995: 59). Of course, farmers see that it does not matter on what type of chip they coordinate, yet there is no way that they can know what the other will choose in advance (see Chapter 8). For this reason they may also choose to randomise their actions (i.e. play mixed strategies) in order to increase their chances of coordination. For example, they may play both options with equal probability.[24] More generally, if farmer I and II meet for the first time in the market, they have no way of knowing the way in which the other will want to trade. Now, suppose that there are N farmers in the economy, and that they are randomly matched in pairs. In period 0 (i.e. the first time they trade) all farmers would face a situation similar to the one described for Farmer I and Farmer II. Assume that all farmers randomise their choices in period 0. In period I, farmers may reconsider their choices given what happened in period 0. This may happen in many ways,[25] but Schotter assumes that all players are able to observe the behaviour of the rest, and that they all observe the same history. Given this assumption in every period, all farmers would revise their expectations about the likelihood of the choice of others given the information about the last period.[26] For example, if the percentage of farmers who play A is 70 per cent in period 0, then in period 1 every farmer would expect others to play A 70 per cent of the time.[27] The logic behind this is the following: 'if it happens in the past, it is likely to occur in the future' (Schotter 1981: 72).

In this model the history of play matters. In period 0, no-one knows what to expect from others, but as time passes the players update their expectations and increase their chances of executing a successful trade. If, eventually, all players expect A (or B) to be played by all the other players all the time, they will play A (or B) all the time and A (or B) will be established as a medium of exchange. Moreover, no-one will have any incentive to do anything else. Both equilibrium points, (A, A) and (B, B), are stable. Note that because of the probabilistic nature of the expectations, updating rule and strategies, the model is not deterministic. Either A or B *may* emerge as a medium of exchange. Both conventions, (A, A) and (B, B), are equally probable. At first sight, it seems that the players would end up coordinating on one of these options, for they may be locked-in to one of the options after some time. But the probabilistic nature of the strategies implies that there is always a chance that some players would choose the less likely option. For example, if the history of play dictates that A is played with 0.9 probability, there will still be players who may play B in the next period. Thus, to assume that the individuals update their expectations does not guarantee that the process will end up in a stable equilibrium – at least in a reasonable period of time.

Schotter makes an important assumption, meant to increase the likelihood that one of the equilibrium pairs is achieved at the end of the process. He assumes that if the state of the model is *close enough* to a state where every individual expects others to play a certain strategy with unit probability (e.g. they believe that 98 per cent of the farmers choose A), every farmer will behave *as if* they have expected every other to play with unit probability.

Normally there are two points, (A, A) and (B, B), where everyone would expect everyone else to play a particular strategy with certainty (see Figure 6.2a). Schotter call these 'absorbing points' because once they are reached no farmer would have an incentive to deviate. The additional assumption made by Schotter (1981: 99–100) tells that every individual would consider points close to (A, A) and (B, B) as absorbing points as well (see Figure 6.2b).[28] That is, players are assumed 'to use a particular pure equilibrium strategy as soon as the players all believe that' a certain 'strategy is very likely to be used' (Schotter 1981: 100). Hence, if the 'absorption area' is large enough, we may expect a medium of exchange to emerge out of this process. Given this assumption, we learn from this model that if market-dependent, economising individuals are able to update their expectations about the others, it is *very likely* that they will coordinate their behaviour concerning the selection of a medium of exchange.

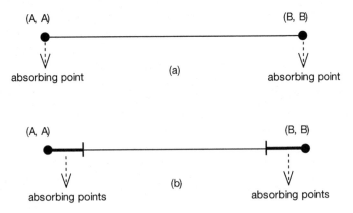

Figure 6.2 Absorbing points in Schotter (1981).

An important difference between Menger's and Schotter's explanations is that Schotter assumes at the outset that trade should be mediated by some intrinsically valueless chips. The agents know that they are supposed to use a medium of exchange. For this reason, the model cannot explain the transformation of a world where the idea of a medium of exchange is non-existent to a world where individuals use a medium of exchange. Schotter only shows that *coordination is possible*. Yet he does not explain how the model economy came to a state where all individuals know that they have to use a medium of exchange. Hence, the emergence of a medium of exchange remains unexplained. To put it boldly, if one wants to explain the emergence of a medium of exchange, he or she cannot assume at the outset that individuals know that they are supposed to use a medium of exchange.

It is easy to see that Schotter's model deals with a small part of Menger's story. He does not have a theory of saleableness, he just assumes that plastic chips are used in trade – thus limiting the number of 'candidates' at the outset. He does not picture a process of discovery, but only a limited type of 'inconclusive' learning from past experience. As it is with the equilibrium models, Schotter shows that the existence of money depends on the common belief that it will be used by others in exchange, and that it is a possible state of the model economy. In addition to this, he demonstrates how a common belief may emerge from a trading environment where individuals are supposed to use a medium of exchange. For this reason, he fails to show that money may be considered as an unintended consequence of human action.

His analysis only makes sense as an account of the emergence of money if it is plugged into Menger's account. Menger explains (in the model world) how individuals may start considering using certain goods to acquire their needs through discovery, learning and imitation. Yet it is possible that at a certain stage in this process different groups of individuals use different goods to mediate their exchanges. At such a state individuals may see the advantage of using a medium of exchange, but they cannot expect every other to accept the same good in their exchanges. This (possible) stage can be considered as Schotter's starting point: there are a couple of goods that are known to be accepted by some, but not all. The individuals have to coordinate their behaviour on a certain type of commodity, and Schotter shows that this is possible if they update their strategies. Moreover, the assumption that if individuals are close enough to the absorbing points they will use pure equilibrium strategies can also be justified with Menger's framework by saying that the increase in the use of a particular good in exchange would increase its marketability and, thus, after some time, individuals would consider that very good as a natural 'candidate' to use in their exchanges. When considered as a contribution to Menger's account, Schotter's model supports the idea that commodity money may be an unintended consequence of human action (i.e. H(1) supported). However, plugging Schotter's account into Menger's story would not make sense for the emergence of fiat money. Since individuals would neither hold a valueless good nor consider using it as a medium of exchange, Schotter *fails* to show that fiat money can be considered as an unintended consequence of human action (i.e. H(2) is not supported).

A similar account of coordination is presented by Young (1998: 11–16, 72–73), who uses the following currency game to demonstrate his general approach to coordination problems. Consider the game in Table 6.3 again. Schotter assumes that the players would use the information about the history of play to update their expectations. Yet he also assumes that only the information from the last period is relevant. Young (1998) suggests that it is more plausible to assume that individuals observe and remember more periods, but not all of them. In what follows, we present a highly simplified version of Young's approach to present his main points.

Young assumes that individuals are boundedly rational and they can only observe a fraction of what happens around them. First of all, they have limited memory: if individuals have been trying to coordinate for t periods they could remember the last m periods and base their decisions for the next period on what happened in the last m periods ($t > m$). When there are n individuals, each individual observes what has happened in the last m periods, calculates the frequency distribution of As and Bs for this time period, and chooses a best reply to this distribution – that is, they try to maximise their expected payoffs, given this distribution. Second, they have limited information. That is, in the n-player case every player is able to observe a fraction of actions of the other players: they may observe only s players' actions ($1 < s < n$).[29] Simply, if player i observes that A is more frequently played than B in the last m periods, then player i chooses A in the next period, because choosing A is expected to yield more utility.

Similarly to Schotter, Young shows that equilibrium points (A, A) and (B, B) are absorbing points (see Young 1998: 51) and that they are (stochastically) stable.[30] More precisely, assume that there are two players that can observe the last two periods. Let w denote the action of player I two periods ago, and x denote the action of player I in the last period. Similarly, let y denote the action of player II two periods ago, and z denote the action of player II in the last period. Then each player is able to observe the information string $wxyz$. Let us now assume that both players played A for the last two periods. Then the information string $wxyz$ would look like the following: AAAA. If the players chose a best reply to this information string then they will have to choose A in the next period. This means that they will be responding to the same information string in the next period, and they will choose A again. Thus, when the model evolves to a state where AAAA (or BBBB) is observed, the model will stay in this state forever. Young also shows that from any initial state one of these goods (A or B) emerges as a medium of exchange with probability one.

But, of course, players may make mistakes, or some players may play in an idiosyncratic way, for example, for other reasons which are not considered in the model. In Young's approach this is reflected by an error rate ε (ε > 0), which is small.[31] That is, player i chooses a best reply to the frequency distribution with probability (1−ε), and chooses an option (strategy) at random with probability ε. In this case, in the short run the best reply dynamics (the process of adaptive learning) will take the system to one of the Nash equilibria, (A, A) or (B, B), and it will stay there for a long period of time. Yet because of the random shocks (or the error rate), the system will not stay there forever. That is, if a number of errors occur in a line, the accumulation of these errors will carry the model from one equilibrium to the other. Young (1998: 12) argues that this may be considered as representing 'shifts in economic and social norms' after long periods of inertia (i.e. abiding to one convention for a long period). Briefly, his model shows that money is bound to emerge if individuals learn from experience, but in the long run social institutions may change if individuals change their behaviour for a sufficiently long period of time.[32] Young (1998: 51–54) also shows that if coordinating on (A, A) yields a higher payoff than coordinating on (B, B),[33] and if the error rate is positive, the equilibrium with higher payoffs, that is (A, A), will be selected and A will emerge as a medium of exchange no matter what happens in the initial stages. Briefly, the 'superior' good is more likely to emerge as a medium of exchange.

Young tries to show how individuals who have limited memory and information and are trying to get the 'best trade' at the moment may bring about a social institution, such as money, as an unintended consequence of their action. Yet our comments on Schotter (1981) also apply to Young. Although he does not explicitly assume that there is a marketing institution as Schotter does, he supposes that individuals are willing to use a medium of exchange by assuming that their interests coincide. For this reason, Young only shows that coordination is possible, and his model fails to explain the emergence of a medium of exchange. Yet plugging his model into Menger's account gives us a better idea about the way in which a medium of exchange may be brought about.

Explanatory progress?

Broadly, we may test a model (verbal or analytical) in two ways: by asking whether it is logically sound and by confronting it with the real world. The above models ask whether Menger's (or the classical) intuition about the origin of money is logically sound. They accomplish this in different ways:

1 by focusing on particular aspects of it;
2 by examining it under different conditions/assumptions;
3 by applying different tools, or by using different methods.

Usually, 1, 2 and 3 go hand-in-hand. For example, Kiyotaki and Wright (1989) focus on storage costs and marketability in a three-good economy with three types of agents. They use equilibrium analysis as a tool for examining this environment. Marimon *et al.* (1990) use 'classifier systems' and 'computer simulations' to examine the Kiyotaki–Wright environment with different specifications (i.e. by changing the number of goods, storage costs, etc.). Gintis (1997, 2000) simulates the behaviour of agents in the Kiyotaki–Wright environment, but he uses ideas from evolutionary theory to model the evolution of strategies. Luo (1999) changes the way in which daily trade is executed and by borrowing notions from evolutionary biology, he introduces a mechanism of imitation. Young (1998), on the other hand, focuses on coordination by abstracting from the trading environment and from the complexities of production and consumption, and assumes that agents may learn from experience and change their strategies accordingly.

Presumably, all of these models have the ultimate aim of providing a better explanation of how money has emerged in history. Yet they simply do not go beyond examining or exploring their model worlds.[34] Our examination of these models teaches us that we should not consider each and every model in economics as providing new explanations, for sometimes they are explorations in model worlds. Sometimes, they do not tell us something new, but they just say that our old intuitions are likely to be correct, or incorrect. The motivations of the authors of the models of the emergence of a medium of exchange support this claim:

> The basic goal of this project is to analyse a simple general equilibrium matching model, in which the objects that become media of exchange will be determined endogenously as a part of the non-cooperative equilibrium.
>
> (Kiyotaki and Wright 1989: 928)

> To be perfectly clear, the goal of the present paper is to use the sequential matching model to derive commodity and/or fiat money endogenously.
>
> (Kiyotaki and Wright 1989: 930)

Here, the goal of Kiyotaki and Wright is not to provide *the* explanation of the emergence of money, rather, the authors want to understand whether money may be endogenously created in a certain type of framework: 'The goal here is to capture monetary exchange as an equilibrium phenomenon and not to force it onto the system' (Kiyotaki and Wright 1991: 217). They ask whether it is possible that money is an unintended consequence of human action in this model world. Obviously, Kiyotaki and Wright's models teach us what we may consider as possible (and what we may not) under certain conditions. Yet they do not tell us whether these conditions were present in history or whether there are plausible mechanisms that may bring about this possibility. Similarly, Marimon *et al.* (1990) explore the possibilities in the Kiyotaki–Wright environment.

Consider the changes from the Kiyotaki–Wright (1989) model to Marimon *et al.*'s (1990) simulation. Marimon *et al.* introduce AI agents instead of rational agents. The implicit question behind this change is the worry that Kiyotaki and Wright's model may not hold if the model agents are not fully rational. Marimon *et al.* find that this worry is indeed true and that under the conditions specified by Kiyotaki and Wright, AI agents rarely use 'speculative' strategies and that usually the least costly to store commodity emerges as a medium of exchange. The suggested improvement is obvious: AI agents are better approximations to real agents and it is more likely that Marimon *et al.*'s results hold in real world. Yet most of Marimon *et al.*'s results are inconclusive and are specific to certain model specifications. For this reason, it is hard to argue that there is much progress in understanding how money is brought about as an unintended consequence of human action. Rather, we learn from this model that Kiyotaki and Wright's results do not hold under every condition. If you like, this may be considered as progress in terms of *removing unsound results* from the Kiyotaki–Wright model. On the other hand, Marimon *et al.* demonstrate that even if individuals are not fully rational, commodity money may be brought about in the Kiyotaki–Wright environment. That is, they show that Kiyotaki and Wright's results concerning fundamental equilibrium hold under more plausible assumptions about individual behaviour and that the fundamental equilibrium may be reached via simple learning dynamics.

Figure 6.3 illustrates the relationship between Menger's model world, Kiyotaki–Wright models and Marimon *et al.*'s simulations: at the first level we have the real world, at the second level we have Menger's rich but vague model. At the third level we have more idealised versions of Menger's model world, Models A and B, which are examined by Kiyotaki and Wright, and Models C and D, which are considered by Marimon *et al.* in addition to Models A and B. Marimon *et al.* examines specific versions of these models, A.1.1, A.1.2, etc., which are particular exemplifications of these models. Figure 6.3 shows how Menger's model has been explored in the contemporary literature on emergence of a medium of exchange (also see Table 6.4).

Consider Young's (1998: xi) goal of developing a new framework to examine the emergence and persistence of institutions (e.g. in our case, money). He states that his aim is (i) 'to suggest a reorientation of game theory in which players are not hyper-rational and knowledge is incomplete' and (ii) 'to suggest how this framework can be applied to study of social and economic institutions.' Young tries to get a better idea of what is possible by isolating his model from other problems and introducing a better approximation to the behaviour of real individuals into the model. Yet he does not suggest that real individuals calculate the frequency distribution of the previous actions of a limited number of agents and calculate the best replies to this distribution. He rather suggests that if we may take this characterisation as an approximation to real individuals learning behaviour, then we may argue that individuals *may* be able to coordinate their

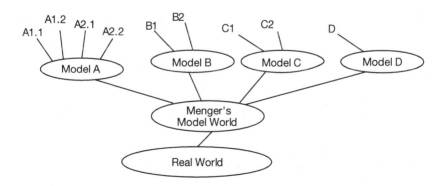

Figure 6.3 Models of emergence of money in relation to Menger's model.

Table 6.4 Models in comparison to Menger (1892a)

	Menger 1892a	Kiyotaki and Wright 1989	Kiyotaki and Wright 1991	Aiyagari and Wallace 1991	Kiyotaki andWright 1993	Starr 1999 I	Starr 1999 II	Williamson and Wright 1994	Marimon et al. 1990	Luo 1999	Schotter 1981	Young 1998	Duffy and Ochs 1999
Economising action	•	•	•	•	•	•	•	•	•	•	•	•	•
Market dependence	•	•	•	•	•	•	•	•	•	•	•	•	•
Absence of double coincidence of wants	•	•	•	•	•	•	•		•				•
Storage costs	•	•		•					•	•		•	•
Transaction costs	•				•	•	•	•					
Marketability	•	•	•		•				•	•	•	•	?
Frequency of use / market traffic	•							•			•	•	?
Imitation / learning	•								•	•	•	•	?
Uncertainty								•					
Rationality — Hyper		•	•	•	•	•	•	•					?
Rationality — Limited	•								•	•	•	•	
Examines[a]	1	1	1	1	1	1	1	b	2	1,2	2	2	3

Notes
a   See the questions in the beginning of this chapter.
b   Introduces a new factor, uncertainty, to examine the fiat money equilibrium.

behaviour. We have seen that his (and Schotter's) model can only be considered as an account of the emergence of a medium of exchange, if considered together with Menger's account. That is, he shows how individuals may be able to coordinate at a later stage in Menger's model world. Figure 6.4 presents the relation between Menger's account and Schotter and Young's.

Schotter and Young present an idealised version of the last stages of Menger's account of the emergence of money. They show how learning from past experience makes coordination possible. We do not learn from these models how money actually emerged in history, rather we learn that it is *plausible to argue* that it may have emerged as an unintended consequence of human action.[35]

After all of these models, we learn nothing new regarding the possible mechanisms that may explain the emergence of money (i.e. in comparison to Menger's account), rather, we have a better understanding of the model worlds from which we may more confidently argue that money may be considered as an unintended

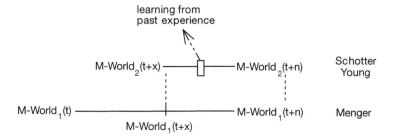

Figure 6.4 Schotter and Young's models in comparison to Menger's.

consequence of human action. While this intuition is 'tested' on logical grounds to a considerable extent, the relation between the model world and the real world has not been examined in a meaningful way. There are few attempts to test these models with real individuals (e.g. Duffy and Ochs's 1999) and no attempts to verify whether the conditions under which money emerges in these model worlds hold in the real world. Unfortunately, it is hard to find information about the motivations of individuals and the particular conditions in which money may have flourished, and lack of this kind of evidence leaves us with partial potential explanations of the emergence of money.

Except Duffy and Ochs's experiment, all models and simulations examined above deal with model worlds that are restricted versions of Menger's model world (see Table 6.4). They examine the conditions under which we may consider money as an unintended consequence of human action, and the mechanisms and factors that may transform a world of direct exchange into a world where trade is mediated by money. For this reason, we may consider them as testing the logical soundness of the intuition that money may be an unintended consequence of human action. They do this by exploring the properties of an abstract world, by focusing on different aspects of an abstract trading environment, by adding elements to, or removing elements from this world. Some examine whether the 'absence of double coincidence of wants' would lead to a medium of exchange, some examine the effect of 'transaction costs', and some others examine whether the beliefs of individuals is important in the process of the emergence of money. Some other models examine what happens in a world where agents learn from past experience, or imitate others' behaviour. But all these efforts are concerned with model worlds, or with variations of a model world.

Alternatively, we may say that these models contribute to the explanation of the emergence of a medium of exchange in these model worlds. Yet these models do not alert us to new explanatory mechanisms.[36] Rather, these models show us that under certain conditions these factors and/or mechanisms may explain the emergence of a medium of exchange in the abstract world of models. They try to explicate how certain mechanisms, in isolation from others, may work together, or whether they are consistent with the existence of the explanandum phenomenon.

By logically testing and exploring the intuition that money may be considered as an unintended consequence of human action, they strengthen the belief in this intuition. They provide firmer grounds for arguing/believing that commodity money may be considered as an unintended consequence of human action. Nevertheless, we do not know – any better than Menger did – whether money was brought about by similar mechanisms in history.

When theoretical explanation is at stake, it is not usually easy to assess whether there is any 'real' explanatory progress unless the model that forms the basis of the explanation is applied to particular cases. As neither any of the above models nor their combination is used to explain particular cases, it is hard to asses the amount of progress in the economics literature on 'the origin of money'. For this reason, it is useful to distinguish between *actual explanatory progress* and *progress in potential explanatory power* to discuss the contribution of these models, simulations and experiments.

Actual explanatory progress occurs when a *particular case* is 'better' explained with a new or improved model.[37] For example, a singular explanation may be better than the other if it provides a better understanding of the actual and effective causal mechanisms and structural relationships that are responsible for the particular fact or event (e.g. the emergence of money at a particular time period and place). These models do not provide such an explanation. For this reason, we may confidently argue that there is no actual explanatory progress.

Yet to have a better explanation of particular cases one may need a better collection of tools, models, theories, etc. Thus, if the research in one particular subject develops in the line of providing better models that may guide us in the search for a good explanation, we may say that there is progress in terms of potential explanatory power. In isolation from each other, the above models, simulations and experiments contribute very little to what Menger has already said in terms of suggesting new possibilities. Yet we may have a better idea of what is accomplished by these models by considering them in combination, as forming the parts of a meta-model (or theory) which is still under construction.[38,39] Or, alternatively we may consider them as constituting the toolbox of economists when they are confronted with the task of explaining particular cases in the real world. Then we may see that the above models are giving more detailed pictures of particular areas of Menger's model world. That is, by the introduction of new models that examine particular aspects of existing model worlds, the meta-model (or theory) of the emergence of money gets more detailed in time. Since particular models examine what may happen under certain conditions and assumptions, we obtain a more detailed picture of what is possible and how it is possible. This, of course, increases the potential applicability of the meta-model and its potential explanatory power.

Concluding remarks

We have said that Menger's intuitions are formulated in the current literature in the following ways: H(1) – commodity money may be an unintended consequence of human action; and H(2) – fiat money may be an unintended consequence of human action. The results of our examination of the literature on 'the emergence of a medium of exchange' may be summarised as follows:

1 The end-state models show some of the conditions under which 'a medium of exchange' is a possible state of 'the model world'. These models show that the existence of 'a medium exchange' is consistent with rational agents pursuing their own interests.
2 The process models show that under certain conditions individual mechanisms, such as 'imitation' and 'learning from past experience', bring about a medium of exchange in 'the model world'.
3 H(1) is confirmed, yet H(2) is not. The models, simulations and experiments that are examined in this chapter *fail* to show that fiat money may be considered as an unintended consequence of human action – a conclusion which is in line with Menger's intuitions (see Chapter 3).
4 Experiments with real individuals show that storing costs may be more important than marketability considerations. Yet the limitations of these experiments suggest that this conclusion may be doubtful. On the other hand, experiments show that under certain conditions, H(1) is confirmed.
5 Different models of 'the emergence of a medium of exchange' examine different aspects of 'the model world', and by way of doing this they contribute to the explanation of the transformation of M-World*(t)* into M-World*(t + n)*. That is, when considered as a collection of models that form a meta-model (or theory) of 'the origin of money', they provide a better explanation of the transformation of 'the model world' from a state of unmediated exchange to a state of mediated exchange. These models increase the potential explanatory power of this meta-model.
6 The relationship between 'the model world' and the real world is not examined. No particular cases are explained. There is no actual explanatory progress in the sense of providing satisfactory singular explanations.

If we consider the above models as a collection we may see that they increase the potential explanatory power of the meta-model (or theory) of the emergence of money by providing epistemic access to a wider range of situations and possibilities. These partial models only make sense if we conceive them as contributing to an (imaginary) meta-model. Otherwise, they have little to say. This interpretation is further supported by the fact that applied research in economics usually starts with a survey of the related theoretical literature. When economists try to explain particular cases they consider most of the available partial models to see which of these fit better to the situation under consideration, and they usually combine insights from different models to explain particular cases. Yet philosophers of social science and critical heterodox economists usually consider these models in isolation from the other related models and from related literature, and find no value in them. The message of this chapter is also important for this reason. It should be added that partial models, such as the ones examined in this chapter, have the potential of stimulating research in related areas. For example, models of the emergence of money have implications for monetary theory and have influenced further models of the emergence of institutions.

Nevertheless, it should again be emphasised that these models do not explain particular cases; they only increase our *chances* of explaining particular cases. Progress in potential explanatory power is necessary for attaining actual explanatory progress, but unless the model is verified, we cannot say that we have improved understanding of the process of the emergence of money; we can only say that we *think* that we have improved understanding. That is, the proof of the theoretical explanation (or model) is in its success in explaining particular cases. The above literature leaves us with conjectures, which may or may not 'really' explain the emergence of money.[40] For this reason, they have to be tested further, analytically, experimentally and historically. Nonetheless, they give us reason to believe that the economising actions of individuals, learning, imitation and saleability of goods may have been important in the process of the emergence of money.

Our examination also suggests that models do not always represent/examine the real world directly, instead, sometimes they are built to examine the results of other models under different conditions. In this sense, some models are akin to *thought experiments*, in that they test certain hypotheses in an abstract world. We have also seen that these models were also tested with real individuals, which takes us closer to the type of experiments that are conducted in physics. Experiments take us closer to the real world and give us a better understanding about the conditions under which our models may work. The relation between different type of models and reality needs more attention and the next chapter examines this relation from a philosophical perspective. Reconsiderations of Schelling's models of residential segregation guide this discussion.

7 Models and representation

Introduction

Explaining the emergence of unintended social phenomena entails the construction and exploration of model worlds. These models are supposed to represent the real world, but they also contain an element of speculation. The construction, examination and exploration of models tell us what may be possible in the real world and show new ways in which we may look at the world, and/or help us evaluate the plausibility of our conjectures about the real world. This chapter discusses these issues in greater detail in light of the relevant philosophical literature. The synopsis of the argument is as follows:

1 The explanations we have considered in this book are partial potential explanations that rest on partial models. In order to understand the nature of these models, it is necessary to understand how partial models are supposed to explain.

2 Models help us explain by way of providing a proper way to conceptualise the real world, and for this reason the representational role of models is important. Sometimes models may represent particular states of the world, but the models of unintended social phenomena are not meant to represent any particular phenomenon. This implies that the relation between the model world and the real world has to be examined further.

3 The relation between the model world and the real world is rather complex. Models usually rest on a web of abstractions and idealisations, and maybe more importantly, their representations are often 'flawed'. Understanding how representation works in these models gives us clues about why they may have a chance of alerting us to real possibilities.

4 The examination of the accepted philosophical views on the relation between models/theories and the real world suggest that the similarity between the model and the real system it represents is crucial for the success of explanation. Yet similarity as such is not a well-defined concept and does not help us in evaluating models of unintended social phenomena.

5 Consideration of models of unintended social phenomena reveals that the similarity between such models and the real world amounts to the existence of certain (known) tendencies (individual mechanisms) in the model world,

and for this reason these models demonstrate some of the possible ways in which these individual mechanisms (tendencies) may interact. That is, models and theories may be conceived as webs of idealisations and abstractions. In virtue of the similarity between the constituents of the model and the parts of the real world they represent, they alert us to possible ways in which certain individual mechanisms in the real world (or certain tendencies) *may interact* – even if some of the assumptions of the model do not hold.

6   Good examples of these models alert us to new possibilities, but exactly for the same reason, it is necessary to test the plausibility of these models, by exploring their premises and implications and by confronting them with the real world. Explorations of Menger's model (see Chapter 6) and the chequerboard model indicate the different ways in which models are explored.

7   The fact that models are explored in many different ways suggests that they are somewhat independent from the real world and theories, and that they may serve different functions. Thus, a model of an unintended social phenomenon cannot (and should not) be evaluated in isolation from other related models and hypotheses about that phenomenon and about other related phenomena. When singular facts are in need of explanation, all relevant models and data about the very fact to be explained may be used and for this reason they may be considered as forming an incomplete theory of that fact.

Briefly, this chapter critically discusses the important philosophical views concerning models and explanation in light of our examination of models and explanations of unintended social phenomena. It develops the argument that models of unintended social phenomena have a chance of alerting us to real causal mechanisms and structural relationships for they employ known individual mechanisms (i.e. tendencies) and explore the ways in which they may interact. Usually, their novelty is the demonstration of this *interaction*. By way of showing how certain familiar mechanisms[1] may interact and bring about the explanandum phenomenon, they serve as eye openers. They expand our mental horizon and increase our chances to explain particular instances of those social phenomena.

Partial potential explanations

The philosophical literature on scientific explanation makes distinctions between (a) singular and theoretical explanation, and (b) complete and incomplete explanation. Singular explanations are explanations of singular facts or events (such as the fact that different ethnic groups are living in different parts of the city in Rotterdam).[2] A singular explanation is expected to explain why a certain fact occurred in a certain way, to provide the causal history of it, or to inform us about the causal and structural relations that produced the explanandum phenomenon. If the phenomenon is fully accounted for, we have a complete singular explanation. A complete theoretical explanation, on the other hand, is supposed to provide all the causal and structural relationships that may explain all instances of the explanandum.

Most of the philosophical discussion is concerned with complete explanations:[3]

> If 'scientific explanation' does not mean 'explanation actually offered in science', the sense of the expression is far from obvious, and needs to be made clear. Many philosophers of explanation use it merely in the sense of 'an ideally complete explanation'.
>
> (Ruben 1990: 19)

Moreover, generally the focus is on complete singular explanations, rather than theoretical explanations. The rationale behind this is that what holds for complete singular explanations also applies to theoretical explanations (e.g. Ruben 1990: 19–23, Ylikoski 2001: 8), and that this abstract discussion is useful for it provides a reference point to evaluate actual explanations. The explanations actually offered by scientists differ more or less from these ideals; that is, they are usually incomplete. Obviously, the explanations we have encountered in this book are not complete explanations.

An important kind of incompleteness that Hempel discusses is partiality. An explanation is *partial* if it does not fully account for its explanandum:

> Often, however, explanatory accounts exhibit a more serious kind of incompleteness.[4] Here, the statements actually included in the explanans, even when supplemented by those which may reasonably be assumed to have been tacitly taken for granted in the given context, account for the specified explanandum only partially.
>
> (Hempel 1965: 415)

For example, a *complete* singular explanation of the emergence of residential segregation in Rotterdam has to provide all the reasons why it emerged in Rotterdam. It has to tell us about the arrival of foreign workers and people from former Dutch colonies; the initial housing decisions of the foreigners; Dutch housing policies; the employment areas for these people; their income levels compared to Dutch citizens; the immigration policies of the Dutch government; the racial preferences of the individuals; and all the events and facts that contributed to the emergence of residential segregation in Rotterdam. Obviously, such complete singular explanations are rare. Usually, one is content with providing the most important causal and structural factors. Hence, we usually face incomplete, or partial, singular explanations in real life. On the other hand, a complete theoretical explanation of the emergence of residential segregation should be able to account for all particular exemplifications of residential segregation in different times and places. Again, such complete theoretical explanations are rare, at least in social sciences. It is usually the case that general models account for some of the factors that may explain the emergence of residential segregation. Actual theoretical explanations are usually partial.

We have seen that Menger and Schelling's explanations are partial in the sense that they only take into account some factors that may explain the explanandum phenomenon and for this reason they cannot fully account for the explanandum phenomenon in Hempel's sense. They ignore some of the factors known to be relevant for explaining the emergence of money and segregation (e.g. the fact that social and economic factors bring about segregation). Yet we should emphasise that partiality is not a special feature of explanations of unintended social phenomena; rather, most of the explanations offered by scientists are more or less partial:

> many explanatory accounts offered in the literature of empirical science have the formal characteristics of partial explanations, and that as a consequence, they overstate the extent to which they explain a given phenomenon.
>
> (Hempel 1965: 417)[5]

If we know that the explanans of a partial explanation is true, then having a partial explanation is better than having nothing – for we can develop the explanation by plugging in the other necessary elements. Yet some explanations may also fail to satisfy this truth condition, or the empirical adequacy condition. (Hempel and Oppenheim 1948: 248) The empirical adequacy condition states that 'the sentences constituting the *explanans* must be true.' This means that the statements of the antecedent conditions and the general regularities utilised by the explanation must be true. If this does not hold, then we have a *potential explanation*, instead of a true or correct explanation (Hempel 1965: 338). We have seen that this condition is not satisfied by the explanations we have examined in this book, because the descriptions of the antecedent conditions are rather conjectural or fictional. Particularly, Menger and Schelling's explanations are potential (theoretical) explanations in Hempel's sense, for we cannot guarantee the truth of their premises.

We have noted that most of the philosophical literature on explanation is concerned with ideal explanations. In such a view, complete theoretical explanations are explanations of laws and generalisations with other laws and generalisations,[6] and for this reason they are assumed to rest on an idea of well-established theory (e.g. Hempel 1965). A well-established theory may be defined as having all the conceptual (and methodological) resources for explaining the particular exemplifications of the type of phenomenon it is concerned with. However, we have seen that Menger and Schelling's explanations are not based on well-established theories, and they do not entirely rest on generally accepted generalisations and/or laws. Rather, these explanations are partial and based on partial models. Our examination of the contemporary literature on the emergence of a medium of exchange showed that partial models are important tools in the explanations of unintended consequences of human action, and that models may contribute in different ways to these explanations. The assumption of a well-established theory cannot help us understand how partial theoretical explanations work and how they could help us explain particular cases. A better understanding of the actual explanations of unintended social consequences rests on a better understanding of theoretical explanations and their relation to partial models.

Models and explanations

In previous chapters, we have discussed two alternative (but complementary) interpretations of the invisible hand: the process interpretation and the end-state interpretation. Explanations that subscribe to the process interpretation of the invisible hand may be considered as causal explanations for they focus on the mechanisms that may bring about certain unintended social phenomena. The causal view of explanation[7] suggests that an explanation has to inform us about the way in which entities in the real world are causally related to each other, or about the causes of the explanandum phenomenon (Salmon 1984, 1990).[8] Yet a mere statement of the causes is not enough. A proper explanation has to inform us about the way in which causes are connected to their effects and to explicate the causal mechanisms that produce the phenomenon under study.[9] Unfortunately, the causal view of explanation is mostly concerned with singular explanations; that is, with explanations of particular facts.[10] For this reason, it does not help much to have a better understanding of the type of partial potential theoretical explanations we have encountered in this book. Given the requirements of the causal view, we may not argue that a proper explanation is provided unless a particular case is fully explained and/or the existence of the proposed mechanisms is supported by 'objective evidence' (Salmon 1998: 90). What we may take from the causal view is that the knowledge of (particular and/or general) causal mechanisms is important.

What about the end-state models? Since they do not seem to provide the causal mechanisms responsible for the origination of the end-state, they cannot be considered as causal explanations – at least, not of the sort we have discussed above.[11] But it may be that not all explanations are causal. Explaining some property of a matter A by referring to its structure can also be considered as an explanation. Since it is an open question whether structural explanations are causal or not, we need not constrain our view of explanation to the causal view or to a specific type of causation. Rather, we may take it that a proper explanation should inform us about the structural and causal relationships that constitute or produce the phenomenon under consideration. If we accept this view we may consider the end-state models as informing us about the possible ways in which certain social phenomena may be constituted, thus providing certain bits of information that may help us in explaining particular cases of the social phenomenon under consideration (see Chapter 6). Although we will be talking mainly about the process interpretation, it seems wiser to have a view of explanation that is much more flexible than the causal view:[12]

> explanations work in virtue of something determining or being responsible for something. [. . .] we explain something by showing what makes it or what is responsible for it. The fault of causal theory of explanation was to overlook the fact that there are more ways of *making* something what it is or being *responsible* for it than by causing it.
>
> (Ruben 1990: 231)

Mäki (1994: 159) proposes a flexible account of explanation that is consistent with this view, and which takes into account the fact that theories and explanations are, or may be, partial. He thinks that explanation can basically be characterised as re-description. According to this account, 'explanation [. . .] involves redescription of explananda' and re-description should be understood as re-description of what has been empirically described before (Mäki 1990a: 320). An important presumption of this view is the following:

> it is primarily the task of scientific theory to do the explanatory work. Singular phenomena cannot be explained by deriving their descriptions from empirical generalizations. [. . .] It is rather the case that theories account empirical facts directly. It is only by means of the conceptual resources of a theory – not being reducible to the observational language of empirical facts and generalizations – that empirical facts can be redescribed in a way which reveals what those facts are really are.
>
> (Mäki 1990a: 321)

Here, Mäki is talking about singular explanations, and argues that by using the conceptual resources of a theory we may explain particular facts (i.e. provide singular explanations). Re-description of particular facts depends on theories. But this view also entails that theories and models represent parts of the real world. It is with the help of these representations that we are able to re-describe particular facts and provide singular explanations. Clearly, when we try to spell out the relation between theories, models that represent parts of the real world and redescriptions (singular explanations) that rest on these representations, the meaning of 're-description' gets blurred. Considering explanation as re-description is appealing, but what re-description really amounts to is not very clear. Consider Schelling's chequerboard model, for example. Is the chequerboard model supposed to re-describe the way in which segregation emerges, or to represent the way in which it emerges and help us re-describe particular cases of segregation? But on what kind of prior empirical description of emergence of segregation does this re-description rest? A proper prior empirical description (whatever it really is) of emergence of segregation seems to be non-existent in Schelling's case. Mäki may be right in that all explanations rest on the existing state of knowledge about the phenomenon under study, and this knowledge rests on empirical descriptions of that phenomenon. But characterising explanation as re-description does not tell much about the way in which the chequerboard model (or Menger's model) is constructed and is supposed to explain. There seems to be a more complex relationship between theories, models and explanation – at least for the cases we have examined in this book.

Moreover, the term 're-description' wrongly suggests that models represent particular phenomena. For example, concerning the chequerboard model, Sugden challenges the re-description view (particularly Mäki's account of isolation and representation) by arguing that the chequerboard model does not represent any particular city:

it does not seem right to say that the chequerboard model isolates some aspects of real cities by sealing of various other factors which operate in reality: just what do we have to seal off to make a real city – say Norwich – become like a checkerboard? Notice that, in order to arrive at the checkerboard plan, it is not enough just to suppose that all locations are identical with one another (that is, to use a 'generic' concept of location): we need to use a *particular form* of generic location. So, I suggest, it is more natural to say that the checkerboard plan is something that Schelling has *constructed* for himself. If we think that Schelling's results are sufficiently robust to changes in the checkerboard assumption, that assumption may be justified even if it is not an isolation.

(Sugden 2000: 22)

Sugden is right in that the chequerboard model is constructed and is not a model of a particular city. Yet this does not contradict with the statement that models represent and are isolated from other complexities of the real world. It is probably the terminology of the explanation-as-re-description view that wrongly suggests that models represent particular phenomena. As will be seen in the rest of this chapter, the relation between models, theories and explanations is complex. For this reason, we need a better picture of how models and theories represent and help us to explain.

To prevent other possible misunderstandings caused by the notion of 're-description', and to mediate the complexity of the act of representation and explaining, we may argue that explanation entails discovering ways to *conceptualise* the phenomenon at stake:

Whether a fact as normally understood explains or is explained depends at least in part on the way in which the properties involved are conceptualised: relative to the conceptualisation of a property in one way, the fact may be explanatory, relative to a different conceptualisation of the same property, the fact is not explanatory.

(Ruben 1990: 177)

Explaining a complex phenomenon entails discovering how to conceptualise it, that is, to discover the way in which the world works, or may be working in producing a certain phenomenon. Given the importance of partial models in explanations of unintended social phenomena, and the conjectural character of the reasoning involved, we may further argue that novel partial potential theoretical explanations (e.g. such as Schelling's) involve 'creative conceptualisations' that are partly discovered through the construction of models and conveyed by way of presenting them. It is on this complex task of representation that explanation rests. The next section examines the way in which models represent.

Models and representation

Models have many uses in everyday life, in science and in philosophy of science. Models might be material, visual, formal or representational, theoretical, or imaginary. Mäki (2001a) argues that there is one thing common to all these different types of models: 'they represent something beyond themselves'.[13] In this sense, we may consider the chequerboard model of residential segregation as a representation of the way in which segregation emerges in the real world. But what is the nature of this representation? To see this let us start by examining how we can represent an already known particular phenomenon, and gradually introduce the types of issues that are relevant in modelling social phenomena.

Isomorphisms

Basically, any 'environment can in principle be described by a set of states and a transition function (or next state function) that specifies how the states change overtime' (Holland *et al.*, 1989: 30).[14] Thus, a model characterising an environment like a city with a focus on segregation patterns may describe the city to be in a state where there is no segregation, and in another state where there is segregation. A transition function, then, can be characterised with the elements in the model that carry the city from the first state to the other. Thus, the transition function(s) (of the model, T' in Figure 7.1) represents the causal mechanisms that bring about segregation.

Assume now, for the sake of the argument, that we have all sorts of information about a particular city (City A) concerning its states prior to and after the emergence of segregation. We also have a considerable amount of information about individuals' motivations and discriminatory preferences and about the way in which two different ethnic groups are distributed in City A. Yet we do not know why and how segregation emerged. That is, we have all sorts of information about R-World *(t)* and R-World *(t + n)*, but we do not know the way in which R-World *(t)* was transformed into R-World *(t + n)*. We want to know the mechanisms (T) that have contributed to the emergence of segregation.

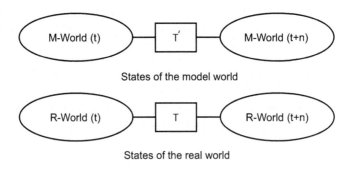

Figure 7.1 Model world and the real world (adapted from Holland *et al.* 1989).[15]

What would a model of this transition look like? We may start with a detailed description of R-World *(t)* and R-World *(t + n)* (i.e. describing every individual, neighbourhood, action, etc.). We may also try to present a detailed picture of every state of City A from no-segregation to segregation.[16] But a description where every aspect of the city finds a place would be like the city itself and would be too complex to tell us something about the way in which segregation emerged in this city. That is, it would be hard to build a model that is *isomorphic* to the city it represents (i.e. where there is a one-to-one relation between the model city and the real city). Simply, a complete description of what we know about the city is not very informative, or it is too complex to tell us something.[17]

Abstractions and q-morphisms

One important concern about how to represent the transition of a real city from one state to the other is that the real world is usually too complex to be modelled in this way. Thus, it is 'unreasonable to expect' models 'to be isomorphisms in which each unique state of the world maps onto a unique state in the model' (Holland *et al.* 1989: 31). A fundamental component of thinking about the world is abstraction. Holyoak and Thagard (1996: 19–20) remind us of a story of Jorge Luis Borges, *Funes the Memorious*, in which the fictional character Ireneo Funes has an exceptionally good memory to remember every detail in his life. But his alertness to details makes him incapable of abstract reasoning.

> Not only it was difficult for him to comprehend that the generic symbol 'dog' embraces so many unlike individuals of diverse size and form; it bothered him that the dog at three fourteen (seen from side) should have the same name as the dog at three fifteen (seen from the front).[18]

Borges's story suggests that abstraction is extremely useful for our reasoning and it guides our inferences, which usually go beyond our direct observations and mere descriptions of the real world.[19]

Neither ordinary human beings nor scientists can reason without abstraction. The model builder is then forced to form general concepts and to simplify the model by ignoring unnecessary or irrelevant components of the world for the specific task at hand. In daily life we use such categorisations and abstractions quite naturally without much thinking, yet to build a model we also need conscious thinking about the way in which certain entities should be characterised, categorised, etc. For example, for the case of segregation, we may categorise individuals into different categories, such as racist individuals, individuals who prefer a mixed neighbourhood, individuals who cannot tolerate a certain minority status, etc. Or, we may categorise individuals into different groups, such as individuals who have above-average income, average income, low income, etc. Note also that 'segregation' itself is an abstraction and that such categorisations are not independent from our existent state of knowledge.

Having formed certain categories and generalisations, we may use them in our model. But this time the relation between our model city and the real city is not isomorphic (in the sense defined above). Rather it is a many-to-one – that is, *homomorphic* – relation. 'A faithful model based on categories, in which the mapping from elements of the world to elements of the [model] is many-to-one, is called homomorphisms' (Holland *et al.* 1989: 31). Indeed, this is what happens in many economic models: certain concepts in our models, such as the consumer and the producer, represent many individuals.

However, it is also true that our categorisations and generalisations are not exceptionless. For example, if we represent several individuals who prefer a mixed neighbourhood with a few model agents, and homogenise their preferences concerning their neighbourhoods, we may overlook the fact that some of the (real) individuals may be more tolerant to a minority status than others. Our representation of the states of City A would most probably be incomplete, subject to exceptions. Generally, our models of the world are typically far from perfect, that is, the mapping between the model world and the real world is flawed. In Holland *et al.*'s terminology, such mappings are called *quasi-homomorphisms* (q-morphisims). Considering the case of segregation, this means that our model of the transformation of R-World *(t)* to R-World *(t + n)* is likely to be incomplete, and the relation between R-World and M-World would be q-morphic.

Isolation

The fact that our models of the real world are usually q-morphic implies that they are deformed images of the real world. Yet such deformations are necessary for dealing with the complex real world, that is, for understanding, for living in, etc.

> Faced with the essential complexity of the world, every science is compelled to employ methods of modifying and deforming it so as to make it or the image of it theoretically manageable and comprehensible.
>
> (Mäki 1992a: 317)

We have seen that one of the tools that helps us represent the real world in a comprehensible manner is abstraction and that it is a subspecies of isolation (see Chapter 4, note 10). We use the method of isolation in everyday life, consciously or unconsciously. We use abstractions and we also idealise the environment we are living in. Likewise, scientists isolate; they focus on some factors rather than the others, given their initial hypothesis about the way in which a certain fact may be explained. As Mäki would agree, one-to-one representations of the real world phenomena are intractable. This is one reason to utilise the method of isolation. But more importantly, the ability to isolate is the key to understanding the world.[20]

Analogical thinking

But how do we choose the factors that are relevant, that is, those to be isolated from the influence of others? Until now we have implicitly assumed that it is possible to model the transformation of City A from a state with no segregation to a state with segregation. Yet the real problem in modelling this transformation is to discover a way to represent it, a way that would tell us something about the real world. If we do not have any idea about the way in which segregation may have emerged, it would be practically impossible to present a model of the emergence of segregation. Having an initial hypothesis about the way in which segregation may have emerged is certainly necessary for building a model of this process. Generally, scientists do not start their investigation from scratch. They have certain ideas about what may have caused a certain phenomenon, or what may be relevant in explaining a certain fact. Obviously, not all the available information about the states of City A would be relevant in explaining the emergence of segregation in this particular city. For example, we may ignore the data about the shoe sizes of the individuals, or the type of armchairs they use as being irrelevant to, or having negligible impact on, the emergence of segregation. Selecting the seemingly relevant factors is not independent from our existing state of knowledge about the world, about the things we know about segregation in general, and about other aspects of social life. What we know about the real world helps us conjecture about the factors that may be relevant, and to formulate an initial hypothesis about the way in which segregation may have emerged. When we try to build models, we usually go beyond simple statements of bits of information, or mere descriptions of what we observe, and our model depends on, or starts from, our initial (i.e. prior to model building) knowledge about the world. For this reason, model building certainly involves some type of analogical or metaphorical thinking.

One fundamental purpose of analogy is 'to gain understanding that goes beyond the information we receive from our senses' (Holyoak and Thagard 1996: 9). Human beings recognise similarities between what they observe and what they already know about other things, and usually if they find enough (or some) similarities between two different phenomena, they use these similarities to reason about the relatively less-known phenomenon. If we are allowed to use 'analogy' in a loose way, we may say that all we know about our neighbours, how they react to or interact with another ethnic group, or the things we know about social behaviour, individuals' reactions to social differences, etc. may help us in forming an idea about the way in which segregation may have emerged in City A. Of course, analogical or metaphorical thinking cannot be the definite source of our knowledge. But it may help us form our initial hypothesis about the emergence of segregation. That is, 'analogy is a source of plausible conjectures, not guaranteed conclusions' (Holyoak and Thagard 1996: 30).

Simple analogical thinking involves the discovery of the similarities and differences between objects, and of relations in two different domains. Most of our everyday explanations make use of similarities between causal relationships in two different domains. Moreover, scientific explanations involve more complex high-order relational and causal mappings, as well as the use of metaphors.[21] The ability to map what we observe onto what we already know is an important mechanism that makes us generate thoughts about what we are trying to explain. We find out what we believe to be the relevant relationships about a phenomenon by conceptualising it in light of our existing knowledge together with our observations about the phenomenon to be explained. In short, what we know about the real world helps us generate fallible hypotheses or conjectures about the way in which a certain phenomenon may be explained.[22]

Representation

Now, let us assume that what we know about segregation is the following: we know that there are people who have strong discriminatory preferences, and that there are welfare differences between two ethnic groups. We also know that existing accounts of the emergence of segregation suggest that these two factors explain segregation in other cities. Then, other models of residential segregation may help us single out the relevant factors for our own model, and study their relation in isolation from other factors. That is, we may infer that what is true for other segregated cities may also be true for City A, and start our modelling from here. This helps us categorise and organise the available information and try to see whether such and such is really the case for City A.

Briefly, given what we know about segregation, we may isolate a couple of factors, strong discriminatory preferences and welfare differences, and try to see whether emergence of segregation in City A can be explained by these factors. In this case, our task is to show how these factors interact in bringing about segregation in City A. The only thing we have to do is to check whether our initial hypothesis is confirmed by what is known about City A. If this is the case, we would have a singular explanation of the emergence of segregation in City A. Yet the reader may object that we hardly have a theoretical model of segregation in this case, rather, what we did is to use the existing models / theories of segregation to explain another instance of segregation. This is true. We have explained segregation in City A with the help of previous 'theories' of the emergence of segregation. Models of this type do not provide new insights about general phenomena; rather, they confirm previous insights and generalisations, and help us explain new particular cases. As we have seen in the previous chapters, this is not what happens in explaining unintended social consequences.

Until now, we have assumed that modelling starts from particulars and models represent particulars. Remember that 'in an isolation, something, a set X of entities, is "sealed off" from the involvement or influence of everything else, a set Y of entities; together X and Y comprise the universe'. The elements of the set of X entities need not be particulars. Yet we have seen that 'explanation as redescription' is sometimes wrongly interpreted as suggesting that models represent particular phenomena.[23] The practice of modelling does not always start from particulars, rather scientists try to model a general phenomenon (e.g. the emergence of segregation in general) and consider a set of other generalisations and concepts in isolation from others to be able to model segregation. To see this, consider the case of the emergence of segregation again.

Figure 7.2 pictures a possible way in which we may conceptualise segregation prior to our modelling efforts. What we know is that there are many residentially

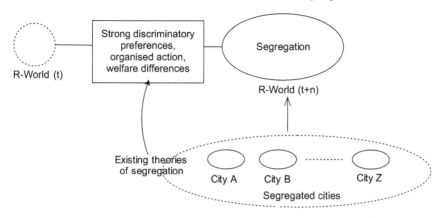

Figure 7.2 A possible way to conceptualise segregation.

segregated cities, and that segregation in general is explained with strong discriminatory preferences, organised action and welfare differences. That is, our initial conception of segregation can be considered as a general (mental) model about segregation. We start our examination of residential segregation from this general conception of segregation, that is, given what we know about segregation. This means that what we have characterised as the real world (what has been characterised as R-World before) may be considered as a 'model' itself; it is our conception of the real world, which is characterised by many generalisations and isolations. We rarely build theoretical models of particular phenomena from scratch; we usually have a mental image of them and start from there. Scientists build theoretical and empirical models to *refine* this image.

Let us assume now that our knowledge of residential segregation has other aspects; particularly that we also know that there have been many attempts to reduce the degree of segregation in different cities. Known policy interventions against segregation focused on reducing welfare differences among different ethnic groups, and banned and controlled possible types of organised action that may prevent the formation of integrated neighbourhoods (e.g. public policies for monitoring housing agencies). Yet these policies were unsuccessful in reducing the amount of residential segregation, and residential segregation persisted despite these acts and controls.[24] Given this knowledge, we may suspect that there may be other causes of segregation; that is, other than strong discriminatory preferences, organised action and welfare differences. In particular, we may suspect that individuals may be segregated because many individuals enjoy living with similar others, and living as an extreme minority is not something that everyone would prefer. Of course such an intuition does not come from nowhere. Our existing knowledge of social sciences helps us in forming this initial hypothesis. For example, we may know that there are unintended social consequences, and that social psychology suggests that individuals do not like living in an extreme minority status.[25] The question is whether there is another mechanism of segregation rooted at the individual level that may produce segregation as an unintended consequence of human action. The problem is that we do not know any city where these other factors were absent and we know very little about the actual preferences of individuals. In general, we do not know enough about R-World *(t)* to show that segregation may indeed emerge as an unintended consequence of human action.

What we need to do is study the conditions under which segregation may emerge, and in order to be able to focus on the effects of mild discriminatory preferences we may isolate them (in our model) from other known causes of segregation (e.g. from organised action). Note that when we look at aggregate statistics about segregated cities, we cannot deduce the type of preferences individuals have. Even if we have some data about the preferences of individuals, we know that individuals are diverse and their preferences range from being highly tolerant about other ethnic groups to zero-tolerance. This may make our task very complex. On the other hand, if we start our analysis from a particular city, where segregation is somewhat simultaneous with the arrival of a new ethnic group, it may be difficult to understand the extent to which mild discriminatory preferences may be responsible for segregation. Thus, it may be a good idea to build a model where we can study the emergence of segregation in isolation from the complexities of the real world. Now consider Schelling's effort to show that mild discriminatory preferences may be responsible for segregation. His preference was to 'postulate some mechanisms and see whether they produce segregation'. The chequerboard city is a model that represents cities that have at least two different ethnic groups. But it is not in any way a representation of the emergence of segregation in any particular city. Rather, it is *constructed* to test the initial hypothesis that segregation may be an unintended consequence of human action. It is constructed, yet it is based on knowledge of how things work (or may work) in the real world.

To build the chequerboard model we start from the picture presented in Figure 7.2 and we conjecture about the ways in which segregation *may* emerge in the absence of organised action and welfare differences (see Figure 7.3). It is true that the chequerboard model represents the emergence of segregation and that the chequerboard city represents real cities, but in a special way. Not all the characteristics of the chequerboard model and what happens therein have been previously observed. The model is rather constructed with some modification and rearrangement of what we know about segregation and, most importantly, it is constructed to present a what-if scenario. Several types of isolation, as well as some conjectural construction, are at work here. The view that considers models merely as representations or explanations as re-descriptions fails to convey this idea, and for this reason it is open to misunderstanding.

Representations need not start from concrete phenomena and in economics they usually do not. Economists and other social scientists usually start modelling from our everyday conceptualisations of phenomena (see Mäki 1996: 434). Yet what makes economics different from folk views concerning economic phenomena is that economists 'rearrange' and 'modify' these everyday (mental)

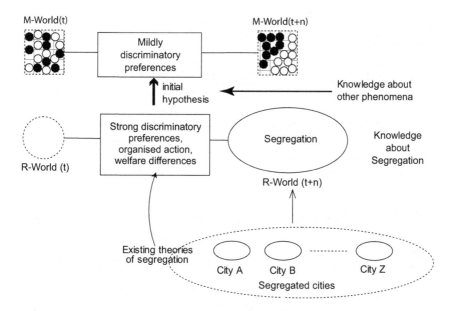

Figure 7.3 The chequerboard model.

representations to gain a better understanding of the general facts to be explained (Mäki 1996).[26] Models in this sense are webs of isolations (i.e. idealisations and abstractions). They are constructed by way of rearranging and modifying the available conceptualisations of relevant and related phenomena. Moreover, we have seen that rearrangement and modification have a conjectural component in models of unintended social consequences. The combination of the isolated elements in the model is essential. For example, we have not seen something like the chequerboard city before. By way of combining certain individual mechanisms (see Chapter 4), Schelling suggests that his initial intuition may be true. It is due to the familiarity of the individual mechanisms presented in the model that we tend to think what happens in the chequerboard city may happen in the real world. The initial hypothesis that residential segregation may be an unintended consequence is a conjecture and the chequerboard model shows that it is a plausible conjecture. More generally, models of unintended social phenomena embody familiar elements, what we have called individual mechanisms, and it is the *interaction* of these mechanisms that produce the unintended consequence.[27] The similarity between these models and the real world consists in depicting some familiar mechanisms in isolation from others. Economising action and discriminatory preferences are such mechanisms. We know that these mechanisms exist. Models that characterise social phenomena as unintended consequences demonstrate a possible way in which they may interact and may produce certain results under certain conditions. If you like, we may call these individual mechanisms 'tendencies' or 'capacities' (Cartwright 1999). These models show us how those tendencies may work together (see the next section).

We have seen that two conditions must be satisfied by explanations: the explanation has to be valid in the model world in order to render the initial hypothesis plausible, and the explanatory model has to be similar to the real world in certain respects. When these two conditions are met we get plausible conjectures about the way in which certain individual mechanisms may interact in the real world. These models alert us to *some* of the possible ways in which these mechanisms may interact in the real world. Let us explore these ideas further by way of discussing some of the prominent views on models and theories.

Models and theories

There are two prominent philosophical accounts of theories: the syntactic view and the semantic view. The syntactic view rests on an assumption of well-established theories that can be presented as axiomatic systems, and emphasises the linguistic and logical structure of theories. The syntactic view does not pay attention to the role of (partial) models in science.[28] Thus, it cannot be used as an account of the models of unintended social phenomena.[29] The alternative view, the semantic conception of theories,[30] emphasises the role of models in science.[31] It holds that theories are basically collections of models. The semantic view of theories has three distinct interpretations that differ in the way in which they characterise the relation between the models (or theory) and the real system. The first interpretation considers models as isomorphisms, that is, as being isomorphic to the particular part of the real world they represent. In this interpretation there is a one-to-one mapping between the theory and the observable part of the real world it represents (e.g. van Fraassen 1980). The second interpretation suggests that models describe an idealised and abstracted version of the portion of the real world they represent, and hence they are not necessarily isomorphic (e.g. Suppe 1989). The third interpretation conceives the relation between models and the real world in a more flexible manner, in terms of similarity (e.g. Giere 1988a).

According to van Fraassen,

> To present a theory is to specify a family of structures, its *models*, and secondly, to specify certain parts of those models (the *empirical substructures*) as candidates for the direct representation of observable phenomena. The structures which can be described in experimental and measurement reports we can call *appearances*: the theory is empirically adequate if it has some model such that all appearances are *isomorphic* to empirical substructures of that model.
>
> (van Fraassen 1980: 64, last emphasis added)

Van Fraassen holds that if a theory is isomorphic to the observable part of the real world it represents (i.e. empirically adequate), then we may believe in it for it helps us to hold a coherent view of what we have observed.[32] In the previous section we have seen that the condition of empirical adequacy (i.e. isomorphism) does not hold. The empirical adequacy condition is too strong and cannot be expected to hold in all models.[33] This interpretation considers adequate models as mere descriptions of the real world. This is not the case for many scientific models.[34] Other proponents of the semantic view criticise van Fraassen's interpretation for similar reasons. Particularly, Suppe (1989: 102) argues that the 'process of abstraction carries no guarantee that any of the theory's models or substructures will be isomorphic to any actual phenomenal systems; hence there is no guarantee that there will be any models such that all appearances are isomorphic to empirical substructures of the model.' Or, Giere (1988a, 2000) suggests that isomorphism is a strong relation and usually a weaker relation (e.g. similarity) holds between theories (or models) and real systems.

Suppe's (1989) alternative is to consider models as descriptions of an abstractly conceived world (i.e. physical systems). He characterises scientific theories as descriptions of 'the behaviour of physical systems, which are idealised replicas of actual phenomena' (Suppe 1989: 67). Under this interpretation, a theory is a collection (or a cluster) of models, which characterises possible 'physical systems' (i.e. possible model worlds). Theories, in this sense, are about the behaviour of abstract systems, 'what the theory does is to directly describe the behaviour of abstract systems, known as physical systems, whose behaviours depend only on the selected parameters' (Suppe 1989: 83). For this reason, theories and models suggest how things *could be* in the real system they represent:

> the theory does not characterise the actual phenomena, but rather characterises the contribution of the selected parameters to the actual phenomena, describing what the phenomena *would have been had* the abstracted parameters been the only parameters influencing them.
>
> (Suppe 1989: 82–83)

Suppe's conception of theories suggests that laws only hold *ceteris paribus*.[35] The argument is that if the assumptions of a certain theoretical model hold in the real world then the results derived from the model would hold in the real world. The statement of a model (or a theory) under this conception is that if such and such were the case in the real world such and such results would hold. Yet we have seen that some of the assumptions of the models of unintended social phenomena do not hold in the real world. For example, consider the Kiyotaki–Wright (1989) model. Under this version of the semantic conception, the Kiyotaki–Wright model is supposed to say that if there were three types of rational agents and three different commodities with different storage costs (and if certain other conditions hold) then money would emerge as an unintended consequence of human action in the real world. Or consider the chequerboard model: if there were two types of agents that can recognise each other's types, if they all had similar preferences regarding the other type, and if no other factors were influencing their preferences (and if the other assumptions about neighbourhoods hold), then segregation would emerge in a real city (as an unintended consequence).[36] We would be doing injustice to the authors of these models if we were to accept this interpretation. The models of unintended social phenomena have *more to say*. They suggest that the proposed mechanisms may be working in the real world even if the assumptions of these models do not strictly hold in the real world.

A more flexible version of the semantic view suggests that models may tell us something about the real-world systems in virtue of certain similarities between the proposed models and the real-world systems they represent. We have seen that the semantic conception conceives theories as descriptions of abstract systems, thus, they are true about the systems they represent, and they do not claim something about the real world as such. But theories and models make claims about the real world with a theoretical hypothesis (Giere 1988a, 2000). They suggest something about the world through a hypothesis about the real world. But how can we say that the model shows that the hypothesis is defensible, that is, that it holds in the real world? Giere's argument is that the *similarity* between the model and the real system suggests that the theoretical hypothesis may hold for the real system. Giere's account of the relation between models and theories seems to be the best fit among the alternatives we have discussed above. However, there is still an important unresolved issue in his account: similarity is a vague concept and it is not clear what it amounts to in Giere's account.[37]

Models, similarity and tendencies

In contrast to van Fraassen's interpretation, Suppe and Giere's versions of the semantic view suggest that models and theories represent (or have a chance to represent) the way the world works, or the connecting principles of nature and society (see Chapter 5). But we have seen that if we conceive their general claims as laws that hold *ceteris paribus*, it would be difficult to conceive models of unintended social phenomena as meaningful models. Giere's notion of similarity, on the other hand, is not well defined, and it does not tell us much about how models explain. An alternative suggestion is that 'the laws that hold *ceteris paribus*', or the results we obtain by way of studying the behaviour of models worlds show us how certain tendencies and capacities are realised in the model world. Or, they inform us about those tendencies and capacities that exist independently of the assumptions of the model. Cartwright (1999) presents a recent defence of this view.

Cartwright conceives models as 'nomological machines'.[38] She argues that the relations that hold in the model only hold under certain conditions (i.e. model's assumptions), 'they obtain just when a nomological machine is at work' (Cartwright 1999: 25).

> Models in economics do not usually begin from a set of fundamental regularities from which some further regularity to be explained can be deduced as a special case. Rather they are more appropriately represented as a design for a socio-economic machine which, if implemented, should give rise to the behaviour to be explained.
>
> (Cartwright 1999: 139)

The generalisations (i.e. 'laws') that are employed by the model (e.g. every individual would prefer a less-costly to store object), or the ones deduced from the model (e.g. actions of the self-interested economising individuals bring about a medium of exchange) are only valid in the model world. Yet she proposes these 'laws' indicate tendencies, or capacities, that work in the real world.[39] Particularly, she suggests that the models of economics 'provide us with a set of components and their arrangement. The theory tells us how capacities are exercised together' (Cartwright 1999: 53).[40] For example, concerning game theory, she argues:

> In game theory various concepts of equilibrium describe what is supposed to happen when the capacities of different agents are all deployed at once.
> (Cartwright 1999: 55)

Cartwright (1999: 57) argues that to build a theory one needs 'parts described by special concepts' and a 'special arrangement'. Moreover, she suggests that nomological machines (i.e. models) need 'shielding' to work, coming close to our argument that models isolate. But do Cartwright's suggestions answer our main concern about the relation of abstract model worlds to the real world? The suggestion is that in virtue of certain similarities between the model world and the real world, models depict certain *tendencies* in the real world.

Cartwright (1983, 1989, 1999) suggests that models inform us about certain tendencies, or capacities that exist in the real world even if the assumptions of the model do not hold. To see how models of unintended social phenomena may alert us to real tendencies, we have to realise that there is a two-way relation between them and the real world. To be able to point out real tendencies, models have to utilise some other real tendencies. Models depict or represent certain tendencies (e.g. individual mechanisms) in isolation from other factors, and then suggest possible ways in which these individual tendencies may interact. Considering the (process) models we have examined above, we may say that models alert us to possible aggregate tendencies. For this reason, Cartwright's suggestion is correct, but somewhat incomplete: Models are based on our knowledge of certain tendencies or capacities. For example, we know that people have discriminatory preferences, or that individuals try to economise or decrease their transaction costs when possible. Models of unintended social phenomena portray these tendencies in isolation from the factors that may prevent them from being realised, and then demonstrate the ways in which they may interact under certain conditions. They point out possible 'aggregate' tendencies: a possible way in which those individual mechanisms (i.e. tendencies) may interact in bringing about a certain social phenomenon. It is because of our knowledge of, and familiarity with, these tendencies, or individual mechanisms, that we consider the model world to be similar to the real world. For this reason, we may argue that models of unintended social phenomena alert us to a certain way in which these mechanisms may interact in the real world.[41]

Individual mechanisms that are embodied in a model need not necessarily be common sense elements or things we know from a previous body of (scientific) thought about the subject matter. But good examples of invisible-hand models have this property. The novelty of these models come from showing how these individual mechanisms may interact. And in the best cases (e.g. the chequerboard model) the type of interaction (aggregate mechanism) suggested by the model is a novel one, that is, it is a previously unnoticed (and maybe counterintuitive) type of aggregate mechanism. The way the model connects with the world and with our cognition is the familiarity of the individual mechanisms, and with the help of this familiarity the model may convince us that a previously unnoticed type of interaction among these individual mechanisms is possible, and that this aggregate mechanism may be responsible from the phenomenon under investigation. Using familiar elements do not at all prevent novelty. By creatively conjecturing about the way in which those familiar things (e.g. economising action, different goods with different saleability, etc.) may interact in bringing about a certain phenomenon, one may find out new (previously unnoticed) aggregate mechanisms. For example, no-one would disagree if you say that 'some people have mild discriminatory preferences', but one may disagree if you say 'mild discriminatory preferences may cause residential segregation'. However, if you show how those mild discriminatory preferences are connected to the aggregate phenomenon of segregation in the model world, then you have demonstrated some of the 'connecting principles' and a possible aggregate mechanism that may explain particular cases of residential segregation.

Knowledge of the existence of individual mechanisms that are isolated in the model is an important factor for assessing the similarity between these models and the part of the real world they represent. Yet we cannot accept these models merely because they represent certain tendencies that we know about. The speculative element in the model, the suggestion that those mechanisms may interact in a certain manner, necessitates that certain other external (i.e. external to the model) constraints hold for us to consider their conjectures plausible. As every model is unique, we can only cite some general issues here. The first thing that comes to mind is that if the model contradicts already known facts about the phenomenon, it would be hard to accept. That is, the consistency of the model or the explanation with what we already know about the real world is an important criterion. If the model is consistent and coherent with what we already know about the phenomenon (e.g. segregation) and about related phenomena (e.g. social psychology) then it is easier to conceive the possibilities that may be generated in the model world as possibilities of the real world. Moreover, complementarity with other available accounts of the same phenomenon is also important and preferable. A partial potential explanation (which is based on the model) that contradicts available accepted explanations would have a limited chance of survival. The reader may object that the requirements of complementarity and coherence are not 'real' or firm criteria for evaluating a model. This is true. But the point here is that if a model or explanation that is partial and suggests certain possibilities departs radically from what we know, we would have a hard time fitting it to our world picture. Moreover, as we have seen, models of unintended social phenomena do not generally challenge the previously accepted causal mechanisms as explanatory factors. They should not be interpreted as rejecting the idea that other mechanisms may be at work. Unless one explains a particular case by pointing out that the proposed mechanisms are indeed working in the suggested way, it better not contradict most of what we know about the world. Thus, logical plausibility,[42] coherence and consistency with what we know determine more or less the strength of our beliefs in the suggested possibilities. By virtue of presenting what we know in a novel way, *some* models of unintended social phenomena suggest new ways to look at the world. Some models, on the other hand, test other models in terms of logical plausibility, coherence and consistency with the facts of the real world.

Nevertheless, it is not possible to define a priori the amount of 'acceptable realisticness' for a model. Concerning theoretical models of the sort we are interested in, the only real 'hard' criterion seems to be the success in explaining particular cases. Explanations of unintended social phenomena are rarely singular explanations, and their claims should remain claims about what may be possible in the real world, how certain tendencies may possibly interact, etc. The similarity between the model world and the real world is important for it constrains the range of possibilities generated by the model, but we cannot predetermine a degree of acceptable realisticness. This, however, should not prevent us from testing the plausibility of our hypotheses further.

Models and exploration

It is a fact of economics that models are explored. A quick look at any survey article on a certain topic would reveal this fact. Generally, an influential model built on observations about the phenomenon under consideration, on knowledge of other areas of research and on knowledge available tools suggests something new. Other economists, then, go on to study this model. Some check whether the same conclusion holds under different assumptions, some check whether the model is in agreement with data, or whether the model works when other disturbing factors are introduced, some others instead may test the model with real individuals (i.e. conduct an experiment). It seems that there is no single way, or methodology, to study a model and its implications. For models of unintended social phenomena, availability of data and historical records restrict the possible ways in which one can explore and test a model. Yet we have seen in Chapter 6 that this does not prevent researchers from testing the existing models or hypotheses concerning unintended consequences of human action.

It has been argued that a good invisible-hand explanation should suggest a novel way to look at the real world. Our story of exploration can then begin from there. Consider the chequerboard model. It suggests that mild discriminatory preferences may bring about segregation. More properly, the chequerboard model tests this hypothesis and suggests that it is plausible, and that we may expect to see the depicted individual mechanisms to interact in the real world as they do in the model. Of course, we have every right to be suspicious about this claim, or at least about the possible range of conditions under which it may hold. And if we are, we may test it. For example, we may ask whether the hypothesis holds under extreme conditions, such as when everyone strictly prefers a mixed neighbourhood to any other mixture. Pancs and Vriend (2003) present an example of this. They test the chequerboard model under the conditions of strict preference[43] for perfect integration and conclude that Schelling's results hold even if this assumption is made. What they do is not to increase the realisticness of the model, or to test it with available data. They simply question the strength of Schelling's conclusions. They partially test the chequerboard model in the abstract.

Alternatively, we may introduce new factors to the chequerboard model. We may change the definition of the neighbourhoods, the number of agents, the way in which individuals are initially distributed, we may add certain other specifications to the agents and all that. There are numerous possibilities, and all can help us to understand the strength of Schelling's argument, and to have a better idea of the implications of the chequerboard model. Two examples of this approach are presented by Epstein and Axtell (1996) and Zhang (2004a). They simulate the chequerboard city by way of changing the number of agents and by setting a certain limit on the lifetime of agents. Again, these simulations are not any closer to the real world than the chequerboard model. They test it under different conditions, or with specific parameters. Both Epstein and Axtell (1996) and Zhang (2004a) show that Schelling's initial hypothesis holds under these conditions for a variety of initial starting points.

Another possibility is to integrate some factual information to the chequerboard model and hence see whether it holds when some external factual constraints are introduced. For example, Sander *et al.* (2000a,b) and Zhang (2000) use survey data to determine the preferences of different types of agents in the chequerboard city. That is, their assumptions are consistent with real individuals' preferences. The survey data suggest that individuals are tolerant to mixed neighbourhoods, but that whites are less tolerant than blacks. By way of integrating this information, Sander *et al.* (2000a,b) and Zhang (2000) demonstrate that Schelling's insights hold.

It should be noted that the survey data by themselves suggest that Schelling's insights may be true and that they need further examination. Bobo and Zubrinsky (1996) and Farley (1997) argue that many individuals are highly tolerant to mixed neighbourhoods. This indicates that the individual mechanisms (i.e. tendencies) depicted in Schelling's model actually exist. Maybe more importantly, they suggest that strong discriminatory preferences alone cannot explain residential segregation. This should give us reasons to examine how tolerant individuals contribute to the emergence (or persistence) of segregation in particular cities, that is, examining the way in which individual mechanisms interact with each other as well as with others (e.g. economic mechanisms). We have argued that Schelling's model is valuable because it suggests new explanatory mechanisms (i.e. the interaction of individual mechanisms). To be able to utilise this suggestion we need to examine the way in which these mechanisms interact with others in particular cases.

Introducing into the chequerboard world the factors that were isolated is another way in which we may explore and test it. In fact, Sander *et al.* (2000a,b) pursue this strategy in combination with others. They assume larger neighbourhoods and integrate housing costs and costs of moving into the model. Moreover, they use survey data to give shape to the preference functions of the individuals. The actual discriminatory preferences of blacks and whites are represented in the model by defining, consistently with survey data, three types of agents of each group. Sander *et al.* then simulate the model under some 'what-if' scenarios to see how different factors (e.g. housing costs, moving costs, discriminatory preferences) may be related to each other. What we have in this case is confrontation with data as well as further conjectural scenarios. Yet another thing to do is to confront a certain aspect of Schelling's models with statistical data. This is exemplified by Clark (1991). By studying statistics for certain particular segregated cities, he confirms that integrated equilibria (i.e. mixed neighbourhoods) are not stable.

Of course, there are other ways to explore and test Schelling's model. Appendix III presents a brief survey of the literature concerning the chequerboard model in order to give a better idea of the different ways in which models may be explored and tested. Nevertheless, exploration is not (and cannot be) a well-defined concept. There is no single way in which we can examine the plausibility of a certain hypothesis or its implications. Economists (and other scientists) test the available hypotheses in different ways. They sometimes test them partially, sometimes test their implications, sometimes their premises. But the important idea is that they explore the existing models to assess their plausibility. Moreover, they also entertain new hypotheses and test their plausibility in different ways by way of building models (Figure 7.4). In this sense, models are like thought experiments:[44] when models are confronted with the real world, they come close to 'real' experiments.

The relation between thought experiments and 'real' experiments can be captured by distinguishing between two types of isolation. Mäki (1992a) identifies two subspecies of isolation: material and theoretical. In material experiments the laboratory environment is materially isolated to test a certain hypothesis. 'Theoretical isolation is based on "thought experiments" instead of laboratory experiments: isolation takes place in one's ideas, not in the real world' (Mäki 1992a: 325). Morgan (2000) and Boumans and Morgan (2001) suggest that we may distinguish between different types of experiments by way of studying the materiality of the intervention. In a thought experiment, the intervention is immaterial. We

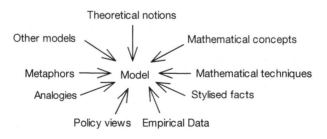

Figure 7.4 Conceptual tools for model building (adapted from Boumans 1999: 93).

simply make assumptions about the model entities (e.g. about the way in which they change). In a material experiment, however, we physically intervene in the process under consideration. For example, if we want to know how changes in X (e.g. temperature) affect Y (e.g. a certain material), we materially intervene to change the physical temperature of the laboratory environment. Morgan (2000) suggests that there are also quasi-material experiments where the intervention is only partially material.

If we are allowed to use this terminology, we may say that models of unintended social phenomena may also be tested by way of conducting quasi-material experiments, if not with material experiments. For example, we may test whether real individuals behave in the way the model theoretic agents behave under certain conditions. Most economic experiments are considered to fall under this category. There may be different reasons for conducting such experiments. We may be interested in seeing whether individuals behave according to the predictions of the model, or we may want to check whether our hypothesis is valid if we let real individuals 'play the game'. Such experiments bring the model close to the real world for they connect the model world with the real world. We have seen an example of this in Duffy and Ochs (1999). They have created a laboratory environment within which they let real individuals behave (see Chapter 6). The environment was artificial in that real individuals had to assume they were behaving in an imaginary world, yet Duffy and Ochs have controlled the environment by way of changing some of the parameters.[45]

All the complex ways in which new models are constructed and explored cannot be pictured easily. Some of the possible ways to explore and test existing models on a certain topic or about a certain phenomenon are listed below. Note that this list merely indicates some of the possibilities.

Assume that Model A suggests and supports a certain hypothesis, H(0). H(0), or its premises [H(a0), H(b0), etc.] or its implications [H(0a), H(0b), etc.] may be tested:

1　By using a specific version of Model A, and by examining what happens under certain parameter values (e.g. by changing the number of agents, goods or neighbourhoods).
2　By trying to construct a more coherent/consistent/robust model with restricted (i.e. more unrealistic) assumptions (i.e. by way of applying further isolations to Model A).
3　By trying to construct a more coherent/consistent/robust model with some relaxed and some restricted assumptions (i.e. partially more realistic and partially more unrealistic compared to Model A).
4　By trying to construct a more realistic model with relaxed assumptions.
5　By adding some other relevant factors to the model and by examining how they interact.
6　By testing the existing model against real-world data, or by conducting (quasi-)material experiments.

Most of the models we have seen in the previous chapters may be considered as thought experiments concerning a certain hypothesis. When the experiment results are positive, then the thought experiment renders our initial hypothesis (or a modified version of it) a plausible conjecture about the real world. Like material experiments, thought experiments may be conducted under different conditions. The rationale behind this is that this may help us see how plausible our initial conjecture is. Abstract models help us to check the logical coherence / consistency of certain hypotheses and the robustness of our models. Quasi-material experiments help us connect the model with the real world. If they support the hypothesis, we gain more confidence about its plausibility. Yet these models cannot go beyond alerting us to certain possibilities unless the suggested individual mechanisms are found to interact in the way the model suggests for particular cases.

Models as mediators

Finally, let us consider the implications of 'exploration' for our view of the relation between models and theories. The semantic view helps us see one important fact about models: a collection of models of the same phenomenon (e.g. emergence of money) can be considered as supplying the resources for explaining singular cases if they form a more or less consistent body, that is, if the models that emphasise different explanatory factors are more or less complementary. But the semantic view holds that models and theories are related in a certain way: models describe possible abstract systems and theory is a collection of these models.

Our examination of the models of unintended social phenomena and their exploration suggests that the relation between models and theories may be more complex. Models may derive from existing theories, from observation, as well as from other models. For this reason, they may be characterised as somewhat independent objects, or tools that *mediate* between theory and the real world, and between models and other models.

A recent defence of the view that models are mediators is presented in *Models as Mediators* (Morgan and Morrison 1999). It argues that models may be seen as 'autonomous agents' that serve as 'instruments of investigation' (Morgan and Morrison 1999: 10). The main idea is that the construction of models is a complex matter and that models do not have a defined relationship with theory or data.[46] For example, by studying economic models, Boumans (1999) argues that these models use a rich pool of conceptual resources.

On one hand, model construction is a complex process and scientists use varieties of resources to be able to reason about the phenomenon under question. On the other hand, models are usually partial and cannot alone account for particular cases. Explaining particular exemplifications of the explanandum phenomenon requires the use of other resources, such as other models of the same phenomenon or of similar phenomena. For this reason, abstract models should not be evaluated in isolation from the rest of the body of thought developed on a certain topic. Models serve different purposes. Some of them provide partial potential theoretical explanations, some others test other models, some help us conduct experiments, some help us measure and some help us explore the implications of our thoughts. But if considered all together, distinct but related models of a certain phenomenon help us explain its particular exemplifications. As a collection, they form something like an incomplete theory, serving the conceptual resources to provide singular explanations. For example, Schelling's model of residential segregation, its reconsiderations, and the other models and explanations of segregation that emphasise organised action and economic factors may be considered as forming an incomplete, partial theory or a meta-model of segregation. This 'theory' supplies the conceptual resources of a social scientist when confronted with particular cases. Similarly, Menger's verbal model, its formal reconsiderations and historical/anthropological accounts of money may be considered as forming an 'incomplete theory' of the emergence of money.

Considering models as mediators also helps us to understand the nature of models of unintended social phenomena and of many models in economics. Models constitute the toolbox of many economists and, hence, models that do not explain any particular case are useful: they prepare the base for explaining particular cases and alert us to certain types of mechanisms. Any economist will be familiar with the fact that most of the theoretical journals contain what seem to be extremely abstract models with little or no explanation of particular real-life situations. An economist will be able to publish a model which is a modification of another (known) model, if he or she can show that under such and such (different) conditions such and such (different) results are obtained, or if he or she could prove that the same results hold under a wide variety of conditions. Many people, including philosophers, have criticised this practice. However, the fact that no particular singular explanation is provided cannot immediately be used against these models, because these models are usually parts of the preparation process prior to a singular explanation. Many of these models, as exemplified by models of unintended social phenomena, provide partial potential theoretical explanations that alert us to certain ways in which the real world may be working. Such models are mediators between our theories, hypotheses and explanations. They are further explored to have a better understanding of the real world.

Concluding remarks

It has been argued in this chapter that models that characterise social phenomena as unintended consequences are based on 'known' individual mechanisms and that they explicate the way in which these mechanisms may interact in bringing about the social phenomenon under investigation. The novelty of good invisible-hand explanations, such as Schelling's, is to suggest a previously unrecognised aggregate mechanism by way of explicating the interactions of known individual mechanisms. Such explanations are appealing because we are familiar with the individual mechanisms and surprising for we did not think about a certain way in which they may interact in bringing about the explanandum phenomenon. To prevent misunderstandings, it should be emphasised here that this does not imply that every such model is based on such 'known' mechanisms. It is entirely possible that a model would suggest new mechanisms both at the individual and aggregate level. Yet if no evidence were provided, it would be hard to accept such a model. That is, it would be hard to establish the similarity between the model world and the real world.

By way of suggesting some of the possible ways in which certain individual mechanisms may interact, models of unintended social phenomena are important tools for getting a better picture of society. They may be abstract and ahistorical, but they should not be criticised merely for this reason, because what they suggest may be integrated into historical research and historical research may be enriched by way of searching for the existence of the possible mechanisms they entertain. Furthermore, even if their suggestions may prove wrong, by way of expanding our mental horizon they may facilitate further research and may lead to the discovery of some other aspects of social phenomena.

Many models in economics (particularly models of unintended social consequences) are not complete representations of a particular phenomenon; rather, they are incomplete, partial representations of types of phenomena. These representations are necessarily 'flawed' in certain aspects. Because of these 'flaws', 'the process of model construction can be viewed as the progressive refinement of q-morphisms' (Holland *et al.* 1989: 34) – where refinement does not always mean increasing realisticness. Models may be refined in terms of logical integrity, as well as representative power. Both help us in getting a better idea about the world. Representation is never done at once, but it is a continuous process. Economists explore their models to get a better picture of the model world, which in turn is expected to provide a better picture of the real world. Analogies, metaphors, representation, isolation, speculation, construction, experimentation and exploration are all important aspects of these models. This implies that the relation between models and the real world is complex and for this reason we should always be alert to the ways in which these models are (may be) tested logically and empirically.

We have also seen that models are explored in different ways to get a better picture of reality. For this reason, models of similar phenomena cannot be evaluated independently from each other. This is important because examining a model in isolation from other related models would give us a completely misleading picture of what is going on in economics. In the previous pages we have seen the relation between different models of emergence of money and segregation; the following chapter examines some of the game-theoretic models of coordination conventions in light of the arguments of this chapter.

8   Game theory and conventions

Introduction

Thus far, we have examined two prominent examples of explaining unintended
social phenomena and several models that reconsider these explanations with the
modern tools of economics. This final chapter examines game-theoretic models of
conventions from the perspective set out in the previous chapters. Modern models
of institutions are built with the help of a wide variety of tools and methods. Con-
ventional and evolutionary game theory, stochastic–dynamic games, agent-based
computer simulations, computer tournaments and laboratory experiments are a
few of the labels that come into mind when talking about these models. Hence,
there is a rich variety of models that characterise institutions as unintended social
phenomena. It would be beyond the limits of this book to examine this variety in
detail. Hence, this chapter only focuses on the emergence of coordination conventions and on the use of game theory in this literature. The rationale of this choice
is that 'coordination conventions' more or less underlie what we have been talking about in this book. In particular, we have seen in Chapter 6 that emergence of
money has been modelled as a coordination game among the market-dependent
members of a direct exchange economy. In fact, it is generally argued that coordination problems lie at the heart of many institutions and that many institutions
can be considered as solutions to coordination problems.

   David Lewis's attempt 'to render the notion of convention independent of any
fact or fiction of convening' (Quine, in Lewis 1969: xii) is a corner stone in the
history of explaining the emergence of institutions as unintended consequences
of human action. Many attempts had been previously made to show that institutions might be considered as unintended consequences. Lewis combined these
ideas, especially those of David Hume and Thomas Schelling, in order to present
a convincing argument along these lines. He followed Schelling's (1958: 208)
suggestion that 'the coordination game probably lies behind the stability of institutions and traditions'. His analysis of convention led to a research area known as
'economics of convention'. It is in this area that conventions are usually regarded
as solutions to coordination problems.[1] Yet for non-trivial cases, such as the coordination problem represented by the money game (see Chapter 6), there are
multiple solutions to a single coordination problem: the use of gold as a medium

of exchange solves the problem of double coincidence of wants, as does the use of silver. While authors such as Schelling and Lewis think that which of the two solutions is established as a convention depends on the particularities of the environment, modern (evolutionary) game theory abstracts from these particularities. That is, in contrast to Schelling's and Lewis's ideas, research in this area has been predominantly unempirical. This has been one of the prominent lines of criticism directed at these models. Many authors, such as Sugden (1998a,b, 2001), have suggested that the study of convention should be empirical, in the sense that more attention should be paid to the particularities of existent institutions. We have seen similar disputes in previous chapters and here it is argued that the difference of opinions regarding the study of conventions can be resolved by interpreting game-theoretic models of conventions as fulfilling diverse tasks in the process of explaining the emergence of conventions. While some of these models provide partial potential explanations, others examine the conditions under which a certain outcome is plausible. In fact, different models of convention fit each other in a way that allow us to see them as providing a good framework for understanding the emergence of particular conventions and for empirical research concerning conventions.

The plan of the chapter is as follows: the first section introduces the idea that conventions are solutions to coordination problems and points out the importance of history and existing institutions in the explanation of the emergence of conventions. The worry that abstract game-theoretical analysis may not explain the emergence of conventions is also introduced here. The second section discusses the possibility of coordination in the model worlds created by game theorists. First, standard static games and problems such as 'equilibrium selection' and 'justifying Nash equilibria' are discussed. It is argued that static models cannot explain the emergence of coordination (and, hence, conventions), even in the model world. The best interpretation of these models is that they study the conditions under which coordination is possible. The examination of these conditions suggests that history and existing institutions are important in the process of the emergence of institutions. Second, the possibility of coordination in a dynamic setting is discussed; in particular, learning and models with boundedly rational agents are discussed. It is argued that learning models explicate some of the ways in which coordination may be brought about in the model world. Nevertheless, consideration of these models fortifies the point that history and existing institutions are crucial for explaining the emergence of conventions in the real world. The third section discusses these issues in a general setting and presents a general evaluation of models of coordination and coordination conventions. The fourth section concludes the chapter.

Coordination conventions

Let us say that we want to explain the emergence of the 'rules of the road' and to show how such a convention could emerge merely from the interactions of individuals who do not intend to bring it about. As we want to explain the emergence of the convention, we should start from a situation where no such convention exists. If there are no such conventions, when two people approach each other from opposite directions, they have to make a choice. They may drive on the left, or they may drive on the right. It does not matter which side of the road they choose as far as the other person chooses the same side. This is the only thing that matters, because if they fail to coordinate they may confront hazardous situations. In the terminology of game theory this is a coordination game, as presented in Table 8.1.

In the driving game, 'Left' and 'Right' are the possible options for players A and B. The letters, a and b, in the table represent the payoffs to their actions. If both player A and player B choose 'Left', they have positive payoffs. Similarly, if they both choose 'Right' they have positive payoffs. If they fail to coordinate, they do not have positive payoffs. For simplicity, henceforth, we will assume that $b = 0$. This table tells us that it is in their interest to coordinate. Let us suppose that they have to make their choices simultaneously without any communication. Additionally, assume that individuals A and B are rational, that they both know the rules of the game, and that Table 8.1 presents all the information available to them. Given the payoff structure of the game, each player has an incentive to predict what the other will do. The question is, 'how?'

The driving game is a sample coordination problem[2] and, according to Lewis, if a population of agents expects each other to choose a certain action all the time, we may talk about the existence of a driving convention. More properly,

> A regularity R in the behaviour of members of a population P when they are agents in a recurrent situation S is a convention if and only if it is true that, and it is common knowledge in P that, in any instance of S among members of P, (i) everyone conforms to R; (ii) everyone expects everyone else to conform to R; (iii) everyone prefers to conform to R on condition that the others do, since S is a coordination problem and uniform conformity to R is a coordination equilibrium in S.
>
> (Lewis 1969: 58)

That is, if in a society everyone drives on the right; expects everyone else to drive on the right; and prefers to drive on the right on condition that the others do, then we may say that the driving convention in this society is driving on the right-hand side of the road. Briefly, in the driving game, both (right, right) and

Table 8.1 The driving game

		Player B	
		Left	Right
Player A	Left	a, a	b, b
	Right	b, b	a, a

Note
$a > 0$ and $b = 0$

(left, left) are candidates for becoming a driving convention. Basically, (right, right) and (left, left) are Nash equilibria of this game, or coordination equilibria, in Lewis's terms.[3] An important aspect of this coordination game is that there is no guarantee that the agents will be successful in reaching one of the equilibria (Lewis 1969: 24). In order to explain how conventions emerge, one has to show that coordination is possible and how concordant mutual expectations arise.

Now let us consider the case of two players in order to focus our attention on this point. If two drivers are driving in the middle of the road and have no other information than what is available from the above game-theoretic presentation, they will actually have no way to tell rationally what the other will do. In the terminology of game theory, the two equilibrium points (left-left and right-right), which represent the alternative conventions, are formally indistinguishable and the problem facing the players is known as an *equilibrium selection problem*. Yet standard game theory suggests that there is also a mixed-strategy equilibrium. In a mixed-strategy equilibrium players randomise their choices according to the payoffs attached to the alternatives. Since the payoffs in this game are symmetric, the mixed-strategy equilibrium of the game consists of the situation where agents choose one of the two alternatives with equal probability (50 per cent, 50 per cent). However, such an equilibrium could not form the basis of a convention. To see this, consider the case for many players: there is no guarantee for success even if all agents mutually expect the others to use a mixed strategy.[4] When everybody uses mixed strategies no-one would expect others to conform to a certain pure strategy equilibrium all the time. Agents who continuously use mixed strategies cannot bring about a driving convention.

Moreover, it has been suggested that instead of playing mixed strategies, agents would search for clues for successful coordination. For example, Schelling (1960) argues that existing conventions, norms, personal history, imagination and analogy help individuals to single out one of the many equilibria and help them solve novel coordination problems. Lewis expresses the same idea by arguing that salience and precedence are two important means of creating concordant mutual expectations (see Appendix IV.1 for some examples). In the language of game theory, the equilibrium that stands out among others as a salient option is called a *'focal-point* equilibrium'. Schelling (1960) reported a series of informal experiments (see Appendix IV.2) where real individuals were much more successful in coordination than their model counterparts. Lewis and Schelling's intuition and informal experiments have been confirmed by formal experiments (e.g. Mehta *et al.* 1992, 1994a,b) that revealed that ('boundedly rational') real individuals were much more successful in solving coordination problems than their (hyper)rational model theoretic counterparts. This suggests that real individuals who are confronted with the driving game base their expectations about others on the particularities of their environment and on history; rather than using mixed strategies.

In general, experiments point out a gap between the 'predictions' of standard game theory and the actual behaviour of individuals, a gap that has been confirmed by a number of other experiments concerning some other games, such as the ultimatum game (e.g. Güth *et al.* 1982; Henrich *et al.* 2001, 2005; see Appendix IV.3).[5] In sum, it is suggested that in these games real individuals were not doing what the theory predicted them to do and that history, cultural and personal traits, experience and analogies with previous situations are relevant for explaining how real individuals behave and how institutions emerge. If this is true then there appears to be an important disparity between the model worlds created by game theorists and the world in which real individuals live.

These experiments support those authors (e.g. Sugden 1998a,b, 2001) who demand that explanations of the emergence of institutions should be more empirical. Of course, the idea that 'institutions and history matter' is neither new nor surprising. Many institutional economists have demonstrated (e.g. North 1990, Greif 1998) that they do.[6] What is surprising may be that many economists and game theorists are convinced that institution-free and history-independent models, such as the driving game, may provide insights about socio-economic phenomena, despite the existence of intense criticism and evidence on the contrary. If so many economists are confident that these abstract, ahistorical models have something to contribute, one is tempted to give them the benefit of doubt. It may be that there is some serious misunderstanding between economists who use abstract models and their critics. In order to investigate whether there is such a puzzle, we need to attend to some of the difficulties presented by equilibrium selection and game theory in general.

Coordination in the world of models

The driving game presented in Table 8.1 is a one-shot game. It may be considered as a representation of the state of affairs when two individuals face the aforementioned coordination problem for the first time. Moreover, it serves as a representation of the possible outcomes they may reach after making their choices. As we have mentioned, two pure strategy equilibria, (left, left) and (right, right), are considered as states where individuals have no intention to deviate, that is, change their strategies. Moreover, if many individuals are involved, these equilibrium points *represent alternative conventions*. In order to explain the emergence of conventions, or how individuals coordinate, one has to explain how concordant mutual expectations emerge.

For a certain equilibrium (e.g. left, left) to get established and maintained, agents need to 'know' what the others will be doing. In other words, every individual should know that every other has a good reason to play a certain strategy (e.g. left) and that this is common knowledge:[7]

> So if a convention, in particular, holds as an item of common knowledge, then to belong to the population in which that convention holds – to be party to it – is to know, in some sense, that it holds. If a regularity R is a convention in population P, then it must be true, and common knowledge in P, that R satisfies the defining conditions for a convention. If it is common knowledge that R satisfies them, then everyone in P has a reason to believe that it is true, and common knowledge in P, that R satisfies them; which is to say that everyone in P must have a reason to believe that R is a convention.
>
> (Lewis 1969: 61)

We have argued that given that the Nash equilibria of the driving game are formally indistinguishable from each other, it is not possible to explain how individuals would rationally succeed in reaching one of them unless they succeed by chance using mixed strategies. Moreover, the reasoning behind the use of a mixed strategy does not allow us to argue that if an equilibrium point is reached it may be maintained if the game is repeated. That is, if individuals have no clue about what to expect from the other player and do not update their expectations with the information they have acquired in previous plays, we cannot explain how they may expect others to play a certain pure equilibrium strategy. Hence, we cannot explain how concordant mutual expectations are established.[8]

The problem of explaining why individuals play a certain equilibrium strategy is based on a deeper problem in game theory: it is commonly argued that Nash equilibrium does not follow from the assumption of rationality of the players, but it is a consequence of the additional assumptions imposed on the players. It is more generally argued that the notion of Nash equilibrium is based on the assumption that players are able to anticipate others' actions (Bernheim 1984).[9]

Aumann argues:

> Nash equilibrium does make sense if one starts by assuming that, for some specified reason, each player knows which strategies the other players are using.
>
> (Aumann 1987: 2)

Justifying Nash equilibrium, or explaining how it gets established, has been an important problem for game theorists that has led to the literature known as the *refinements* literature. Although this problem is usually considered as being distinct from that of equilibrium selection, it is closely related (Harsanyi and Selten 1988; Samuelson 1998).[10] Both questions are relevant if we wish to explain the emergence of conventions. In the refinements literature, many suggestions have been made on how to render an equilibrium rational without relying on the assumption of common knowledge. The orthodox justifications[11] (which are based on static games) fail to explain how and why individuals would play certain equilibrium strategies (Colman and Bacharach 1997; Crawford 1997: 210–211; Kandori, Mailath and Rob 1993: 29; Janssen 1998a: 12). For example, Aumann and Brandenburger (1995) and Brandenburger (1992) assume that individuals have coordinated expectations,[12] yet, as Janssen (1998a: 9) argues, the justification of the Nash equilibrium in this context requires an explanation of how individuals acquire coordinated expectations. In fact, in order to justify Nash equilibrium in the context of a game one has to make assumptions about agents' expectations or knowledge about others and every such assumption would be in need of further explanation.

Generally, the problem of justifying the Nash equilibrium and explaining why and how agents would choose a salient strategy are similar problems. In order to justify the Nash equilibrium one has to explain, in a sensible manner, why agent I would expect agent II to choose the Nash strategy, and expect agent II to expect himself (agent I) to expect agent II to play the Nash strategy, and so on. In order to explain why a salient strategy is chosen one has to explain why agent I would expect agent II to choose the salient strategy, and expect agent II to expect himself (agent I) to expect agent II to play the salient strategy, and so on. For example, if two game theorists play a game with a unique Nash equilibrium, they would consider the Nash equilibrium as a salient option because they would mutually expect the other to play his part in the Nash solution, given that they know that their co-player is a game theorist as well.[13] This is because 'the salience of any particular mode of behaviour depends critically upon whether that salience is universally recognised' (Bernheim 1984: 1010). Yet if two individuals who do not have knowledge of game theory play the same game, we would have no good reason to believe that they would play their Nash strategies. Neither the emergence of the Nash outcome nor the selection of a salient equilibrium among multiple equilibria can be explained without explaining how agents *come to believe* that the others will behave in a particular way. A satisfactory explanation of how a certain Nash equilibrium gets established seems to require a model of learning or of how agents form and update their expectations. Particularly, one has to explain why agents would consider a certain equilibrium as being focal or salient.

Focal points

Schelling's (1960) focal point argument and Lewis's concept of salience has been interpreted by game theorists in different ways. Game theorists have tried to integrate the idea that individuals may be successful in coordination games if a certain strategy stands out as an obvious option into the formal structure of the game. Two well-known standard examples of this approach are the arguments from Pareto dominance and risk dominance (see Appendix IV.4). It has been argued that if one of the Nash equilibria, Pareto, dominates others then one may argue that agents might use this as a coordination device (Harsanyi and Selten 1988).[14] That is, Pareto-dominant equilibrium may be considered as a focal-point equilibrium. Yet, in some games Pareto-dominant equilibrium is not always an obvious solution. For example, in the stag hunt game (see Appendix IV.4) agents may perceive the Pareto-dominant equilibrium as a risky option and try to play their parts for the risk-dominant equilibrium (Carlsson and van Damme 1993ab[15], cf. Harsanyi and Selten 1988). Nevertheless, unless agents are assumed to know how the other player thinks, neither Pareto dominance nor risk dominance seems to be a compelling argument for equilibrium selection.[16]

Another line of research in the line of justifying or rationalising focal points focuses on the attributes of the different alternatives that are present in the game set-up.[17] For example, suppose that you are a participant of a select-a-ball experiment. You and your co-player are located in different rooms and you are asked to choose one ball among three balls. Two of the balls are red, one is green, and the red balls are indistinguishable. If you are able to coordinate on the same ball you will get fifty euros, if not you get nothing.[18] It is argued that since individuals cannot discriminate between the red balls (*principle of insufficient reason*) they should choose the green ball (see Bacharach 1991; Janssen 1998a: 15, 2001a,b). This approach tells us that individuals may rationally play their part in a focal-point equilibrium, given that there is a unique strategy which is Pareto optimal and that the individuals are able to cluster the indistinguishable alternatives together (*principle of coordination*).[19]

There were some attempts to study the importance of labelling and framing in coordination games. Sugden (1995) distinguishes between the strategic structure of the game and the way in which the game is described or labelled for players. The result is that the particular way in which the game is described (or perceived) influences the outcome of a coordination game. Yet he assumes that the labelling procedure is common knowledge among the agents and it remains to be explained how the labelling procedure becomes common knowledge.[20] Bacharach and Bernasconi (1997), on the other hand, try to formalise the different ways in which strategies may be framed. They generalise Bacharach's (1991) idea that players' options are acts under descriptions and they are distinguished by the concepts the players use to specify them. This model permits to conceptualise the possible differences in agents' perceptions and for this reason it is a step further in understanding how these differences may influence the outcome of a coordination game. Like Janssen (2001b), this model focuses on the attributes of the alternative strategies and how players of the game perceive these attributes. Yet, unfortunately, the model is only able to 'predict' the outcome of simple coordination games (see Appendix IV.5).

Formal models of focal points commonly focus on whether one may explain the selection of the focal-point equilibrium without assuming common knowledge, or common history. In these models, the modeller predetermines the salient or focal option. For example, in the select-a-ball game a 'focal' alternative, green, was embedded in the game set-up.[21] What remains to be explicated in this context is the conditions under which rational individuals would choose the focal alternative. These conditions are expressed as principles, such as the principles of insufficient reason, coordination, rarity preference, etc. Although formal approaches to focal points justify the intuitive idea that rational individuals would choose the green object (the odd one) in the three ball version of the select-a-ball game, they do not help us solve the equilibrium selection problem presented by games with two Pareto non-comparable Nash equilibria, such as the driving game in Table 8.1.

Consider a version of the select-a-ball game where there are only two balls: one red and one green. In this game, there is no unique strategy that is Pareto optimal. For this reason, the alternative strategies remain formally indistinguishable from the point of view of the models examined above. Yet remember Schelling's argument that real individuals are more successful in coordination than the model theoretic agents in games. Then, according to him, connotations of green and red, as well as culture, history and experience of the players might influence the way in which this coordination problem would be solved in the real world. For example, given the existing conventions concerning the colours on warning tags and traffic symbols, if both agents think that the colour 'red' is more prominent than 'green' and expect the other to do the same they might be able to coordinate by selecting the red ball.[22] That is, the existing conventions might influence the way individuals play this game. (A similar argument is developed for Bacharach and Bernasconi's model in Appendix IV.5.)

In sum, static models (discussed above) *study the conditions under which coordination may be possible*, rather than focusing on the mechanisms of coordination. This is closely related with our distinction between the end-state and process interpretations of the invisible hand (see Chapter 5). Standard game theory, refinements and models of focal points introduce the conditions under which a certain equilibrium is plausible, but the emergence of conventions remains unexplained.[23] We learn from these models that successful coordination is a plausible outcome of a coordination game if conditions such as 'common knowledge', 'correlated expectations', or 'shared frames' hold. Successful coordination is compatible with rationality only under these conditions. If we believe that individuals have a tendency to behave rationally *ceteris paribus* we should take these results seriously. This framework implies that explaining the emergence of a particular convention requires the introduction of further factors (e.g. existing conventions, history, etc.) because the structure of the game does not necessarily tell us how such an equilibrium may be reached. Bernheim makes a similar point:

> The economist's predilection for equilibria frequently arises from the belief that some underlying dynamic process (often suppressed in formal models) moves a system to a point from which it moves no further. However, where there are no equilibrating forces, equilibrium in this sense is not a relevant concept. Since each strategic choice is resolved for all time at a specific point during the play of a game, the game itself provides no dynamic for equilibration.
>
> (Bernheim 1984: 1008)[24]

As a model of conventions, the driving game in Table 8.1 does not explain how conventions may emerge, but merely provides a framework for analysing some of the properties of conventions. The explanation of the emergence of convention appears to require that we bring in more ingredients to this model and consider the process of emergence of conventions. The next section discusses whether 'learning' may explain the possibility of coordination.[25]

Rationality and learning

The driving game presented in Table 8.1 is a one-shot game. It may be considered as a representation of the state of affairs when two ('clueless') individuals face the aforementioned coordination problem for the first time. A two-player one-shot game cannot be a good model of the emergence of driving conventions. At most, it describes the relevant coordination problem for two individuals, but not how it is solved. Coordination of two drivers is not enough to create a convention: for a convention to exist there should be many drivers who are coordinating on one of the equilibrium points and who are expecting the others to do the same. Hence, the relevant game-theoretic concept here would be a multi-player repeated game, for example, where *n* individuals repeatedly meet in pairs and play the one-shot driving game in Table 8.1 (which is called the 'stage game' in this context). However, even if we present the game in this form, standard (non-evolutionary) game theory is not very helpful in explaining the emergence of conventions, or in showing how one of the two possible equilibrium points is reached. The perfectly rational model players who are able to reason about all possibilities in the repeated game would fail to bring about a convention for they would have no clue about what exactly to expect from others given the structure of the game (also see Bernheim 1984: 1008–1009). Obviously, random play (e.g. playing a mixed strategy) of all agents would not bring about a convention. Moreover, even if all (or most of) the agents would be able to coordinate on one of the equilibria by chance, this equilibrium would not be stable and would not constitute a convention. That is, unless agents update their expectations or learn to play in a certain way as they repeatedly play the game, we cannot explain how a certain equilibrium point would be self-supporting.

Given the concepts of salience, precedence and focal points, explanation of individuals' success in coordination and emergence of conventions necessitates the study of learning in coordination games. The mechanisms of imitation, reinforcement and best reply dynamics have been employed in various forms to study the consequences of individual learning behaviour in coordination games.[26] There is a large number of models that use different assumptions concerning how individuals learn in repeated games. It is not necessary to give a full account of this literature here.[27] It will suffice to examine some of the important ideas in order to give a flavour of the models that focus on learning and evolution in the context of a coordination problem.

Standard justifications of the Nash solution concept and solutions to the equilibrium selection problem fail to explicate why rational individuals would play their part in the Nash equilibrium or choose a certain Nash equilibrium among many. Assuming 'common knowledge', 'correlated expectations' or 'shared frames' does not help us explain how real-world agents are able to coordinate, and individual dynamics of coordination have to be examined to explain the emergence of conventions as unintended consequences of human action. An important question is whether rational individuals who learn from experience might arrive at a coordination equilibrium. Or whether we could explain the emergence of a convention without restricting individuals' expectations and learning behaviour with a certain form of common knowledge assumption. Goyal and Janssen (1996), who study similar questions, argue that rationality alone does not suffice to explain coordination even if individuals are able to learn.[28] The idea behind this argument is the following: in order to ensure coordination in the next period, every agent has to take into account the previous plays of other players. However, since every player knows that the other players are using the information gathered in previous plays to form their expectations for the next period, in order to ensure coordination every player has to know how the others are forming their expectations. The problem is that the outcome of the previous encounters does not restrict the type of hypotheses they might entertain about each other. In other words, as Goyal and Janssen argue, at any point in time one may entertain an infinite number of hypotheses about others, which are consistent with their information. Thus, unless the modeller restricts the number of these hypotheses, rationality does not ensure that players learn how to coordinate.

Crawford and Haller (1990) and Kalai and Lehrer (1993a,b), on the other hand, argue that rational individuals can learn to coordinate. While Crawford and Haller assume that there are optimal rules for learning, Kalai and Lehrer put certain restrictions on individuals' prior beliefs. Yet these assumptions (restrictions) imply that rationality alone cannot ensure coordination. More specifically, Goyal and Janssen (1996) argue that even if there may be optimal rules for learning how to coordinate, these rules are not unique. That is, if agents' learning behaviour is not coordinated at the outset they might not be able to coordinate. Similarly, Kalai and Lehrer's model indicate that agents' prior beliefs have to be coordinated to ensure their success in coordination. Both models imply that pre-existent conventions are necessary for individuals' success in coordination. Goyal and Janssen's argument is consistent with our interpretation of the literature on refinements and focal points: in the model world, coordination is only possible (i.e. individuals might be able to learn to coordinate) if conditions such as common knowledge, shared background or correlated expectations hold. Again, from the perspective of explaining real world coordination problems, this means that the knowledge of pre-existing conventions is necessary to explain how coordination is achieved.

Note that Goyal and Janssen's argument supports Lewis, Schelling and others who argue that explanation of the emergence of conventions is an empirical matter. Yet one may still argue that models of learning point at certain dynamics in the explanation of successful coordination. Schotter's model (see Chapter 6) shows that under some conditions simple learning might bring about conventions. Crawford and Haller's model indicates that if we can ensure that individuals are using a certain type of learning rule they may be able to coordinate. Kalai and Lehrer's model can be interpreted as saying that agents drawn from a population with shared conventions may learn to coordinate and bring about conventions. That is, a learning mechanism may bring about conventions under certain conditions. Of course, this alone does not explain the emergence of any particular convention. These models suggest certain possibilities.

Nevertheless, many argue that rationality is not a good criterion either. It is argued that if we want to understand how real individuals achieve coordination we should consider more realistic models of real individuals (e.g. Marimon 1997: 278–282; Young 1998). A typical model that follows this suggestion has been examined in Chapter 6. Let us return to the driving game in order to recall the results of models with bounded rationality.

Bounded rationality and learning

In Chapter 6 we examined Young's (1998) model of emergence of money that assumes bounded rationality. In brief, his model involves a dynamic known as *fictitious play,*[29] in which each player constructs a simple statistical model of what the other people are doing based on fragmentary information on what they did in the past. The idea is roughly as follows: each player observes the actions that the others have chosen up to a given time *t*. Then player i computes the observed frequency distribution for his sample size and chooses a best reply to this distribution. The outcome of this process is that after some time individuals coordinate on either (left, left) or (right, right). Both of these equilibrium pairs may be considered as conventions because when individuals reach a state where everyone chooses the same strategy, their best reply to this state of affairs would be to continue playing the same strategy. The model also says that the outcome of this process depends on its initial states, that is, it is non-ergodic. Similar results apply if one of the equilibrium pairs Pareto-dominates the other; for example, when driving on the right yields a higher payoff than driving on the left.[30]

The above model says that it is possible in this set-up that one of the alternative conventions emerges. Yet real individuals make mistakes and this may be incorporated into this model by introducing small persistent stochastic shocks. These shocks represent the 'mistakes' of the players and/or other reasons they may choose an action other than the one indicated by the history of play. In our case we may simply assume that every player chooses his best reply strategy with a high probability $(1 - \varepsilon)$ and with probability ε she chooses another strategy (Young 1998). Foster and Young (1990) argue that when there are shocks of this sort the dynamic system spends most of its time in certain (Nash) equilibria than in others. They have called such an equilibrium a *stochastically stable equilibrium.*[31] The introduction of persistent stochastic shocks changes the results of the model.[32] Because of the mistakes (or mutations, if you wish), now there is a positive probability that the system might move from one Nash equilibrium to the other. That is, conventions emerge but they do not stay forever. In the long run, the 'society' occasionally switches between alternative conventions. When both conventions are equally desirable, the model cannot tell which of the two conventions will emerge. However, if one of the conventions is better than the other, then the system spends most of its time in the Pareto-optimal equilibrium (which is also risk dominant). That is, in this model mistakes ensure that the better convention is followed most of the time. In cases where the Pareto-optimal equilibrium is not also the risk-dominant equilibrium (as in the stag-hunt game), risk-dominant equilibrium is the stochastically stable outcome (Foster and Young 1990; Kandori, Mailath and Rob 1993; Young 1993a). Hence, the model solves the equilibrium selection problem and the risk-dominant equilibrium gets selected.

Ellison (1993) points out an important issue concerning these models. It is argued that the model converges to the risk-dominant equilibrium in the long run. But how long is the long run? If we assume for a moment that the assumptions of the model hold for a particular society in the real world, could we expect to observe the emergence of a convention in a reasonable period of time? Ellison examines the nature of convergence and argues that if individuals interact locally (i.e. if individuals mostly interact with their neighbours) then the dynamics introduced by the above model may be plausible for large populations.[33] In brief, the final result of these models is that boundedly rational agents who interact locally might bring about conventions.[34]

Yet it seems to be somewhat puzzling that while rational individuals could not learn to coordinate, myopic individuals can. In fact, bounded rationality assumption is a way of constraining individual behaviour. When individuals are myopic and base their decisions on fragmentary information one may dispense with the common knowledge assumption in a dynamic setting. Yet even if individuals are not fully rational they need to form expectations about others. Indeed, they are implicitly assumed to expect others to continue doing what they did in the past. Moreover, they are assumed to form their expectations in a certain manner, for example by constructing a simple statistical model of what the other people are doing based on fragmentary information. When considered from the perspective of the real world, these assumptions are in need of further explanation. One has to justify in one way or the other that this is a plausible assumption about individual behaviour. Briefly, while these models dispense with the common knowledge assumption they constrain individual behaviour in another way. It should also be noted that there is no guarantee that real individuals would conceive the problem situation as described in the model. They may consider alternative strategies or entertain different hypotheses about other individuals' future behaviour. Thus, assuming bounded rationality does not help us avoid the questions concerning existing institutions and conventions. Or in other words, assumptions concerning learning behaviour are problematic in a similar manner to assumptions concerning common knowledge. In the latter one has to explain how common knowledge (or common priors, correlated expectations) is acquired, in the former one has to explain why individuals form their expectations in that particular manner. Yet a single model cannot explain everything at the same time. Learning models study how concordant mutual expectations may emerge and for this reason they fare better than static models with respect to explicating the mechanisms that may bring about conventions.

One may also ask whether the agents in these models learn anything at all. For example, Fudenberg and Levine (1998: 143) argue that agents in these models have information about the current state and this is the only thing they care about. They respond to this information, yet they do not learn anything about others at all. Fudenberg and Levine suggest that the assumptions concerning the agents' learning behaviour can be viewed as an approximation to a model where individuals are less perfectly informed. More properly, these models ask what would happen if individuals respond to fragmentary information concerning the history of play. Note that these models define an individual mechanism that may be called a *best reply mechanism*, or a *fictitious play mechanism*, and show that their interaction may bring about conventions. The interaction of the individual level mechanisms forms an aggregate level mechanism that may be called the mechanisms of the *accumulation of the precedent*. Thus, these models explicate how 'precedence' may help individuals solve coordination problems. Although precedents may accumulate in different ways in the real world, these models suggest a particular way in which they may relate to individual mechanisms.

Interpretation

This section discusses various implications of our discussion of game-theoretic models of coordination and convention in light of the arguments developed in the previous chapters of this book. It should be noted here that game theory poses many difficult philosophical and methodological questions. Concepts such as 'rationality' and 'utility' are especially prone to deep philosophical criticism. This section merely aims at discussing the nature of the explanations provided by models of coordination and convention. The philosophical discussion of 'rationality' and 'utility' are beyond the limits of this book.

End-state models

In previous sections we have examined the possibility of coordination in the model world. We have learned that coordination is possible in the model world but only under specific assumptions concerning common knowledge and rationality. The necessity of such assumptions may be interpreted in two ways. On one hand, one may argue that game theory cannot explain how a certain Nash equilibrium emerges out of a coordination game and hence dismiss such models. On the other hand, one may argue that the necessity of these assumptions implies something about the real world: rationality alone is not sufficient for successful coordination; other conditions need to be satisfied. The former interpretation is based on a correct observation. It is true that these models do not explain how successful coordination emerges. Yet it does not follow from this that these models should be dismissed. Our discussion of models and explanation suggests that it is not the task of these models to explain how and why coordination and conventions emerge. Rather, static models of coordination are partial models; they do not take into account every relevant aspect of coordination. They study whether successful coordination is consistent with rationality and if so, under what conditions. Thus, they should not be considered as ready explanations of particular real-world cases. This would be expecting more than these models could offer. They tell us what is possible under certain conditions and suggest further lines of research for the development of general models of coordination and emergence of conventions. They bring forth the idea that rationality alone is not sufficient for successful coordination and that other conditions have to be satisfied.[35] It is then argued here that static games should be considered as partial models that test the plausibility of certain hypotheses in the abstract. They should not be expected to provide an explanation of why and how successful coordination and conventions emerge. These models are end-state models (see Chapter 5) and their role should not be overestimated.[36]

Process models

The models that study the process of emergence of coordination and conventions (e.g. Young 1998) are more concerned with explicating the mechanisms that may bring about coordination and conventions. These models capture certain tendencies concerning individual behaviour in isolation: for example, we know that some people imitate others, or individuals take their decisions based on fragmentary information concerning their environment. These models study these 'known' tendencies in isolation from other factors to see whether these tendencies, or individual mechanisms, bring about coordination in the world of models. They demonstrate that the interaction of such individual mechanisms may bring about coordination and conventions, and that it might be possible in the real world that such mechanisms may be working behind the development of conventions. Nevertheless, they provide partial potential theoretical explanations. They do not provide ready-made explanations for particular cases; rather, they alert us to the idea that mechanisms, such as learning and imitation, may be used to explain particular cases.

Conventions as unintended consequences

These models study whether conventions could be unintended consequences of human action and argue that they might be. This does not amount to holding that this is the only way conventions may arise. Researchers in this particular field are in fact commonly conscious of the fact that conventions may also be brought about intentionally. Lewis notes three means of producing concordant mutual expectations: agreement, salience and precedence.[37] Neither Lewis nor other game theorists deny that conventions may be brought about by agreement and that they might be intended. Nevertheless, they put this possibility aside and examine whether they may also be brought about as unintended consequences. For example, Young argues:

> We have shown how aggregate patterns of behaviour at the societal level can emerge from many decentralized decisions at the individual level. Of course, it would be absurd to claim that this is the only way in which such patterns arise.
>
> (Young 2001: 150–151)

These remarks are well in line with the interpretation that these models are partial. Partiality of these models suggests that the explanation of particular cases *might* require taking into account the mechanisms of learning and imitation, and explicit agreement together. Although these models rule out the possibility that conventions might be partially or fully intended, this does not imply that conventions cannot be intended. Rather, they suggest that the mechanism they postulate might be playing an important role in the process of the emergence of conventions. In particular cases, one or more of the postulated mechanisms may be working separately or in combination. Moreover, the suggested mechanisms (e.g. learning) might be important even though a particular convention has been imposed by agreement.

Necessity of empirical research

It has been suggested throughout this chapter that existing institutions and history are relevant if we want to explain the emergence of particular conventions. The same idea can also be expressed in a more general way: in Menger's terms, these models provide an exact understanding of coordination and convention. As we have noted in Chapter 3, such an exact understanding concerns only a small part of the real world in isolation from all other (possibly relevant) influences. Since these models are concerned merely with a 'special side of phenomena of human activity', other models focusing on other aspects of the real world as well as empirical research are needed in order to get a better understanding of how individuals coordinate their activities and how conventions emerge.

Remember that in the beginning of this chapter we noted an important criticism concerning game-theoretic models of conventions. These models are criticised because they are ahistorical and too abstract. It is not suggested here that this criticism is without foundation; rather, it is argued that this criticism should be re-evaluated. It is one of the arguments of this chapter that explanation of particular cases necessitates empirical research – that is, empirical research is needed for explaining particular cases. One has to bring in the relevant historical factors and the peculiarities of the specific environment under study in order to explain why one particular convention rather than another is established at a particular place and time. Yet this does not imply that general models of coordination and convention should be given up. These general models portray the possible outcomes of individual interaction and explicate the ways in which they might be brought about in the real world. They are useful for empirical research exactly for this reason. Note that the formal framework of game theory allows scientists to study individual behaviour in specific settings and to learn more about individual behaviour. From this perspective, although standard game theory cannot explain the mechanisms that bring about coordination, it played (and still plays) an important role in the process of understanding how individuals may coordinate and bring about conventions. Process models, on the other hand, indicate further lines of empirical research. Since they suggest that specific types of learning behaviour may be important in the process of emergence of conventions, further experiments concerning the learning behaviour of individuals are necessary.[38] To sum up, the suggestion here is that the formal apparatus of game theory should not be expected to provide full explanations of particular real-world cases. While such partial models and explanations necessitate empirical research, empirical research is also in need of a general guiding framework.[39]

A framework for understanding coordination

We may further argue that different models of coordination (e.g. with different assumptions concerning individual rationality and learning) can be considered as providing an understanding of the different aspects of coordination and convention in the model world. Hence, the totality of these models can be considered as providing a general, albeit incomplete, framework for empirical research and for explanations of particular cases. Individual models study what is possible under different conditions. While one model studies whether rational individuals who adopt a simple learning rule may bring about conventions, another model studies this problem with boundedly rational individuals. Some models use uniform interaction where everyone has an equal chance of interacting with every other, another model studies the same dynamics when individuals are more likely to interact with their neighbours. In some models there is no place for mistakes, in others individual mistakes are formulated in different ways. Thus, different models study different parts of the real world and they make more sense if they are considered together. The collection of different models of conventions can be considered as forming a general, albeit incomplete, model or theory of conventions. Hence, the collection of these models can be considered as a general framework within which particular cases may be examined.

Interpretation of game theory

Our discussion of game-theoretic models of coordination and convention also has implications concerning the interpretation of game-theoretic models in general. These implications are discussed in what follows. Yet it should be noted that this is a preliminary discussion, only meant to suggest other research questions concerning game theory. An examination of other research areas that utilise the tools of game theory, as well as concepts such as utility and payoffs, is needed in order to develop a good interpretation of game theory. Since this is not one of the tasks of this book, we will only suggest what appears to be the most plausible interpretation, given our discussion in the previous chapters.

The classical interpretation of game theory is that games should represent the physical and institutional rules of the game in the real world. Yet we have seen in the previous pages that models of conventions and coordination do not reflect the physical and institutional rules in the real world, rather the rules of the game are usually the invention of the theorist. That is, these models do not provide a description of the environment within which a particular convention has emerged; rather, they abstract from such factors.[40] An alternative interpretation is that 'to make sense a game should present the way in which individuals (players of the game) conceive the situation' (Rubinstein 1991). Of course, game-theoretic models portray the way in which *model agents* perceive the hypothetical scenario described by the theorist. Yet they do not represent the way in which real individuals perceive the problem situation in the real world. Rather, most of them represent the way in which (hyper or boundedly) rational agents *may* perceive the conjectured situation. Moreover, most of evolutionary game theory (e.g. replicator dynamics) portrays individual agents as pre-programmed machines. Hence, the perceptions of real agents have no role in these models. Generally, the theorist presumes that agents would perceive the situation in a certain manner and then examine the results of this presumption. For example, in the money game we have seen that the theorist assumes that there are two types of commodities that agents consider as candidates for a medium of exchange. Real individuals, of course, do not know the 'candidate' goods in advance, and they have to decide which goods may be considered as good candidates to serve as media of exchange.[41] Another example is the driving game. The game considers only two options: left and right. Yet there is no guarantee that real individuals would conceive the situation in a similar manner in the case where driving conventions are nonexistent (see, for example, Sugden 1998a). In brief, the interpretation that games represent how agents perceive the situation does not apply.

Another interpretation of game theory suggests that it provides a framework for analysis (e.g. Schelling 1984a; Binmore *et al.* 1993a). Schelling argues that anyone who tries to deal with the complex real world needs to isolate his model from some of these complexities. He suggests that game theory does not describe 'how people make decisions but a deductive theory about the conditions that their decisions would have to meet in order to be considered rational' (Schelling 1984a: 215) it may be 'valuable not as "instant theory" just waiting to be applied but as a framework' (Schelling 1984a: 241). Similarly, according to Binmore *et al.* (1993a: 8), game theory is a tool of investigation. It is like thought experiments in that it helps us conjecture about the type of theorem that might be true. By analogy to models in, for example, cosmology and evolution they suggest (1993a: 5) that game theory involves the 'construction of models [. . .] that make no claim at being demonstrably correct. Their purpose is to show only that a particular type of explanation is viable, in the sense that it can be expressed in a logically coherent manner.' Considering the types of evolutionary models we have examined in this chapter and in Chapter 6, Binmore *et al.* suggest that

> Such explanations are not testable in any real sense. They only provide *possible stylised explanations* of how things might have come about. [. . .] But this can be a valuable insight, since the key to breaking out from the preconceptions that imprison our thought is often nothing more than the realisation that other ways of looking at things is intellectually respectable.
>
> (Binmore *et al.* 1993a: 5–6)

The interpretation of game theory as a framework for analysis, or as providing stylised explanations, is well in line with our analysis of models of the emergence of money and segregation.[42] Under this interpretation the collection of diverse game-theoretic models constitutes a framework for studying diverse issues in the real world; and a collection of the different game-theoretic models of coordination and convention may be considered as a framework for the analysis and explanation of particular conventions and for empirical research.

Concluding remarks

In sum, the following arguments have been made in this chapter:

1 Static models of coordination (and convention) are concerned with examining the conditions under which certain outcomes are plausible, rather than

explaining why and how such outcomes are brought about. Hence, such models are in line with the end-state interpretation of the invisible hand.

2   Dynamic models of coordination provide partial potential (theoretical) explanations of the emergence of coordination and conventions. Hence, such models are in line with the process interpretation of the invisible hand.

3   Though these models examine whether successful coordination and conventions may emerge as unintended consequences of human action, this does not amount to a denial that conventions may be brought about intentionally. The interpretation of these models as providing partial potential explanations is well in line with this remark.

4   Explaining particular cases (e.g. explaining the emergence of a particular convention) necessitates empirical research. General models of coordination and conventions, however, need not be empirical or historical.

5   The collection of different models of coordination and conventions may be considered as providing a general framework for empirical research and for providing singular explanations.

6   In general, game-theoretic models may be interpreted as a framework for analysis, rather than providing ultimate explanations concerning social phenomena and individual behaviour.

The overall suggestion of this chapter is that, rather than seeing distinct models and accounts of institutions as alternatives to each other, one should try to see what may be gained by looking at the overall picture presented by their collection. Moreover, the accounts that present institutions as intended consequences need not be in conflict with the models that portray them as unintended consequences. The apparent conflict in these accounts disappears when we realise that the real world is complex and that social phenomena may unfold in various ways. The models we have examined in this chapter consider only a few aspects of the real world in isolation from others. It would be far-fetched to argue that what is possible in these small model worlds exhausts the possibilities in the real world. Thus, we have concluded that these (process) models alert us to some of the possible ways in which conventions may emerge. On the other hand, the argument that 'institutions are intended' is commonly based on historical accounts of particular institutions. It would, again, be far-fetched to argue that what seems to be true for some particular institutions is true for all. Menger made us aware of the fact that 'exact knowledge' and 'historical knowledge' contribute in different ways to our understanding of phenomena. He also told us that both abstract and historical analysis is necessary for a good understanding of the real world. While it is true that more empirical and historical research is needed, this does not imply that abstract models of institutions are valueless. These models are necessary components of the research regarding institutions and they help us entertain and test our hypotheses concerning the emergence of institutions. Understanding what these models could accomplish and what they could not is important both for methodologists and practicing scientists. Knowing the limits of these models would prevent unnecessary debates and facilitate constructive criticism.

9   Concluding remarks

There is a general trend in philosophy and in methodology of economics (and of social sciences in general) that bases the discussion concerning economic models on the available theories in philosophy of science without paying sufficient attention to the content of these models. This type of research has the potential danger of alienating practising economists to the philosophical literature concerning economic models. In fact, very few economists pay attention to what philosophers of economics say about economics and it may be argued that the predominant trend in philosophy of economics is precisely what prevents a potentially mutually rewarding dialogue between these research domains. This book followed another path and focused on the content of the models it is concerned with, basing the philosophical discussion on an examination of the characteristics of these models.

One of the main themes of this book is that models that characterise institutions as unintended consequences of human action are partial models and that they serve diverse purposes in the process of explaining social phenomena. It has been argued that models that focus on the process of emergence of social phenomena are partial potential theoretical explanations. This point cannot be overemphasised because many authors interpret these models literally as suggesting that institutions are exclusively unintended and/or believe that these models provide the ultimate explanations concerning the phenomenon under investigation. Many overvalue these models for this reason, and interestingly, for this same reason, many others undervalue these models. This book has tried to identify what these models can and cannot offer.

Once we see that process models suggest how certain individual mechanisms may possibly interact and bring about a consequence at the societal level, we may understand that these models can work as eye-openers. Some of these models suggest a new way to look at the real world and one should not ignore these messages if one wishes to have an (approximately) correct answer concerning our questions about the real world. Other models, which do not suggest new mechanisms, test the plausibility of the related models and mechanisms. We have seen that these tests vary in their methods. Some of them test the plausibility of existent hypotheses in the abstract; others confront them with the real world. For this reason, they contribute in different ways both to our understanding of the existing world of models and of the real world.

Another theme in the book was that we should not evaluate extant models in isolation from other related models; rather, we should try to see how these models fit together. Each of the models we have examined tries to answer tiny questions concerning the real world. Since separate models deal with diverse questions, their individual contributions may seem insignificant. Yet, when evaluated together we may see them as contributing to the answer of a larger question: how do institutions emerge and persist? These models propose answers to their minor questions concerning the emergence of macro-social phenomena. In turn, their answers yield further questions and by the 'accumulation of precedents' they may provide a better picture of what may be possible in the real world.

The aim of philosophical and methodological inquiry concerning social sciences should be taking part in the process of clarifying and improving this overall picture. Understanding how the world of models fit together is an important task because it is with these models we try to understand how things relate to each other in the real world. Many criticise economists for over-specialising on minor issues (it is true that they do); they ask tiny questions about the real world and / or about other existing models. Yet, it should be the task of other scientists and philosophers to think how these small model worlds fit to each other and how they relate to the real world. Questions about how this picture may be improved might only follow after having a clear view of what has been done before. That is, in order to improve the existing approaches to social phenomena, one has to have a good understanding of them. In this book, I have tried to give a better picture of the models that characterise institutions as unintended consequences of human action. Hopefully, this picture will facilitate more research concerning the nature of these models and the ways in which they may be improved.

In particular, this book is an attempt to examine how models concerning similar phenomena are related to each other. What remains to be done is to go further on one particular topic and try to see how our understanding of that area can be improved. It is the easy way out to argue that more empirical and historical research is needed. Instead, one should try to develop ideas about how such research can be integrated into what we already know and catalyse the process of cognitive improvement concerning social phenomena. For this reason, an important area of research for philosophers and methodologists of economics appears to be the inquiry into the nature of the 'tests' concerning the proposed mechanisms in particular areas of research. As it has been argued in this book, there are several ways in which one may test a certain hypothesis, and the nature of these tests should be further examined.

Unfortunately, few authors consider questions concerning the 'nature of abstraction' and 'experiments in economics' as belonging to a continuum of research activities that are tightly connected with each other. Both abstract models and experiments test the plausibility of our hypotheses concerning the real world. On one hand, we have 'thought experiments' that test these hypotheses in the abstract, on the other we have experiments that confront them with the real world.

In between these two, we have a wide variety of tools that integrate real-world data into the models at different degrees. Moreover, experiments in economics are not as distinct from abstract models as they may be in physics because abstract models are usually an important part of the laboratory environment in economic experiments. Investigating the different ways in which these seemingly distinct research areas are related to each other is an important task for philosophers and methodologists of economics.

One particular interesting area of research for philosophers of economics is the inquiry into the nature of agent-based computer simulations.[1] While there has been very little effort on the part of philosophers and methodologists of economics, researchers in the field of agent-based simulations have discussed the pros and cons of their simulations extensively. This suggests that both parties may benefit from an inquiry into the methodology of these simulations. Agent-based simulations are interesting for they lend themselves for integration of real-world data in a way mathematical models do not. Moreover, the task of 'exploration' appears to be easier with computer simulations, in that one is able to change several features of the model and get the results for these specifications quite easily – of course, once the model is explicitly specified. Yet, as we have mentioned in the previous pages, results of such simulations are usually specific to the chosen parameters. Agent-based simulations in economics present interesting and difficult questions for philosophers of economics.

For example, an interesting issue is whether these models have something to offer concerning the workings of the 'invisible hand'. People who apply agent-based computer models think that society is a complex evolving system (e.g. Holland 1995, 1998; Arthur *et al.* 1989, 1990; Anderson, Arrow and Pines 1988). Tesfatsion (2000: 1) presents agent-based computational economics (ACE)[2] 'as the computational study of economies modelled as evolving systems of autonomous interacting agents'. He then goes on to state that these simulations offer an explication of the invisible hand. Complexity theory,[3] although it is not yet a well-defined domain (Rosser 1999), examines the systems of highly *interconnected agents* (e.g. neurons, individuals, etc.) that bring about aggregate behaviour. To understand and explain the phenomena at hand, they start with an abstract representation of the system (such as neural networks, social networks, etc.) and then try to demonstrate what emergent properties result from different properties at the lower level. 'For such explorations', Holland (1998: 119) argues 'computer based models provide a halfway house between theory and experiment'. These remarks suggest that the relation between the invisible hand, conjectural history and scientific thought experiments appears to be another interesting research subject. Finally, some of these researchers who adopt ideas from complexity theory argue that causes in the social realm may be chemically composed (e.g. Arthur *et al.* 1989, 1990; Anderson, Arrow and Pines 1988; cf. Mill 1843: 533). Similarly, some other economists like Sugden (2000: 21) find it implausible that all economic phenomena are mechanical. It is interesting to see whether agent-based simulations suggest different forms of causal connections than the ones offered in other parts of economic theory.

There is a rich variety of models that characterise macro-social phenomena as unintended consequences of human action, and further examinations of these models will teach us more about the nature both of these models and of macro-social phenomena. Yet such fruitful research can only emerge from a detailed examination of models in economics. Because, as John Casti (2000: 4) suggests, 'using the scientific terms without their content is like using soap without water – it doesn't clean up the situation but it further muddies it'. Hopefully, this book has been successful at getting some of the mud out of your hands.

Appendix I

Smith, Jevons and Mises on money

Smith on money

A similar story of the origin of money to that of Menger's can be found in Adam Smith's *The Wealth of Nations*. Smith builds the story of money on his ideas about the emergence of the division of labour:

> This division of labour, from which so many advantages are derived, is not originally the effect of any human wisdom, which foresees and intends that general opulence to which it gives occasion. It is the necessary, though very slow and gradual, consequence of a certain propensity in human nature which has in view no such extensive utility; the *propensity to truck, barter, and exchange one thing for another.*
>
> (Smith 1789: I.2.2, emphasis added)

(Note that Smith sees division of labour as an *unintended consequence* of human action. See Chapter 5 of this book on Smith's 'invisible hand' and 'unintended consequences'.)

According to Smith, it is the human propensity to exchange goods that opens the way to the division of labour:

> And thus *the certainty of being able to exchange* all that surplus part of the produce of his own labour, which is over and above his own consumption, for such parts of the produce of other men's labour as he may have occasion for, encourages every man to apply himself to a particular occupation, and to cultivate and bring to perfection whatever talent or genius he may possess for that particular species of business.
>
> (Smith 1789: I.2.3, emphasis added)

He argues that while the disposition to exchange goods stimulates the division of labour, it is the 'extent of the market' that limits this effect:

> When the market is very small, no person can have any encouragement to dedicate himself entirely to one employment, for want of the power to ex-

change all that surplus part of the produce of his own labour, which is over and above his own consumption, for such parts of the produce of other men's labour as he has occasion for.

(Smith 1789: I.3.1)

Thus, the existence of a market and its size is a precondition of the development of the division of labour; a condition that also finds its place in Menger's exposition of the origin of a medium of exchange. Given that individuals are able to exchange goods at the market, and given that this leads to specialisation, Smith (1789: I.4.1) argues, 'it is but a very small part of man's wants which the produce of his own labour can supply'. For this reason, individuals are forced, even more, to exchange goods at the market to acquire the goods they need. That is, market exchange encourages the division of labour that, in turn, encourages more exchange. Yet, with the increased market traffic, the inconveniencies of direct exchange become more important for every individual who needs to exchange his products in the market to be able to supply his needs:

In order to avoid the inconveniency of such situations, every prudent man in every period of society, after the first establishment of the division of labour, must naturally have endeavoured to manage his affairs in such a manner, as to have at all times by him, besides the peculiar produce of his own industry, a certain quantity of some one commodity or other, *such as he imagined few people would be likely to refuse in exchange for the produce of their industry.*

(Smith 1789: I.4.2, emphasis added)

Although Smith does not have a theory of saleableness, he argues that metals would be good candidates for being a medium of exchange for they have certain characteristics (e.g. durability, easy storage) that would give good reasons to think that other people would be likely to accept them. In Smith's version, money develops out of direct exchange among market-dependent individuals' actions to overcome the inconveniencies of direct exchange. Smith goes further to argue that after the emergence of a medium of exchange, new inconveniencies would follow and they are to be solved centrally by issuing a standardised money.

Jevons on money

Jevons (1876) extensively discusses the early forms of money in *Money and the Mechanism of Exchange*. He starts the book with a discussion of 'barter', where he discusses the inconveniencies of barter. According to him, barter or direct exchange was how early societies supplied their needs. In this stage of society, he argues, it is difficult to find double coincidence of wants, and, moreover, individuals would be in want of a measure of value and a means of subdivision:

three inconveniences attach to the practice of simple barter, namely, the improbability, of coincidence between persons wanting and persons possessing; the complexity of exchanges, which are not made in terms of one single substance; and the need of some means of dividing and distributing valuable articles.

(Jevons 1876: III.1)

According to Jevons, money solves these inconveniencies by serving as a medium of exchange and a common measure of value. He also argues that other functions of money, such as being a standard of value and a store of value, follow these two functions. The qualities of a commodity that is used as material money – such as its utility, portability, indestructrability, homogeneity, divisibility, stability of value and cognisability – are related with these functions. He argues (1876: V.2), 'money requires different properties as regards to its different functions'. He observes that the most important quality of a commodity that would help its acceptance as 'a medium of exchange' and a 'store of value' is its utility:

Certainly, in the early stages of society, the use of money was not based on legal regulations, so that the utility of the substance for other purposes must have been the prior condition of its employment as money. [. . .] We may, therefore, agree with Storch when he says: – 'It is impossible that a substance which has no direct value should be introduced as money, however suitable it may be in other respects for this use.'

(Jevons 1876: V.4)

For example, Jevons (1876: IV.5) argues that in the pastoral stage sheep and cattle were used as money for they were naturally 'the most valuable and negotiable kind of property'. (Note that Jevons presents etymological and historical evidence on this point. See Jevons 1876: IV.7.) Jevons does not have a story of the origin of 'money' in its earlier forms, but based on the above quotation, it is plausible to argue that he believes that the driving forces in the genesis of money were inconveniencies of barter, self-interest and custom. Coined money, however, was issued by central authorities.

Also note that Jevons (1876: VIII.22–25) argues that force of habit and social conventions are important in understanding social phenomena:

Over and over again in the course of history, powerful rulers have endeavoured to put new coins into circulation or to withdraw old ones; but the instincts of self-interest or habit in the people have been too strong for laws and penalties.

(Jevons 1876: VIII.22)

We must notice, in the first place, that the great mass of the population who hold coins have no theories, or general information whatever, upon the subject of money. They are guided entirely by popular report and tradition. The sole question with them on receiving a coin is whether similar coins have been readily accepted by other people.

(Jevons 1876: VIII.23)

Mises on money

In *The Theory of Money and Credit*, Mises's (1954) story of the origin of money is very similar to that of Menger's, thus there is no need to present it here. Below is a representative passage from his version of the story:

> those goods that were originally the most marketable became common media of exchange. [. . .] And as soon as those commodities that were relatively most marketable had become common media of exchange, there was an increase in the difference between their marketability and that of all other commodities, and this in its turn further strengthened and broadened their position as media of exchange.
>
> (Mises 1954: I.1.10)

He makes similar assumptions about the initial stages of the society, as can be seen below:

> Where the free exchange of goods and services is unknown, money is not wanted. [. . .] The phenomenon of money presupposes an economic order in which production is based on division of labor and in which private property consists not only in goods of the first order (consumption goods), but also in goods of higher orders (production goods). In such a society, there is no systematic centralized control of production [. . .] The function of money is to facilitate the business of the market by acting as a common medium of exchange.
>
> (Mises 1954: I.1.1–2)

The idea that money can only exist under free exchange and that the existence of free commercial transactions is a precondition for the existence of money is as controversial as the assumption that human beings have a disposition to exchange. It is these assumptions that reduce the credibility of the explanation of the emergence of money as an unintended consequence of human action. The objections to these assumptions are discussed in Chapter 3.

Appendix II

Models of emergence of money

Kiyotaki and Wright (1989)

The expected discounted lifetime utility is defined as:

$$E\sum_{t=0}^{\infty}\beta^t I_i^U(t)U_i - I_{i*}^D(t)D_i - I_{ij}^c(t)c_{ij}$$

where U_i is the utility from consuming good i, D_{i*} is the disutility from producing good $i*$ ($i* \neq i$), and β is the discount factor. c_{ij} denotes the cost of storing good j for type i. $I_i^U(t)$ is equal to one if the agent consumes good i, and it is zero otherwise. $I_{i*}^D(t)$ is one if the agent produces $i*$ and it is zero otherwise. Likewise, $I_{ij}^c(t)$ if agent i stores any good j, it is zero otherwise.

The direct utility of consuming i for type i and producing $i*$ is: $u_i = U_i - D_i$. In addition to this, there is an indirect utility of storing $i*$. Thus, the expected discounted utility for type i for this occasion may be defined as:

$$V_i(i) = u_i + V_i(i*)$$

The indirect utility of storing good $j \neq i$ is:

$$V_i(j) = c_{ij} + \max \beta(E[V_i(j')|j])$$

where $E[V_i(j')|j]$ is the expectation of V_i at next period's random state j', conditional on j, and the maximisation is over strategies.

Lastly, the distribution of potential matches is characterised by the time path of $P(t) = [\ldots p_{ij}(t) \ldots]$, where $p_{ij}(t)$ is the proportion of type i agents holding good j in inventory at time t. Yet it is assumed that $p(t) = p$ for all t. And finally, $b = \beta/3$. Given these definitions,

Theorem 1

In Model A, under the maintained assumptions,

(a) if $c_{13} - c_{12} > 0.5bu_1$, then there is a unique equilibrium in which all agents use fundamental strategies and good 1 serves as the unique commodity money;

(b) if $c_{13}-c_{12}<(\sqrt{2}-1)bu_1$, then there is a unique equilibrium in which type II and type III agents use fundamental strategies while type I agents speculate, and both goods 1 and 3 serve as commodity monies;

(c) these are the only equilibria.

(Kiyotaki and Wright 1989: 939)

Theorem 2

In model B, under the maintained assumptions, there always exists an equilibrium in which all agents play fundamental, with goods 1 and 2 serving as commodity money; for parameter values implying

$$c_{23}-c_{21}<(\sqrt{2}-1)bu_2 \text{ and } c_{32}-c_{31}<(1-0.5\sqrt{2})bu_3,$$

there also exists an equilibrium in which types II and III speculate while type I agents play fundamental, with goods 2 and 3 serving as commodity money;

these are the only equilibria.

(Kiyotaki and Wright 1989: 941)

Fiat-money equilibrium

Aiyagari and Wallace (1991) generalise the Kiyotaki–Wright (1989) model to N goods + 1 fiat money and show that there always exists an equilibrium where the lowest-storage-cost commodity serves as a medium of exchange. And if 'fiat money' is such an object, it becomes a medium of exchange. Similarly, Kiyotaki and Wright (1991) show that fiat-money equilibrium exists. Briefly, they pick up the unsettled questions about the existence of fiat-money equilibrium in the Kiyotaki–Wright (1989) model. They (1991: 222) inquire whether 'a worthless paper or shell' may serve as a medium of exchange 'merely because' individuals believe that others will accept it in exchange, or not. They show that if agents believe that the others would exchange their commodities for the fiat good, then there exists an equilibrium where an intrinsically useless object is used as a medium of exchange, that is, as fiat money.

Kiyotaki and Wright (1993) provide a simpler and more tractable version of the Kiyotaki–Wright (1991) model, where they discuss the welfare implications of their model in addition to the proof of the existence of pure monetary equilibrium. The Kiyotaki–Wright (1993) model is defined with a large number of infinitely lived agents and a large number of (indivisible) consumption goods, which they call *real commodities*. There is also a good with no intrinsic value, which we may call as fiat money if agents accept it in their exchange. Initial stage of the model economy consists of a fraction M of agents who are endowed with the intrinsically valueless good, and a fraction $(1-M)$ of agents who are endowed with real commodities ($1 > M \geq 0$). Agents who are endowed with real commodities are called 'commodity traders', and those who are endowed with fiat good are called 'money traders'. The fiat good cannot be produced, and real commodities can. In accordance with the Kiyotaki–Wright (1989) model, agents meet randomly at the marketplace, and trade entails one-for-one swap of goods; they cannot consume what they produce, and they cannot produce if they do not consume. In contrast to the Kiyotaki–Wright (1989) model, in Kiyotaki–Wright (1993) commodities can be stored with no costs; yet real commodity barter entails transaction costs (ε), while there are no transaction costs for monetary exchange – that is, for the fiat good. (However, they show that this assumption may be relaxed.) In a similar fashion to the Kiyotaki–Wright (1989) model, 'agents maximize their expected discounted utility from consumption net of transaction costs, given strategies of others' (Kiyotaki and Wright 1993: 66).

Under these conditions, agents always accept a barter offer if one of their consumption goods is offered, and it is assumed that a commodity trader never accepts a real commodity if it is not one of his consumption goods (note here that this assumption rules out the existence of a commodity money in this economy). A parameter x is defined to characterise the level of differentiation of commodities and tastes in this model economy, that is, in terms of what agents produce and consume. x may be considered as the probability that a real good will be accepted in exchange. Accordingly, x^2 may be considered as the probability that a one-for-one swap of real commodities occur. Similarly, let Π denote the probability that a commodity trader A accepts money, which represents the trading strategies of others. Therefore, in the Kiyotaki–Wright (1993) model, whether individuals accept money or not depend on the trading strategies of others (i.e. on whether others will accept the fiat good in exchange, or not) and on the initial endowment of the fiat good in the economy, which is represented by M.

Given these specifications and assumptions, Kiyotaki and Wright (1993) prove that there are three kinds of equilibria: non-monetary equilibrium ($\Pi = 0$), pure-monetary equilibrium ($\Pi = 1$), and mixed-monetary equilibrium ($\Pi = x$). The intuition behind these equilibria is as follows. If the fiat good is accepted with lower probability than a barter offer (i.e. if $\Pi < x$), then it is a good idea to never accept the fiat good in exchange. If the fiat good is accepted with a higher probability than a barter offer (i.e. if $\Pi > x$), then it is a good idea to use fiat good in exchange. And if the probability that fiat good and real commodities can be exchanged is the same (i.e. if $\Pi = x$), then individuals would be indifferent between using fiat good in exchange, or not. In addition to the specification of these equilibria, Kiyotaki and Wright (1993) show that introducing fiat money to a barter economy enhances welfare, and that an economy with multiple currencies is possible (i.e. an economy where several type of intrinsically valueless goods serve as a medium of exchange).

Marimon *et al.* (1990)

Economy A1 (A1.1 and A1.2) converges to a fundamental equilibrium, that is, the lowest-storage-cost commodity emerges as a medium of exchange. But Marimon *et al.* (1990: 359) report that the convergence is lower when 'initial classifiers are randomly generated' – that is, for A1.2.

Economy A2 is similar to Economy A1, but only the utilities of the commodities are higher. Marimon *et al.* (1990: 360) convey that the results for this economy are 'fairly inconclusive'. The simulations for Economy A2.2 do not support Kiyotaki–Wright argument that the economy would converge into the speculative equilibrium. Remember that speculative equilibrium is an equilibrium where some agents use a high-storage-cost commodity as a medium of exchange, for they believe that it is more marketable. They argue that the 'trading patterns were closer to fundamental equilibria'. But it is also argued that fundamental equilibrium only exists if the discount rates are sufficiently low. (Marimon *et al.* (1990: 362) report that Marimon and Miller's (1989) simulations for Economy A2 converged to speculative equilibrium.) Results for Economy B support the idea that fundamental equilibrium is selected. Yet in early stages of the simulation, economy B1 converges to a speculative equilibrium, before moving to the fundamental equilibrium in later stages. Although the Economy B2 does not converge, it is closer to the fundamental equilibrium.

In economy C, a fiat good (good 4) is introduced to the economy, which does not provide any utility. Good 4 has no storage costs, but it decreases the storage capacity of the agents. It is assumed that some agents hold good 4 in period 0. Marimon *et al.* (1990: 366) report that economy C2 converges to the fundamental equilibrium quite fast. Yet economy C1 does not converge probably because of the low storage costs of good 1 (see Table 6.2), which also provides utility. The simulation of this economy implies that if the storage costs of non-fiat commodities are sufficiently high and if at least some of the agents hold no-utility goods at the initial stage, fiat money may be brought about. Yet, note here that there is no reason for the agents to hold no utility goods and accept it in exchange; rather, it is assumed that agents hold fiat goods. Note that the results presented here are for randomly generated classifiers. Marimon *et al.* do not report what happens when agents know all the available strategies. Obviously, if some of the AI agents consider using good 4 in exchange by chance – which corresponds to believing that it will be accepted – then it is possible that fiat money emerges in the end of the process. In Marimon *et al.*'s simulations, whenever the economy converges to fiat-money equilibrium this happens quickly; if it does not converge quickly (i.e. in the first stages), it does not converge at all. This supports the idea that if agents are able to trade with fiat goods accidentally in the first periods, fiat money emerges; if not, the strength of the rule 'trade with the fiat good' decreases and fiat money does not emerge. Of course, this also suggests that when agents are not fully rational they may explore different strategies and create new opportunities. But in general, there seems to be no reason for the agents to accept a fiat good in exchange if they do not believe that it will be accepted by others – hence the common belief assumption. Finally, economy D depicts a more complex economy with no fiat good. Marimon *et al.* report that although the trading patterns are close to a fundamental equilibrium, the simulations are inconclusive. That is, although agents accept a good in exchange if its storage cost is lower than the storage cost of the good in their inventory, no single medium of exchange emerges.

Gintis (1997, 2000)

An alternative to Marimon *et al.* (1990) is presented by Gintis (1997, 2000) (also see Dawid 2000). Gintis (1997: 24) reports that he replicated Marimon *et al.*'s simulations and found that they were highly sensitive to the choice of parameters. (This sensitiveness supports the argument that Marimon *et al.* only show what is possible under specific conditions.) Gintis's simulation is different in that it is based on Darwinian notions, such as natural selection, mutation and adaptation, and dispels AI agents. Every agent is characterised as having a 'genotype', which determines his strategy. That is, he replaces the rational agents of the Kiyotaki–Wright model, or AI agents of Marimon *et al.*, with a description of strategies. To ensure variation, Gintis assumes that genotypes are randomly assigned in period 0. In period 0 there are five goods with different storage costs. As it is in the Kiyotaki–Wright (1989) model, there are different types of agents who cannot consume what they produce. (Hence) there are twenty types of agents. In period 0, every agent has an empty inventory and zero wealth. Beginning from period 1, agents are matched in pairs randomly and whether they trade or not is dictated by their genomes. Successful trade increases the wealth of trading agents. After some time the least fit agents of each type die. That is, the least successful strategies become extinct after some time and new agents are introduced into the economy. New agents are offspring of successful agents and they take over the successful genomes. Yet there is mutation. For this reason, new agents' genomes are different from their parents' genomes.

Gintis shows that usually the lowest-storage-cost commodity emerges as a medium of exchange out of this process (i.e. fundamental equilibrium). He reports that, other than storage costs, frequency of use can be considered as an important factor in the process of the emergence of money. That is, if there is a good that is more likely to be accepted by all other agents in trade than the lowest-storage-cost good, then this good emerges as a medium of exchange (i.e. speculative equilibrium). Gintis also shows that unless we assume at the outset that a very high percentage of agents accept a fiat good, fiat-good equilibrium will not emerge. That is, unless there is a common belief that a fiat good will be accepted by all others, fiat-money equilibrium cannot be reached even if one exists:

> using fiat money involves a self-fulfilling prophecy, in the sense that if enough of the population expects a fiat good to be accepted as money, then everyone will accept it.
>
> (Gintis 2000: 228)

Gintis's results are generally in accordance with the previously examined models. Yet he introduces an evolutionary mechanism that selects the successful strategies, and which does not give much role to rational decisions of the agents. This may suggest that the emergence of a commodity money equilibrium does not need much intelligence or rationality. Yet the introduction of new agents who inherit the successful strategies suggests that there must be some role for the individual learning or imitation.

Appendix III
Explorations of the chequerboard world

It may be argued at the outset that if a model's conjecture is interesting and promising enough it will be subjected to many tests and explored in different ways. The chequerboard model is a good example in this respect. First of all, the chequerboard model is widely cited in sociological and geographical studies concerning residential segregation (e.g. Aaronson 2001; Aberg 2000; Bayer, McMillan and Rueben 2001; Clark 1991; Denton and Massey 1991; Downs 1981; Farley, Fielding and Krysian 1997; Fielding 1997; Friedrichs 1998; Huttman *et al.* 1991; Iceland 2002; Ihlanfeldt and Scafidi 2002; Massey and Denton 1993; Torrens and Benenson 2005; Zhang 2000. There are also many studies on residential segregation that do not mention Schelling's chequerboard model. For example, Gürke (2006) gives an extensive list of the causes of segregation, but does not mention mild discriminatory preferences as a possible cause of segregation).

Second, the chequerboard model has been theoretically and empirically tested several times. This gives us a chance to evaluate Schelling's conjecture further. To see the strength of Schelling's initial conjecture, let us have a look at the different ways in which the chequerboard model has been explored.

Epstein and Axtell (1996) demonstrate that Schelling's initial hypothesis holds under a wide variety of conditions and for a variety of initial starting points. (For an overview of Schelling-type models and related discrete choice models see Meen and Meen (2003). For an overview of interaction-based approach to social science see Blume and Durlauf (2001).)

Pancs and Vriend (2003) test the chequerboard model under the conditions of strict preference for perfect integration. Remember that in Schelling's model individuals who have mild discriminatory preferences do not strictly prefer a mixed neighbourhood. They are content as long as they do not have an extreme minority status. That is, *A*s (or *B*s) do not care whether the neighbourhood is segregated or integrated as long as *A*s (or *B*s) do not have an extreme minority status. The assumption of strict preference for perfect integration implies that individuals prefer a mixed (integrated) neighbourhood to a segregated one. Pancs and Vriend (2003) show that Schelling's results hold even if individuals have a strict preference for perfect integration. Similarly, Zhang (2004a) tests the results of the chequerboard model in an evolutionary game-theoretical framework and shows that its results hold even if all individuals strictly prefer to live in mixed neighbourhoods (also see Benito and Hernandez 2004; Young 1998; Zhang 2004b).

Portugali, Benenson and Omer (1994) introduce a couple of new elements to Schelling's model, such as stochastic household behaviour and heterogeneous agents, and find that Schelling's results hold under these conditions (also see Portugali and Benenson 1997; Portugali, Benenson and Omer 1997).

Another line of exploration focuses on the structure of the neighbourhoods. Fagiolo, Valente and Vriend (2005) and Flache and Hegselmann (2001) test the robustness of the chequerboard model by changing the topological characteristics of the neighbourhoods and find that Schelling's results hold under a wide variety of neighbourhood forms. These studies basically explore and change certain properties of the chequerboard model and lend support to Schelling's conjecture.

However, not all types of explorations support Schelling's results under all conditions. For example, Sethi and Somanathan (2004) assume that individuals are also affected by the affluence of their communities and argue that both high- and low-income disparities lead to residential segregation when combined with mildly discriminatory preferences. Yet, they also show that intermediate levels of income disparity produce multiple equilibria and both integration and segregation becomes possible (also see Somanathan and Sethi 2004). In a similar manner, Benenson (1998) integrates new elements to the chequerboard model. In Benenson's evolutionary model agents may adapt their behaviour to local or global environments and vacant places are scarce. He finds that the tendencies to adapt to the local environment and to the global environment may be in conflict. If agents adapt to their local environment, then they become more neutral to differences and that residential distribution is somewhat random. However, when they adapt to the global environment residential segregation is observed in the long-run. This implies that residential segregation is more likely when individuals care about the ethnic composition of their neighbourhood rather that their immediate neighbours.

Two other studies consider the role of vision in neighbourhood formation. Lauri and Jaggi (2003) argue that when individuals are able to observe the neighbourhood structure of a wider area, integrated neighbourhoods may become stable and Schelling's results do not hold. (Also see Ellen (2000), who emphasises the role of expectations in the process of segregation.) Fosset and Waren (2005), on the other hand, argue that Lauri and Jaggi's (2003) results are caused by the specifications of their model. They argue that residential distribution freezes in their model because individuals move only when they can improve their satisfaction, and because individuals occupy their place forever if they are satisfied. They suggest that these are implausible assumptions and show that increased vision does not lead to a stable integrated neighbourhood when these assumptions are relaxed. Yet, Edmonds and Hales (2005) also show that if the chequerboard is more crowded than that of Schelling's and if intolerance levels are higher, the chequerboard model would suggest that segregation is decreased. They interpret this result by arguing that Schelling's model does not provide a general theory and that it should not be interpreted as such. These studies confirm the incompleteness of the chequerboard model, yet they generally support the main hypothesis that mild discriminatory preferences *may* bring about segregation under some conditions.

Aforementioned studies that explore the chequerboard model do not test it against real-world data. There are a couple of studies that undertake this task. Some survey data suggest that Schelling's insights may be true and that they need further examination. For example, Bobo and Zubrinsky (1996) and Farley (1997) argue that many individuals are highly tolerant of mixed neighbourhoods. This suggests that strong discriminatory preferences alone cannot explain residential segregation. This gives us a reason to believe that the individual mechanisms (i.e. tendencies) depicted in Schelling's model may actually exist.

Another interesting way to explore and test the chequerboard model is to integrate some real-world data to the model. Sander *et al.* (2000a,b) and Zhang (2000) use survey data to determine the preferences of different types of agents in the chequerboard city. That is, they employ assumptions that are consistent with real individuals' preferences. Their survey data suggest that individuals are tolerant to mixed neighbourhoods, but that whites are less tolerant than blacks. By way of integrating this information, Sander *et al.* (2000a,b) and Zhang (2000) demonstrate that Schelling's insights hold.

Sander *et al.* (2000a,b) assume larger neighbourhoods and integrate housing costs and costs of moving into the model. Moreover, they use survey data to give shape to the preference functions of the individuals. The actual discriminatory preferences of blacks and whites are represented in the model by defining, consistently with survey data, three types of agents of each group. Sander *et al.* then simulates the model under some 'what-if' scenarios to see how different factors (e.g. housing costs, moving costs, discriminatory preferences) may be related to each other. What we have in this case is confrontation with data, as well as new conjectural scenarios.

Another way to explore Schelling's model is to confront certain aspects of the model with statistical data. This is exemplified by Clark (1991). By studying statistics for certain particular segregated cities, he confirms that integrated equilibria (i.e. mixed neighbourhoods) are not stable. In another study, Benenson (2004) simulates the residential dynamics of Yaffo (near Tel Aviv) between 1955 and 1995 with a model that is similar to the chequerboard model, and demonstrate that Schelling's insights may be explanatory.

However, not all of the empirical tests support the chequerboard model. Bruch and Mare (2003) find that the shape of the utility functions of individuals influence Schelling's results. In the chequerboard model individuals have a threshold utility function. That is, they consider moving only when a certain threshold of neighbourhood ethnic mixture is exceeded. Bruch and Mare (2003) show that if individuals have a continuous utility function, that is, if they continuously consider moving whenever they find a neighbourhood where they can be more satisfied, then the levels of residential segregation may decrease. They also show that utility functions of real-world individuals are continuous and argue that levels of segregation are lower than those suggested by the chequerboard model (also see Bruch and Mare 2004).

Finally, we should also note that Easterly (2004) tests the macro implications of Schelling's segregation models and finds that data does not support macro-consequences of Schelling's segregation models. For this reason, his study is at odds with the sprit of Schelling's analysis and does not really test Schelling's insights.

These studies give us enough evidence that the chequerboard model has received considerable attention and its results and implications have been explored and tested in different ways. These explorations give us reason to believe that Schelling's insights may be relevant for the real world. On the other hand, they also imply that mildly discriminatory preferences may fail to bring about residential segregation under certain circumstances. More properly, the individual tendencies to avoid minority status may not cause segregation when other tendencies (e.g. the preference to live in a wealthy environment) interfere.

If the argument that the chequerboard model contributes to a meta-model of residential segregation is accepted, explorations of the chequerboard model may be considered as future refinements and expansions of this meta-model. Hence, Schelling's initial hypothesis and his chequerboard model helped researchers to expand their conceptual toolbox to include more explanatory factors. Our view of residential segregation is more refined after the chequerboard model and its explorations, because we now have a better idea of possible interactions among different explanatory factors.

Appendix IV
Focal points and risk dominance

This appendix introduces some of the concepts mentioned in Chapter 8 and presents some of the related issues for the convenience of the reader who is not much acquainted with game theory.

1 Salience and precedence

Let us, for a moment, dispense with the assumptions concerning the driving game of Table 8.1 and think about how real individuals (i.e. in contrast to the model-theoretic individuals who are inclined to use a mixed strategy) would behave. The first thing that comes to mind is to look for the specific pieces of information that the individuals may benefit from. For example, we may speculate that if the driving wheels of the two cars are on the left they may find it convenient to drive on the right-hand side of the road. Or, we may contemplate that since a majority of individuals are right-handed they may use this as a coordination device; that is, they may expect others to use this piece of information in order to increase their chances of coordination. We may also speculate that right and left have different connotations in the culture to which players belong. For example, if they believe that doing something from the left (e.g. getting out of the bed from the left-hand side) causes 'bad' things, then they would not drive on the left. That is, they would consider 'right' as the salient or prominent option. These examples illustrate the way in which *salience* may work: given their particular environment, agents might think that one of the alternatives (e.g. driving on the right) stands out and expect others to use this alternative as a coordination device. 'Salience in general is uniqueness of a coordination equilibrium in a pre-eminently conspicuous respect' (Lewis 1969: 38, also see Sugden 1986: 47–52).

Alternatively, it might be the case that individuals always walk on the right-hand side of the pavement. That is, they avoid hitting other people by walking on the right. If this is the case, then it is possible that they might consider the driving game as being analogous to their 'walking game' and expect others to adopt the convention of walking on the right in the driving game. If they could coordinate by using this analogy then we may say that driving on the right would emerge as a convention by *precedence*. Another form of precedence may be the following:

let us assume that agents had previously been able to coordinate in the driving game many times by driving on the right. Then, when they have to play again, they might choose driving on the right just because they were successful in coordination in the past by driving on the right. It should be noted here that although salience and precedence are different notions, they are closely related. In fact, 'precedence is merely the source of one important kind of salience: conspicuous uniqueness of an equilibrium because reached it last time' (Lewis 1969: 36).

2 Schelling games

In the following games, players will win a prize if they choose the same alternative or do the same thing. They cannot communicate and they make their choices simultaneously (reproduced from Schelling 1960: 56–57).

1 Name 'heads' or 'tails'.
2 Circle one of the numbers listed below:
    7   100   13   261   99   555
3 You are to meet somebody in New York City. You have not been instructed where to meet; you have no prior understanding on where to meet; and you cannot communicate with each other. Where would you go to meet the other?
4 You were told the date (and the meeting place) but not the hour of the meeting in no. 4. At what time will you appear in the meeting place?
5 Write a positive number.
6 Name an amount of money.
7 Divide $100 into two piles.

In Schelling's informal experiments, most preferred answers were:

1 Heads
2 Number 7
3 Grand Central Station
4 12 noon
5 1
6 $1,000,000
7 $50

3 Ultimatum games and predictions of game theory

The gap between the 'predictions' of game theory and the actual behaviour of individuals may be observed in a number of other experiments that concern some other games, such as the ultimatum game. For example, consider that as a participant of a TV show you are asked to divide 100 euros between yourself and your co-player. The other participant is asked to do the same and you may not communicate in any way. If you and your co-player can independently agree on the division, you and your co-player will receive the amount of money you had specified for yourselves. If not, you will get nothing.

In this problem there are many equilibrium points. Every division that sums up to 100 may be an equilibrium. For example, you may want to get 60 per cent of the money and give away 40 per cent to the other participant. If the other participant independently makes the same offer – that is, 40 per cent for herself and 60 per cent for you – you will get the money you specify. There are many options you may choose from and standard game theory will not help you a lot. However, you may just think that a fair division may be the first thing that would occur to your co-player, and for that reason you would divide the money in two. You may be right in that people, at least in Western countries, may follow this 'fairness norm' to coordinate in such games.

But let us make the problem a little bit more interesting. Let us say that in the same game you are going to make the offer and your co-player has the chance to accept or reject it. If she accepts, you get the amounts you have specified in your offer, if she rejects you both get nothing. (This game is known as the *ultimatum game*. There is a huge experimental and theoretical literature concerning the ultimatum game in its different forms. Some of the papers that analyse ultimatum games can be listed as follows: Cameron (1995); Fehr and Tougareva (1995); Forsythe *et al.* (1994); Güth *et al.* (1982); Hoffmann *et al.* (1994); Roth *et al.* (1991).) Game theory predicts that as a rational player your co-player would accept any positive offer. The intuition behind this is that he is getting money out of nowhere and, hence, he should get what he could. Your rational decision should then be to offer the possible minimum amount to your co-player and keep the rest for yourself – that is, you could keep 99 euros and give away 1 euro to the other participant. However, experiments showed that this is not what real people do in such situations, for example, Güth *et al.* (1982) found that a fair offer (50 per cent, 50 per cent) is a good coordinating device among real individuals. That is, existing norms and conventions (e.g. fairness norm) help real individuals solve this coordination problem.

A striking experiment that reports how institutions and peculiarities of the particular environment matter is a recent study on fifteen small-scale societies (Henrich *et al.* 2001, 2005). It shows that predictions of game theory do not hold and that there is a wide variety of ways in which individuals coordinate their behaviour. Henrich *et al.* (2001, 2005) demonstrate that specific characteristics of different communities (e.g. economic organisation, structure of social interactions) influence the way in which individuals act in such games. The point is that existing institutions (i.e. conventions, norms, regularities in behaviour) matter and they are relevant if we want to understand how people coordinate and how new institutions evolve.

Also note that Schelling argues:

> the mathematical structure of the payoff function should not be permitted to dominate the analysis. [. . .] there is a danger in too much abstractness: we change the character of the game when we drastically alter the amount of

contextual detail [. . .]. *It is often contextual detail that can guide the players to the discovery of a stable or, at least, mutually non-destructive outcome.* [. . .] This corner of game theory is inherently *dependent on empirical evidence.*

(Schelling 1958: 252, emphasis added)

4 Pareto dominance and risk dominance

Pareto dominance

In a game where one equilibrium Pareto dominates, other equilibria agents may consider the *Pareto dominant* equilibrium as a focal point (Harsanyi and Selten 1988). Table AIV.1 represents a coordination game where two individuals have to choose an integer between 1 and 100. If they can simultaneously choose the same number, x, they will be paid x euros. If they fail to coordinate, they will get nothing.

In this game there are 100 pure strategy Nash equilibria, that is, every successful coordination counts as one. However, one of them is superior to others. The argument is that individuals prefer 100 euros to other outcomes and since players may expect the other player to reason in a similar fashion, the Pareto-dominant equilibrium (100, 100) may be considered as a focal point of the game. Nevertheless, since individuals do not know how the other player thinks, they cannot really know whether the other player will consider the Pareto-dominant equilibrium as a focal point or not.

Risk dominance

If $a_{11} > a_{21}$, $b_{11} > b_{12}$, $a_{22} > a_{12}$, $b_{22} > b_{21}$ then the game presented in Table AIV.2 is a coordination game with (D, D) and (Q, Q) as pure strategy Nash equilibria.

If the following condition holds we say that the pure strategy Nash equilibrium (D, D) is risk dominant:

Table AIV.1 Coordination game

		Player II			
		100	99	. . .	1
Player I	100	100, 100	0, 0	0, 0	0, 0
	99	0, 0	99, 99	0, 0	0, 0
	. . .	0, 0	0, 0	. . .	0, 0
	1	0, 0	0, 0	0, 0	1, 1

Table AIV.2 Game

		Player II	
		D	Q
Player I	D	a_{11}, b_{11}	a_{12}, b_{12}
	Q	a_{21}, b_{21}	a_{22}, b_{22}

Table AIV.3 Stag hunt game

		Player II	
		D	Q
Player I	D	10, 10	0, 8
	Q	8, 0	7, 7

$$(a_{11}-a_{21})(b_{11}-b_{12}) \geq (a_{22}-a_{12})(b_{22}-b_{21})$$

To see the intuitive idea behind risk dominance, consider the stag hunt game presented in Table AIV.3.

First, let us show that this is a coordination game with two pure strategy Nash equilibria:

$$10 = a_{11} > 8 = a_{21}, \ 10 = b_{11} > 8 = b_{12}, \ 7 = a_{22} > 0 = a_{12}, \ 7 = b_{22} > 0 = b_{21}$$

Let us assume that I expects II to play D and for this reason I plays his part in the (D, D) equilibrium. If I's expectations are correct both players' payoff is 10 euros (assuming that payoffs are expressed in euros). Yet if I's expectation does not hold and II chooses to play Q, then while II gets 8 euros, I receives nothing. That is, by choosing to play D, I takes a risk of losing 10 euros. Now assume that I expects II to choose Q, and for this reason plays his part in the (Q, Q) equilibrium. If his expectation holds, then both players get 8 euros. Yet if I's expectation does not hold and II chooses to play D, then while I gets 8 euros, II gets nothing. That is, if I chooses to play Q he does not loose anything even if his expectations turn out to be incorrect. Since the same argument holds for II as well, we say that equilibrium (Q, Q) is less risky, that is, (Q, Q) is the risk-dominant equilibrium.

That is, since $(10–8)(10–8) < (7–0)(7–0)$, (Q, Q) is the risk-dominant equilibrium. On the other hand, both players prefer (D, D) equilibrium to (Q, Q), and for this reason it is the Pareto-optimal equilibrium.

5 Theory of focal points of Bacharach and Bernasconi

Bacharach and Bernasconi (1997) try to formalise the different ways in which strategies may be framed. They generalise Bacharach's (1991) idea that players' options are acts under descriptions and they are distinguished by the concepts the players use to specify them. This model permits us to conceptualise the possible differences in agents' perceptions and, for this reason, it is a step further in understanding how these differences may influence the outcome of a coordination game. Like Janssen (2001b), this model focuses on the attributes of the alternative strategies and how players of the game perceive these attributes. Yet, unfortunately, the model is only able to 'predict' the outcome of simple coordination games. Consider the games in Figure AIV.1. In these coordination games individuals are supposed to pick the same object from a set of objects. Bacharach and Bernasconi (1997) predict that in Game I individuals would select the black circle. The principle is that individuals (*ceteris paribus*) prefer to pick an object that is rarer (*principle of rarity preference*). It is more difficult to coordinate in Game II. Here, the *principle of symmetry disqualification* is needed. In order to find out the odd alternative, one has to disqualify the symmetrical or similar objects. It is asserted that players will not be able to discriminate among symmetrical alternatives and choose the one with different attributes, that is, 'U', which is the only vowel. Game III is more problematic, as the 'odd' option is not easily available. It is argued that in such cases there is a trade-off between availability and rarity and that, *ceteris paribus,* agents are more inclined to pick an attribute which is more available (*principle of availability preference*). That is, agents would not base their reasoning on an attribute which is less likely to appear to their co-player's mind. In this game position, shape and size are the most obviously available attributes, and according to this principle, agents should limit their thinking to these. If they do, with some effort, they will see that one of the diamonds is slightly smaller than the others and that it should be picked (i.e. according to the principles of rarity preference, symmetry disqualification and availability preference).

Bacharach and Bernasconi capture the idea that 'in the pure co-ordination game, the player's objective is to make contact with the other player through some imaginative process of introspection, of searching of shared clues' (Schelling 1958: 211). However, from the point of view of explaining the emergence of conventions, their analysis is still in its infancy. Consider the picking game in Figure AIV.2. In this game, Bacharach and Bernasconi's model 'predicts' that 'the arrow' should be picked, that is, according to the rarity preference. Yet it seems to be reasonable to expect agents to choose the circle indicated by the arrow because of the connotations of 'the arrow'. In fact, when I ask my students to play this game they choose the circle which is indicated by the arrow in order to coordinate

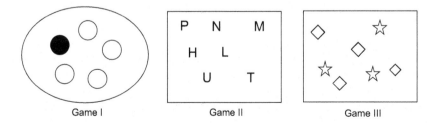

Figure AIV.1 Simple games of Bacharach and Bernasconi (1997).

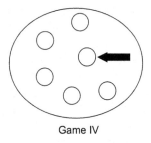

Game IV

Figure AIV.2 A coordination game.

with their co-players. Although 'the arrow' is a simple object, we do not perceive it as such and the conventions concerning the arrow (e.g. the traffic convention that we follow the road which is indicated by the arrow) may influence the way we behave in this particular game. (Bacharach and Bernasconi also mention this problem.) It is evident that physical attributes of the objects cannot be all there is to focal points and coordination. Existing conventions might matter even in the context of simple objects. Then it may be argued that the static models we have examined above fail to explain the possibility of successful coordination.

It should be noted here that from a formal point of view there might be many ways in which a particular set of available strategies may be conceived or framed from the perspective of the agents. For this reason, formalising focal points is not an easy task. It should be evident that the above argument is not meant to degrade the existing models of focal points; rather, it tries to explicate the limits within which these models should be evaluated and criticised.

6 Evolutionary stability and replicator dynamics

Let us, for a moment, assume that the driving game (Table 8.1) is played among deers in a certain area. A large population of deers use a limited number of narrow deer paths. Every day every deer meets at least one other deer coming from the opposite direction. When they meet they have to simultaneously 'decide' whether to use the right- or left-hand side of the path. Since they do not 'want' to reduce speed they have an 'incentive' to coordinate. Simply, they are playing a game similar to the driving game. Should we expect them to bring about a deer-traffic convention?

To examine this question, let us further assume that every deer is predisposed to play a certain pure strategy. That is, a percentage of the deer population always plays 'right', while the rest is predisposed to play 'left'. The deer that are successful in coordination are supposed to have more reproductive success and those who fail to coordinate will become extinct in time. Now consider the following scenario: in time the deer population somehow reaches a state where every deer is predisposed to play the same strategy (e.g. 'right'). Could we say that this equilibrium point (right, right) would be stable? This scenario brings us to the evolutionary analysis of Maynard Smith and Price (1973) and Maynard Smith (1974, 1982) (see Weibull 1995; Mailath 1992, 1998; Michihiro 1997; Friedman 1991; Hofbauer and Sigmund 1988 for a general discussion of evolutionary game theory). They have argued that *evolutionary stability* of an equilibrium of this sort depends on whether the population may be invaded by a mutant strategy or not. An evolutionarily stable strategy (ESS) cannot be invaded by mutant strategies and for this reason evolutionary stable equilibria of a coordination game may be considered as conventions in that they are self-supporting equilibria (for the notion of evolutionary stability see Binmore and Samuelson 1992; Blume, Kim and Sobel 1993; Hofbauer, Schuster and Sigmund 1979; Samuelson and Zhang 1992; Taylor and Jonker 1978; Wärneryd 1991).

In our case, it is easy to see that a mutant deer which is predisposed to play 'left' cannot invade a population of 'right' playing deers, since its average success against other deers will be less than a 'right' playing deer. For this reason, evolutionary game theorists argue that two pure-strategy Nash equilibria of the driving game are evolutionary stable.

(Note that in the standard analysis, it is assumed no individual has an incentive to deviate from his Nash strategy. However, it is still possible that one may be indifferent between his Nash strategy and another strategy, given others' strategies. Given other players' actions, if a player strictly prefers his Nash strategy to other actions then this equilibrium is argued to be a strict Nash equilibrium. Or in other words, if an individual has no alternative strategy that does as good as his equilibrium strategy (i.e. has no alternative best reply) given others' strategies, then the resulting equilibrium is a strict Nash equilibrium. In fact, Nash equilibria of the driving game are strict Nash equilibria, and every strict Nash equilibrium is ESS. See Maynard Smith and Price 1973; Selten 1980.)

However, the notion of ESS does not provide any basis for arguing that the deer population will be able to reach one of these evolutionary stable equilibrium points. It does not help us solve the equilibrium selection in the context of the driving game; rather *it justifies the idea that if a population of agents (human beings, or deer) is able to coordinate their actions somehow, this equilibrium would be self-fulfilling.* Although it tries to capture the dynamics of an evolutionary process, ESS is a static concept and does not explicate why a certain ESS would be selected among others (Mailath 1992: 267). The dynamics behind ESS and its refinements (e.g. stochastically stable strategies) has been studied with different models that vary in the formulations of the evolutionary dynamics. Only a few of these are examined here. First, we will focus on the replicator dynamics that works at the population level and then we will discuss learning models.

Replicator dynamics

The study of natural selection is usually associated with a mechanism known as replicator dynamics. The term 'replicator' comes from Dawkins (1976). For the original statement of 'replicator dynamics' see Taylor and Jonker (1978). The replicator dynamics captures the idea that reproductive success of a certain strategy is a function of its success in games. For example, concerning the deer population, this means that if a certain strategy does better on average than the population average, the type of deers that are predisposed with this strategy will grow at a greater rate than the others. Replicator dynamics has various alternative formulations that need not be examined here (see Fudenberg and Levine 1998: 51–99; Samuelson 1998: 63–75; Weibull 1995: 69–119). Considered from the point of view of pure coordination games, the replicator dynamics produces two interesting results. To discuss these results, consider the telephone game in Table AIV.4.

The telephone game has the following scenario: Player I and Player II are having a telephone conversation. In the middle of their conversation the telephone line is cut off for no specified reason. They have to decide whether to wait for the other to call or to call back. If they both call they cannot communicate for the lines will be busy. If they both wait they cannot continue their conversation either. This is an asymmetric coordination game for they have to choose different strategies for coordination. Now we may ask whether any convention may emerge out of this coordination problem if a finite population of agents are randomly matched to play this game repeatedly. The first interesting result of the replicator dynamics approach is that if the population of agents is homogenous then the only stable equilibrium is the mixed-strategy equilibrium, where agents randomise (Fudenberg and Levine 1998: 56–58). That is, if two players who are randomly matched are from the same population, that is, are identical, then repeated play will not bring about a convention. The second interesting result concerns replicator models with distinct populations. If the players of the game are drawn from distinct populations, then pure strategy Nash equilibria of this game ((wait, call back), (call back, wait)) become stable, but the mixed strategy equilibrium is not stable.

(It should also be noted here that the outcome of the replicator dynamics depend on the initial state of the population in the context of the stag hunt game (see Appendix VI.6). That is, if many agents are playing the Pareto-dominant strategy, initially the replicator dynamics converges to the Pareto-dominant equilibrium. Yet if most of the agents are playing the risk-dominant strategy, then the process converges to the risk-dominant equilibrium. See Samuelson 1998: 79–80.)

The distinct population model may be considered as an approximation to a situation where players can distinguish between 'the caller' and 'the receiver'. Although replicator dynamics is not exactly designed to represent such cases, *it implies that if players can recognise each other in some way (e.g. label each other) then the telephone coordination problem may bring about one of the alternative conventions*: original caller calls back (call back, wait); or receiver calls back (wait, call back). Note that the model remains silent concerning equilibrium selection.

Table AIV.4 Telephone game

		Player II	
		Wait	Call back
Player I	Wait	0, 0	1, 1
	Call back	1, 1	0, 0

Replicator dynamics works at the population level and ostensibly leaves no role to the individual players, as it is suggested by the 'pre-programmed behaviour' assumption. For this reason, replicator dynamics is generally considered as being inappropriate in economic contexts and that the dynamics of an economic process should be based on a model of individual learning (Fudenberg and Levine 1998; Kandori, Mailath and Rob 1993; Young 1998). If successful strategies grow faster, then stable coordination equilibria may be reached and conventions may emerge. Models with replicator dynamics *do not explain* why successful strategies grow faster. Such models study the dynamics at the aggregate level and are silent about which individual mechanisms bring about the consequences at the aggregate level (Fudenberg and Levine 1998: 52). Although we may still argue that replicator dynamics point to the possibility that there may be a certain mechanisms that may bring about conventions, the need for the study of individual mechanisms is obvious. There could be many mechanisms that may be consistent with replicator dynamics, such as learning and imitation. In the following pages we focus on models that study individual learning.

7 Local interaction

Local interaction raises another question about conventions. Since any of the alternative conventions is likely to emerge in the short run, local interaction may result in the emergence of different conventions in different localities. From a real-world perspective, this is quite reasonable for very large populations who are spatially separated, yet it is not always reasonable to argue that alternative conventions are likely to exist in areas such as cities or countries. While countries may abide to different conventions, within a certain country people usually abide to a uniform convention. Yet in some countries different conventions co-exist in different areas. Could these models give any insight concerning the global diversity of conventions while retaining the argument that a single convention is very likely to emerge in local areas?

Young (1996, 2001) analyses the issue of co-existence of conventions by extending his model for the driving game to consider the interactions among different countries who use different driving conventions (e.g. (left, left) or (right, right)). The model is a spatial model similar in some respects to Schelling's residential segregation model (see Chapter 4). Every country has a limited number of neighbour countries and there is consistent traffic on the borders. For this reason, conflicting driving conventions are costly. Let us say that in each time period one country considers whether to switch to the other driving convention by observing the existing conventions of its neighbours. For example, if country A abides to the left-convention and all its neighbours abide to the right-convention it is 'rational' to switch to the right-convention. Yet if its neighbours abide to different conventions then it has to 'assess' the costs of switching. Note here that Young abstracts from the technological costs of switching a convention and argues that this would not change the argument. If we consider the countries as nodes in a network, then this model implies the number of connections one country has with other nodes is the decisive factor given the costs of abiding to different conventions. If, for example, two different parts of the network are densely connected but have a weak link with each other, then the model says that two different conventions may be adopted in these parts of the network. Yet if all countries are densely connected then we may expect a uniform convention. It is also possible to bring in random shocks to this model. Let us assume that one or more countries switch conventions occasionally, whatever the costs (e.g. France switched to the 'right' convention after the revolution and imposed this convention in some of the countries they have occupied; see Young 1996). If there are such idiosyncratic shocks, then in any connected network of countries (i.e. it is possible to drive from any country to any other) it is more likely that all countries adopt the same convention. Moreover, if it exists, the risk-dominant convention is likely to drive other conventions out.

While the above model analyses exclusive conventions in the sense that individuals cannot adopt two or more conventions at a time, not all conventions are like this. Consider the telephone game. It might be that while one country adopts the caller-calls-back convention, another country might adopt the receiver-calls-back convention. It is for these types of conventions that the individuals may switch between conventions with some *costs* (e.g. learning) when they travel. Goyal and Janssen (1997) analyse this type of non-exclusive convention and examine equilibrium selection in such environments. (They use a deterministic model with no learning. In this model individuals form their expectations given the existent local information. Yet they argue that their model is robust to several learning dynamics.) The model implies that, given the costs of adopting both conventions are not very high, both conventions might prevail and co-exist. In a symmetric coordination game the result of their analysis is that both conventions co-exist under all conditions. They further argue that if one convention is better then the other then whatever the costs the Pareto optimal convention is adopted in the long run (note here that this is the result for the case where Pareto-dominant convention is also risk dominant). But if Pareto dominance and risk dominance are in conflict then the following results hold: if the costs are low then Pareto-dominant convention will eventually be adopted in both countries; if costs are high then both countries would eventually adopt the risk-dominant convention. Yet if costs are at an intermediate level then two conventions would co-exist. That is, while different conventions would be adopted in different countries, some individuals would undertake the costs of adopting both conventions. Goyal and Janssen's model indicates that the cost of adopting a certain form of behaviour may be relevant in explaining the emergence and persistence of conventions.

8 Rationality and learning (Goyal and Janssen 1996)

An important question is whether rational individuals who learn from experience might arrive at a coordination equilibrium, or whether we could better explain the emergence of a convention without restricting individuals' expectations and learning behaviour with a certain form of common knowledge assumption. Goyal and Janssen (1996), who study similar questions, argue that rationality alone does not suffice to explain coordination even if individuals are able to learn. Consider the driving game where individuals are randomly paired to play the driving game repeatedly and they are able to observe the whole history of play, including the payoffs. Assume that every time an individual is confronted with another driver he evaluates the previous actions of the other players and bases his expectations on this. The question is whether he and other players could achieve concordant mutual expectation by evaluating the information gathered from previous plays.

A similar model has been examined in Chapter 6. Schotter assumed that when individuals are close to the absorbing points (e.g. 95 per cent plays 'left') they will consider this as an indication that everybody will chose a certain strategy (e.g. 'left') with unit probability. This assumption implies that rational (Bayesian) learning does not guarantee coordination. Additional assumptions concerning learning behaviour have to be imposed on the model. Moreover, agents are assumed to base their expectations on what happened in the past. They expect that what happened in the past is likely to happen in the future. Hence, there is an implicit assumption that says that agents know that the other agents are forming their expectations in a similar way. Otherwise, there is no rational reason for them to expect that what happened in the past would happen in the future. Similar things can be said for other learning models. Goyal and Janssen (1996) discuss Crawford and Haller (1990) and Kalai and Lehrer (1993a,b), who argue that rational individuals can learn to coordinate. While Crawford and Haller assume that there are optimal rules for learning, Kalai and Lehrer put certain restrictions on individuals' prior beliefs. Both assumptions imply that these models put additional restrictions on learning behaviour and, for this reason, rationality alone cannot ensure coordination. More specifically, Goyal and Janssen (1996) argue that even if there may be optimal rules for learning how to coordinate, these rules are not unique. That is, if agents' learning behaviour is not coordinated at the outset they might not be able to coordinate. Similarly, Kalai and Lehrer's model indicate that agents' prior beliefs have to be coordinated to ensure their success in coordination. Both of these models imply that pre-existent conventions are necessary for individuals' success in coordination.

Goyal and Janssen generalise this argument by employing a more sophisticated learning model. The idea behind their model is the following: in order to ensure coordination in the next period, every agent has to take into account the previous plays of other players. However, since every player knows that the other players are using the information gathered in previous plays to form their expectation for the next period, in order to ensure coordination every player has to know how the others are forming their expectations. The problem is that the outcome of the previous encounters does not restrict the type of hypotheses they might entertain about each other. In other words, as Goyal and Janssen argue, at any point in time one may entertain an infinite number of hypotheses about others, which are consistent with their existent information (also see Foster and Young 2001). Thus, unless the modeller restricts the number of these hypotheses, rationality does not ensure players to learn how to coordinate. Again, from the perspective of explaining real world coordination problems, this means that knowledge of pre-existing conventions is necessary to explain how coordination is achieved.

Notes

1 Introduction

1 It is hardly necessary to say that there is only one shortcut from point A to point B. It is assumed here that most of the traffic in the area, where this walking path emerges, is between A and B.

2 Unintended consequences

1 Similarly, Rosenberg (1995: 153) uses 'unintended' interchangeably with 'unexpected'.
2 Merton (1936: 903–904) has another category where he talks about public policy. In this case, the action of the government, say doing Y, to achieve one particular end, X, affects the individuals in the country in such a way that because of the change in individuals' behaviour due to Y, X does not happen. This may be considered as an example of the case where the agent (individual, or government) is ignorant about the fact that Y causes X *ceteris paribus*.
3 He also explicitly says that 'second type would seem to afford a better opportunity for sociological analysis since the very process of formal organization ordinarily involves explicit statement of purpose and procedure' (Merton 1936: 896). In the next chapter it will be clear that this is exactly what this book is not about.
4 By 'social action' I simply mean a number of individuals acting to achieve a social result.
5 Remember that here 'someone having an intention' means 'someone having a purpose', rather than 'someone than having a plan'.
6 Cells 1.2, 1.3, 1.4, 1.6, 1.7, 1.8, 2.2, 2.3, 2.4, 2.6, 2.7, 2.8, 3.2, 3.3, 3.4, 3.6, 3.7 and 3.8 indicate unintended consequences.
7 We may say that (in Merton's terms) we are mostly concerned with imperious immediacy of interests at the individual level.
8 For an example of how these different types of unintended consequences are handled together (or how they are mixed up), see Norton (2002).
9 Note that in Schelling's model individuals are indifferent. They do not care whether the city is segregated or not as long as they are not living as an extreme minority. Yet similar results are obtained even if we assume that individuals strictly prefer a mixed neighbourhood (see Chapters 4, 7 and Appendix III).
10 Note that Collin defines a social fact as follows: 'Social facts are thought to be a product of the very cognition, the very intellectual processes through which they are cognised, explained and classified, in so far as this cognition is a shared, collective one' (Collin 1997: 3). As it stands, with this definition we cannot consider residential segregation as a social fact. Yet I do not see any reason for not considering segrega-

tion as a social fact. If there is racial residential segregation in a particular city and if we state that there is residential segregation in this city, we express the truth about a collection of individuals living together. Thus, 'segregation' should count as a social fact.

3 The origin of money

1 See Chapter 2 for the distinction between 'results' and 'consequences'. Obviously, Menger does not make the distinction, but it is clear that, in our terminology, what he means is 'consequences'.

2 He is not clear about the nature of forces acting in the social realm, but we will not deal with this issue in this book.

3 The explicit consideration of these limitations distinguishes Menger from the organic school, which applies the organism analogy more broadly. An important difference is Menger's individualistic approach. As Hutter (1994: 305) argues, Menger 'defends the "atomistic" paradigm against the dominating historical and organic paradigms'. Thus, he is not using the 'analogy' in its generally accepted form. Also see Stark (1962: 176).

4 Note that Menger (1892a: 239) argues, 'It must not be supposed that the *form* of coin, or document, employed as current-money, constitutes the enigma in this phenomenon [money].' He wants to explain the genesis of the fact that some commodities ('uncoined precious metals', 'cattle', 'skins', etc.) served as a medium of exchange.

5 Menger (1892a: 241) refers to Plato, Aristotle and the Roman jurists, who argued that 'certain commodities, the precious metals in particular, had been exalted into the medium of exchange by general convention or law, in the interest of commonweal'. He states that he is doubtful about this theory and he finds it 'unhistorical'.

6 Some of the conditions that affect the saleableness of goods are institutional conditions, such as the development of the market or social and political restrictions. See Menger (1892a).

7 It is usually argued that the term 'double coincidence of wants' comes from Jevons. It refers to the following difficulty in barter, as stated by Jevons:

> The first difficulty in barter is to find two persons whose disposable possessions mutually suit each other's wants. There may be many people wanting, and many possessing those things wanted; but to allow of an act of barter, there must be a double coincidence, which will rarely happen.
>
> (Jevons 1876: I.5)

8 Also see Horwitz (1999) and Kuniński (1992).

9 See Stenkula (2003) for the argument that Menger was one of the forerunners of the 'Network Theory of Money'.

10 This issue is closely related to Menger's (1892b [2005]) analysis of money as a measure of value. Menger rejects the idea that money bears some quantum of value (also see Campagnolo 2005) and argues that:

> The use of metal as an intermediary made trade easier and offered more accuracy to economic calculation, but it did not change the nature of trade. Still nowadays, the effort of everyone to satisfy their own needs, as much as possible, is the determining cause, not only of the fact of exchange in itself, but also of the formation of prices. The goal of people who do business in the markets is to write down some gain under the income column of their balance sheet and, as to expenses, to provide themselves with as much satisfaction as possible through bartering [*troquer*] money for a commodity. Buying and selling are among the main ways of conveying the universal desire for gain and making one's position

better. Money has become the intermediary of exchange, but if it serves well in measuring prices, it is only in the sense that we have just pointed out. The motive for bartering is profit, but more than that, the quantities that are exchanged for each other get fixed through the subjective advantage of both subjects.

(Menger 1892b [2005]: 260)

11 Figure 3.1 is adopted from Holland *et al.* (1989), where it is used to study the process of induction. In Chapter 7 the rationale behind using this figure is explained. The reader should, for now, merely focus on how it represents Menger's explanation.

12 See also Iwai (1997). Iwai identifies three different views of the emergence of a medium of exchange: spontaneous emergence view, Chartal theory and evolution out of gift (anthropologists' view). We may add to this the Metallist view. Roughly, Metallists argue that only commodities that have intrinsic value may be selected as money with collective acceptance. Chartalists, on the other hand, argue that any object can be used as money, as long as the state guarantees its acceptance. However, from our point of view Metallist and Chartalist views are similar to the design view, for in the first one money is designed by many people, and in the latter by the state. The debate between Metallists (who hold the collective agreement view) and Chartalists is an interesting one, but we will not be dealing with this debate in this chapter. The reader may consult Bell (2001) and the references therein for an overview.

13 These disputed assumptions, in fact, exist in other classical texts of economics (e.g. in Smith, Jevons and Mises). See Appendix I for an exposition.

14 It is common that the 'design view' of money is credited to Knapp (1905), who is one of the advocates of the Chartal theory (see, for example, Hodgson 1992; Hutchison 1962: 143–144; Iwai 1997). However, it can be argued that there is nothing in Knapp that contradicts Menger's account (see, for example, Maclachlan 2003).

15 As could be seen in Appendix I, neither do Smith's, Jevons's and Mises's accounts conflict with the view that coined money was introduced by a central authority.

16 Of course, Polanyi himself builds upon other authors' work, such as Thurnwald (1932).

17 See Quiggin (1949) for a survey of primitive forms of money and Mauss (1930), who presents the gift exchange system with the obligations of give, receive and reciprocate.

18 The reader may wonder whether we may consider Menger's explanation valuable, even if all the historical evidence is against it. The argument here is that Menger points out some possible explanatory factors. This is why it is valuable. But if one is able to show that none of the factors played any role in the emergence of money as a medium of exchange, we should simply dismiss it. Similar issues will be discussed further in Chapters 6 and 7.

19 We may consider empirical forms as the common properties of a set of things, such as tokens modern money.

20 Some believe that Menger defended theoretical economics and disregarded historical economics. Yet, he argues that these different types of inquiries provide different types of knowledge and that they are both necessary. For example, he says, 'The specifically historical understanding of concrete phenomena is also completely adequate for the field of economy' (Menger 1883: 44). With respect to the relation between theory and history, the following quote is relevant:

The understanding of the *concrete* phenomena of economy by *means of the theory*, the application of theoretical economics as *means* for this understanding, the utilization of the theory of economics for the history of economy – all these are [. . .] problems for the *historian*, for whom the social sciences, considered in this way, are *auxiliary sciences*.

(Menger 1883: 46)

21 He also argues (1883: 51) that the degree of the strictness of laws does not affect 'the character of economics as a theoretical science'.

22 The 'realist' in this phrase should not be confused with its accepted philosophical use. Menger is here trying to imply that in this orientation very few abstractions (also see note 24) are made and particular phenomena are examined in their full complexity. In modern language, both the advocates of realistic–empirical and the exact orientations can be considered as realists of sorts. See Mäki (1990b) for an interpretation of Menger as a realist.

23 The emphasis is in the original text.

24 'The laws of theoretical economics are really never laws of nature in the true meaning of the word.' (Menger 1883: 59)

25 Menger also notes that abstraction is inescapable in any type of research: 'An abstraction from certain features of the phenomena in their full empirical reality is unavoidable' (Menger 1883: 79). 'Even the most realistic orientation of theoretical research imaginable must accordingly operate with abstractions' (Menger 1883: 80).

26 ' "The exact theory of political economy" is [. . .] a theory which teaches us to follow and understand in an exact way the manifestations of human self-interest in the efforts of economic humans aimed at the provision of their material needs. [. . .] It has only the task of affording us the understanding of a special side of human life, to be sure, the most important, the economic' (Menger 1883: 87).

27 Menger (1883: 73) argues, '*exact economics by nature has to make us aware of the laws holding for an analytically or abstractly conceived economic world.*'

28 It may be argued that all explanations cannot be characterised like this. The only exception might be structural explanations, or explanations of constitution. That is, we may explain why something has certain properties by referring to its constituents. Yet, this characterisation is appropriate for the examples of this book. See Chapter 7 for further details.

29 For example, Hodgson (1992) argues that Menger ignores the potential quality variation in commodities in his theory of saleability. Yet, Hodgson also argues that for this reason there is room for the state in the development of a medium of exchange. We have seen that this does not contradict with Menger's account. Menger is in no way opposing the idea of intentional refinements of institutions. He explicitly argues that after the establishment of the institutions the need for refinements and/or codification may arise.

30 An exception to this may be the cultures who learned the existence of a medium of exchange from other cultures. Yet, Menger is trying to explain the *origin* of a medium of exchange.

31 It is a common mistake to evaluate Menger's explanation by today's standards. For Menger's explanation is a part of our intellectual heritage – his explanation may seem simple and uninteresting by today's standards.

32 Many would argue that Menger is an Aristotelian of sorts (see, for example, Hutchison 1981, Kauder 1965, Mäki 1990b, Zuidema 2001) and, thus, when he presents the origin of money he proposes to have disclosed the *essence* of money. Even if we assume that this is true, this information does not tell us what is actually accomplished by Menger's account of money. The interpretation introduced here is consistent with Mäki's (1990b: 297) interpretation that Menger is a realist.

4 Segregation

1 The basic idea in Schelling's models is the same, yet it is presented in different ways. The analysis introduced in this chapter is applicable to Schelling's other segregation models.

2 We did not do this for Menger's explanation because the proposed mechanisms in his explanation were not entirely clear.

3   Schelling uses 'blacks' and 'whites' instead of As and Bs.

4   It is also possible to find various simulations of Schelling's model on the web.

5   This also means that in this model individuals who have mild discriminatory preferences do not strictly prefer a mixed neighbourhood. They are content as long as they do not have an extreme minority status, and they are equally happy with other compositions. That is, As (or Bs) do not care whether they have a reasonable minority status or a majority status; or whether the neighbourhood is integrated or not, as long as they do not have an extreme minority status. Strict preference for perfect integration implies that individuals prefer a mixed (integrated) neighbourhood to a segregated one. As we will see in Chapter 7, Pancs and Vriend (2003) test the chequerboard model under the conditions of strict preference for perfect integration and conclude that Schelling's results hold even if this assumption is made (also see Appendix III).

6   It may be argued that preferences cannot be inputs, because they are built-in. Yet the reader should remember that we are talking about how agents are defined in the model. The form of the individual rule is: IF $y \geq x$ THEN move, IF $x > y$ THEN stay. Thus, within the model one can change the preferences of the agents by simply changing the value of y. It is in this way that we consider preferences as inputs.

7   When we interpret the mechanisms as 'IF . . . THEN rules' then reinforcement cannot be considered as a mechanism, it is just a property of the interaction of the IF . . . THEN rules. But, if one wishes, an aggregate mechanism with this property may be called a reinforcement mechanism.

8   Note here the similarity between Schelling's arguments and Menger's characterisation of the realist–empirical orientation (see Chapter 3).

9   See Aydinonat (2005) for an interview with Schelling on the story of the chequerboard model.

10   For Mill's four methods of experimental inquiry see Mill (1843: 222–237).

11   Mäki (1992a: 321) defines isolation in the following way: 'In an isolation, something, a set X of entities, is "sealed off" from the involvement or influence of everything else, a set Y of entities; together X and Y comprise the universe.' Abstraction and idealisation may be considered as subspecies of isolation. In an abstraction we move from a particular case (e.g. Rotterdam, as a city) to a more general representation (chequerboard city, as a model of any city). So to say, abstraction is vertical isolation – that is, the type of isolation where the level of abstraction changes. By contrast, in what we may call horizontal isolation (idealisation), the level of abstraction does not change (see Mäki 1992a: 322–325). 'Idealizations are formulated in terms of limiting concepts designated or designatable by variables with the value 0 or $|\infty|$' (Mäki 1992a: 324). For example, in an idealisation we explicitly neglect some factors (e.g. people have no expectations).

12   Note that both homophily and discriminatory preferences require that individuals are able to discriminate between the group to which they belong and the other group. While 'homophily' emphasises *association* with one's own group, 'discriminatory preference' emphasises the *consequences* of this association in terms of where one wants to live, have lunch, be educated, etc.

13   The relevant concept here may be 'phase transition' (see Kauffman 1995: 56–58; Batten 2001).

14   Note that if there is to be a covering theory A, it must specify all the conditions for this kind of self-organisation. Schelling's models make us understand that there is this phase transition, but they do not specify the conditions for it. For example, when the relative number of individuals to the number of housing possibilities is smaller than what we have seen in the chequerboard model, the individuals may be so detached that segregation will not emerge – or it may be meaningless to talk about segregation. In other words, conditions like the necessary number of connections among individual mechanisms must be specified for a more comprehensive theory.

5 The invisible hand

1 To see the wide variety of references to the invisible hand and its uses consider the following examples: for invisible hand arguments in *philosophy of science* see Kitcher (1993), Hands (1995), Ylikoski (1995), Hull (1988, 1997), Mirowski (1997), Solomon (1994), Wray (2000) and Leonard (2002). For references in *political philosophy* see Holmes (1977), Nozick (1974), Lind (1989) and Postema (1980). For the use of invisible hand in *politics and economic policy* see Sayigh (1961), Bates (1974), Stiglitz (1991: 32) and Frye and Shleifer (1996). For *gender studies* and invisible hand see Feiner and Roberts (1990). For *ethics* and the invisible hand see McMahon (1981), Shepard (1995) and Evensky (1993). For invisible hand in *modern economics* see Stiglitz (1991), Sagoff (1994), T. Smith (1995a,b), Shepard (1995), Coase (1992), Durlauf (1991), Hahn (1970, 1981), Marris and Mueller (1980), Feiner and Roberts (1990), Leibenstein (1982) and Maskin (1994).

2 Thornton (2006) argues that the use of invisible hand in HA does not conflict with its uses in *WN* and *TMS*. He suggests that the invisible hand in HA is concerned with the methodological foundation of Smith's approach. Here we will be investigating what that methodological foundation is and how it is related to the use of invisible hand in *WN* and *TMS*. Thornton (2006) also argues that Smith's economic applications of the invisible hand (in *WN* and *TMS*) were influenced by Cantillon's model of the isolated state.

3 Also see Brewer (2006) who argues that the use of invisible hand in *WN* and *TMS* is not contradictory yet they address different questions.

4 In her survey of the literature Brown (1997) shows the variety of the interpretations of Smith's arguments. She ironically uses Smith's (1795) phrase 'mere inventions of the imagination' to emphasise the implications of the radical differences in these interpretations. Grampp (2000) makes a similar point. He identifies nine different interpretations of the invisible hand and argues that there is no convincing argument in the literature to show that the three invisible hands were indeed related.

5 The misunderstandings that are pointed out in this chapter can also be found in other works about the invisible hand. Thus, there is nothing special about Rothschild's argument in this respect. It was chosen because it is a recent interpretation and because Rothschild powerfully makes her point.

6 Also see Bridel and Salvat (2004: 134–138), who state that Rothschild's argument is not convincing. However, they do not discuss the relation between unintended consequences and the invisible hand and they do not really explain why it is unconvincing. Similarly, Eltis (2004: 155) state that 'Rothschild's attempt to marginalize the invisible hand has no basis' without substantiating this argument.

7 See Davis (1990) and Ingrao (1998) for a discussion of the invisible hand in HA.

8 Fiori (2001) interprets the invisible hand in HA as the invisible chain of events that coordinates natural phenomena. Accordingly, invisible hands in *WN* and *TMS* are related to the invisible chain of events that coordinate social phenomena. We have seen that, according to Smith, these apparently invisible chains of events have to be explicated in order to explain phenomena. Note that Fiori (2001) pairs visible order with concrete facts and invisible hand with abstract analysis. It is true that invisible-hand explanations are generally based on abstract models not on concrete facts. (Remember our discussion of Menger's conception of exact understanding and historical understanding in Chapter 3.) However, it is possible, in principle, to provide singular invisible-hand explanations based on theoretical invisible-hand models (see Chapter 7).

9 It is generally believed that Smith followed Newton in adopting a mechanistic–atomistic view of nature. Montes (2003, 2006) nicely argues that neither Smith nor Newton held a mechanistic–atomistic view of nature. Montes (2003: 731) argues that mechanistic interpretation of Newton and thus Smith owes its origins to French

Enlightenment. See the discussion under the section on modern conceptions of the invisible hand in the present chapter.

10 Note the similarity between this interpretation and van Fraassen's (1980) 'constructive empiricism'. D. D. Raphael and A. S. Skinner emphasise a similar idea in their introduction to the *Essays on Philosophical Subjects*. But their interpretation, as well, contains a minimal realist reading of Smith – as presented in the second possibility above. On Smith's philosophy of science also see Berry (2006), Fleischacker (2004) and Montes (2004).

11 Another possibility is offered by Lindgren (1969). He (1969: 899) argues, 'Smith did not entertain realistic epistemological views' but a conventionalistic view. Lindgren emphasises Smith's views on analogy and his argument that we understand new phenomena with what is familiar to us. And he (1969: 907) argues for 'all associations' – which are necessary for analogical thinking – 'are governed by habit or custom. We must conclude that the systems of signs developed by imitative arts' – to which inquiry is an example – 'are all governed by convention'. Yet there is nothing about conventions, habit or custom that necessarily prevents Smith having a realistic view about the world, or prevents him entertaining 'constructive empiricist' thoughts. In fact, Smith's statements about irregularities and newly observed phenomena suggest that scientific theories change in the light of these phenomena – similarly, habits, customs and conventions change with time. In fact, in HA, Smith demonstrates how theories in astronomy changed through time.

12 He gives Descartes' system as an example here.

13 Davis (1989: 65) argues that the invisible hand 'makes it the responsibility of deity to reconcile self-interest and the social good' and recourses to 'an extra-human, extra-social device'. However, he also points out that Smith is consciously criticising polytheism in HA (Davis 1989: 59). My interpretation is obviously different from that of Davis and from many others who see the invisible hand as a mystical power. Smith's criticism of individuals who explain natural phenomena with reference to invisible powers, and his ideas on philosophy of science, show that Smith's conception of a good explanation is an explanation which uncovers the apparently invisible causal factors in nature of society. This interpretation is further fortified by Smith's explicit attempts, in *WN* and *TMS*, to explicate the processes which seem to be work of invisible hands. One may argue that Smith explanations are not satisfactory, but one cannot argue that he is talking about an extra-human, extra-social device. Smith is at pains to show how people who are following their own interests (intentions targeted at the individual level) bring about unintended social consequences. It is the interaction of individual mechanisms (i.e. individuals pursuing their own interests) that bring about unintended social consequences. There is nothing extra-human or extra-social.

14 Evensky (2005) binds together Smith's apparently conflicting ideas by considering them as attempts to understand how different parts of the world are connected to each other. Under this interpretation the invisible hand may be considered as a metaphor that represents the need for uncovering the connecting principles of nature and society.

15 She mentions that 'caecus', the Latin word which translates as 'invisible', literally means 'blind'.

16 A similar argument can be made for *TMS* as well.

17 In fact, these two forms of blindness may be considered as resulting from 'uncertainties' that individuals may face: 'On the one hand they may not know the exact mechanism by which an outcome (consequence) is brought about by a certain action. On the other hand, a specific outcome often depends not only on the action chosen by a particular agent, but also on the actions chosen by others' (Janssen 1993: 12).

18 Note that 'local environment' represents actions of other individuals in that environment, and the consequences of these actions.

19 One may argue that they are also partially blind in this respect, for there may be unintended consequences at the individual level as well (see the discussion in Chapter 2). Yet they relatively know better than others.

20 Rothschild (2001: 124) argues that 'Smith's three uses of the phrase have in common that the individuals concerned – the people who fail to see the invisible hand – are quite undignified; they are silly polytheists, rapacious proprietors, disingenuous merchants'. In HA Smith literally criticises those who invoke the invisible hand of Jupiter to explain natural phenomena. Although, it does not seem that the proprietors are rapacious (in *TMS*), and that merchants disingenuous (in *WN*) for Smith most probably considers the 'blindness' of individuals as a fact of life. Note that we have already granted that Rothschild is right in that the use of invisible hand is ironic.

21 It should be added that self-regarding or self-interested action should not be understood as necessarily being selfish, given the Smithian notions such as 'emphaty', 'sympathy', 'justice', 'fellow feeling', etc. (see Morrow 1923, Sugden 2002, Werhane 1989, 1991) These should rather be understood as intentions directed to the individual level. It is of course true that in *WN*, Smith mostly talks about economic interests – which may be regarded as selfish – but we would be doing injustice to Smith if we say that all his thought is based on selfish individuals. Also see Rosenberg (1960) on institutional aspects of Smith's thought, especially in *WN*.

22 Schelefer (1998: 16) points out that Smith does not argue that invisible hand would always promote the interest of society. In *WN* Smith uses the phrase 'as in many other cases', not 'in all cases'.

23 In her discussion of invisible-hand explanations, Ullmann-Margalit makes a similar assumption: 'Individuals do not have the overall pattern in mind, neither on the level of intentions *nor even in the level of foresight and awareness*' (Ullmann-Margalit 1978: 271, emphasis added).

24 Irony means 'the use of words to express something other than and especially the opposite of the literal meaning' (as defined in Merriam-Webster's dictionary). Thus, it should be possible that Smith uses the phrase 'invisible hand' that has religious connotations, to imply something irreligious. Coase (1976) makes the following quote from Smith 'Superstition first attempted to satisfy this curiosity, by referring all those wonderful appearances to the immediate agency of the gods. Philosophy afterwards endeavoured to account for them from more familiar causes, or from such as mankind were better acquainted with, than the agency of the gods' (Smith 1789: V.1.152). Then he points out (1976: 19) that 'this is hardly a remark which would have been made by a strong, or even mild, deist.' And he goes on to argue:

> Since Adam Smith could only sense that there was some alternative explanation, the right response was suspended belief, and his position seems to have come close to this. Today we would explain such a harmony in human nature as a result of natural selection, the particular combination of psychological characteristics being that likely to lead to survival. In fact, Adam Smith saw very clearly in certain areas the relation between those characteristics which nature seems to have chosen and those which increase the likelihood of survival.
>
> (Coase 1976: 19)

Coase's interpretation of Smith is well in line with our remarks.

25 Also see Barry (1985: 139), who argues that 'in economics the point of the Invisible Hand theorem is to show how there can be order without a designing mind and without anyone intending specifically to produce such an order'.

26 Barry (1985: 137) argues that unrealistic assumptions of the general equilibrium approach does not prove that the invisible hand is wrong for he considers general equilibrium theory as a self-contained logical exercise. As we will see in the next

chapter, end-state models in economics should not be considered as inconclusive logical exercises, rather they are logical tests concerning our intuitions concerning real phenomena.

27 See Janssen (1993) for the relation between aggregate and individual phenomena in economics.

28 For similar arguments see Evensky (1993), Holcombe (1999), Khalil (2000b: 50) and Knudsen (1993: 149–150).

29 Also note that the common presumption that Adam Smith's invisible hand is the predecessor of general equilibrium theory is problematic on many grounds. For example, there is a common assumption that Smith was influenced by Newton and that they both hold a mechanistic view of nature. Montes (2003, 2006) nicely demonstrates that while the first part of this assumption is correct, the latter part is not. According to Montes (2003: 729), 'Newton repeatedly criticised "mechanical philosophy", as he considered that mechanical principles were inadequate to explain all phenomena.' Montes (2003: 731) argues that mechanistic interpretation of Newton (and thus Smith) owes its origins to French Enlightenment. 'In political economy, the Physiocrats followed this pseudo-Newtonian tradition, which was later adopted and adapted by Walras and played an important part in the subsequent development of general equilibrium methodology' (Montes 2003: 731). Montes (2003: 733) also argues that Smith's conception of individuals were not mechanistic or atomistic either. Rather, Smith emphasised the need to discover the connecting principles of nature and society without endorsing a mechanistic, atomistic or deductive methodology. Moreover, as we have seen, he emphasised the process rather than the end-state.

30 See Barry (1982), Brennan and Pettit (1993), Demeny (1986), Evensky (1993), Hamowy (1987), Heath (1998), Kuniński (1992), Nadeau (1998), C. Smith (2006a,b), Ullmann-Margalit (1978) and Vanberg (1994) for the relation between invisible hand and spontaneous order.

31 For the demonstration of Austrian economics as explaining with causal processes, see Mäki (1990a, 1992b).

32 Hayek (1946a [1949]: 4) considers Smith as one of the founders of 'true individualism' and Menger as the first modern follower of Smith.

33 Also see Barry (1985: 141), who argues that 'the interesting question concerns not the existence of perfect coordination in abstract equilibrium, but the nature of the coordinating process that the Invisible Hand generates'.

34 On the Knowledge problem in Hayek see Caldwell (1997), Vanberg (1994: 79) and Zappia (1995).

35 Also see Barry (1985) and Fiori (2001), who emphasise the aspect of coordination in the invisible hand. We will discuss the emergence of coordination in Chapter 8.

36 Yet Hayek (1945 [1949]: 87) also argues that the 'adjustments are probably never "perfect" in the sense which the economist conceives of them in his equilibrium analysis'.

37 Nozick (1974) considers this mechanism of coordination of economic activity as an example of an invisible-hand process.

38 See Vanberg (1986, 1994) and Vromen (1995) for a discussion of Hayek's ideas about cultural evolution and its relation to individual behaviour.

39 See Berry (1974), Keller (1994), Land (1977) and Otteson (2002b) for interpretations of Smith's explanation of emergence of language.

40 Ferguson (1767: 11) argues:

> [W]e are obliged to observe, that men have always appeared among animals a distinct and a superior race; that neither the possession of similar organs, nor the approximation of shape, nor the use of the hand, [. . .] has enabled any other species to blend their nature or inventions with his; that in the rudest state, he is found to be above them [. . .]. He is, in short, a man in every cognition; and we can learn nothing of his nature from the analogy of other animals.

(Our interest here is in Ferguson's criticism that 'state of the nature' theorists do not get the facts right about human nature – not in the validity of his argument.)

41 C. Smith (2006a: 165, 2006b) analyses the relation between spontaneous order and the invisible hand. He argues that 'spontaneous order theorists believe that they are engaged in a descriptive scientific project that aims at an accurate understanding of the social world'. Yet he also argues that their method is conjectural history. Whether this is a correct interpretation of Smith is debatable. My reading is that Smith aims at unrevealing the connecting principles rather than giving an accurate description of social facts. This distinction will become clearer in Chapter 7.

42 See Klein (1997) for a discussion of the differences between the general idea of spontaneous social order and explaining certain social phenomena as unintended consequences of human action. Otteson (2002b) distinguishes between unintended order and spontaneous order. He argues that the concept of spontaneous order may imply that it is not possible to account of its emergence. What Otteson implies is probably something similar to our distinction between an overall unintended social order and an unintended social phenomenon. As I have noted, it is practically very difficult to explain the emergence of an overall social order as an unintended consequence.

43 Smith's 'invisible hand' has also been associated with theories of biological evolution; with Darwinian evolutionary theory (Carey 1998). The argument is that Smith saw society as an evolutionary process (see, for example, Clark 1990) or that the invisible hand is the invisible hand of natural selection (see, for example, Cosmides and Tooby 1994). We may consider this interpretation alongside the process of interpretation.

44 For invisible-hand explanations see Brennan and Pettit (1993), Pettit (1998), Carey (1998), Curren (1987), Karlson (1993), Keller (1994), Koppl (1992, 1994), Kühne (1997), Mäki (1990a,b), Nozick (1974, 1994), Ullmann-Margalit (1978, 1997, 1998) and Ylikoski (1995).

6 The origin of money reconsidered

1 Throughout the chapter 'model' is sometimes used to refer also to the simulations and experiments that are examined in this chapter for the sake of easy presentation.

2 Fiat money is an object that does not have the intrinsic property of providing utility to the agents, but that nevertheless serves as a medium of exchange.

3 The aim of this chapter is to demonstrate how Menger's story is further explored, not to present these models in full technical detail. Thus, when it is said that something is proved in a model, the proof is not provided. The interested reader should examine the original versions of these models for further details.

4 Some of the other related work may be listed as follows: Alchian (1977), Brunner and Meltzer (1971), H. Dawid (2000), Engineer and Shi (1998, 2000), Iway (1996), Kiyotaki and Wright (1992), Ostroy and Starr (1990), Selgin and White (1994), Selgin and Klein (2000), Townsend (1980) and Trejos and Wright (1995). Particularly, note that Alchian (1977) shows that money may be brought about in the absence of the problem of double coincidence of wants. Rather, he assumes that there are recognition costs (i.e. cost of determining the quality of the commodity), and shows that low-recognition-cost commodity may emerge as a medium of exchange (also see Baird 2000). Alchian's model may easily be integrated to our analysis (i.e. as one of the possible ways to explore Menger's model world). In fact, in what follows we will analyse the Williamson and Wright (1994) model, which utilises Alchian's insights. Also see Cheng (1999), who argues that division of labour is one of the main forces behind the emergence of money. Cheng's analysis extends Adam Smith's explanation of the origin of money (see Appendix I). Another relevant paper is Dowd (1999), which sketches an explanation of the evolution of the monetary system from emergence of medium of exchange to the emergence of the banking system, using Menger's insights.

5 Marketability and saleableness are related concepts. For example, Mises (1954) uses marketability as a substitute for saleableness (see Appendix II). However, this is not true for Kiyotaki and Wright. In their analysis, 'marketability' merely refers to agents' expectations about the acceptability of a good in exchange. Henceforth, marketability is used in a similar fashion to Kiyotaki and Wright.

6 One may argue that it remains a possibility that one or a few factors identified by Menger do not affect the emergence of money and, hence, we cannot consider Menger's set of factors as a subset of the factors in the real world. It is, of course, true that individual mechanisms proposed by Menger may not have played any role in the process of the emergence of money. Yet there seems to be no problem in considering them as a subset of the factors in the real world. The more sceptic reader may consider the figure as a rough representation, because our focus is on the relation between Menger's model and its reconsiderations.

7 See Mäki (2004) for an account of how questions about explanatory power are related to the changes in the sets of *explanantia* and *explananda*.

8 See Appendix II for the theorems about model A and B.

9 This is, indeed, the general criticism to such existence proofs or to the end-state interpretations of the 'invisible hand' (see, for example, Schmitz 2002, Lapavitsas 2005).

10 See Dowd (2001) for a similar argument.

11 The exception to this is the Aiyagari–Wallace (1991) model. They start the proof of the existence of an equilibrium with fiat money with the assumption that every agent accepts money. Later on they show that this is not necessary. In their model the fact that fiat money has low storage costs explains the acceptance of money. Yet they do not explain why individuals should accept an object that provides no utility only because it is less costly to store compared to other commodities.

12 In this sense, these models may be considered as contributing to the explanation of the persistence of fiat money. See Rosenberg (1989: 183), who argues that existence proofs have the quality of explaining maintenance.

13 This model is built on the insights of Akerlof (1970) and Alchian (1977).

14 According to Hodgson (2001: 86–90), acceptance of the need for state intervention contradicts Menger's analysis. However, it is entirely possible that the need for state intervention arises after the emergence of commodity money that serves as a medium of exchange. There is no logical contradiction between Menger's explanation and his acceptance of the need for state intervention at a later stage of the process of emergence of money. Hodgson is right in that money tokens that were used/issued in history cannot be entirely unintended. However, this does not contradict Menger's explanation (see Chapter 3). In fact, models of existence of commodity money and fiat money show that in the process of transformation of commodity money to fiat money some intentional intervention (i.e. intention targeted to the social level) was necessary.

15 This simplified presentation does not reflect all the aspects of the classifier systems as defined by Holland (1975). Yet it is enough to present the simulation in a coherent way.

16 See Appendix II for a detailed presentation of these results.

17 Note that it is often argued that this type of agent-based simulation is different from a standard mathematical model as it does not provide an analytical solution, implying that it is not deductive and not expressed in terms of equations. Yet this argument is not correct. Epstein (2005) nicely demonstrates that agent-based simulations are analytical in the sense that they can be expressed in terms of equations and that they are deductive.

18 It should also be noted here that Marimon *et al.* interpret the inconclusiveness of some of the results as indicating a possible improvement in the specification of the agents (i.e. the way in which they learn). Yet it may also be true that in the absence of other factors (i.e. the factors omitted in the model, such as the influence of increased

market traffic) the emergence of a medium of exchange is not possible under some conditions.

19 For another simulation see Dawid (2000).

20 Brown (1996) conducts an experiment only for 'economy A' and reports that individuals failed to play speculative strategies even if it was rational to do this.

21 See Alvarez (2004) on Menger's theory of imitation. This article contains a brief survey of the related literature.

22 It is assumed that the marketing agent knows the equilibrium price.

23 It is assumed that the price of the plastic chips is exogenously determined and that they can be used as a *numéraire* in the economy.

24 If Farmer I chooses a mixed strategy he should play A with probability α and B with $(1-\alpha)$; that is his strategy is: $\sigma = \alpha A + (1-\alpha)B$. Similarly, Farmer II mixed strategy is: $\tau = \beta A + (1-\beta)B$. Since farmers get the same payoff a from coordination and do not get anything if they fail to coordinate, they should play both options with equal probability ($\alpha a = (1-\alpha)a \rightarrow \alpha = \frac{1}{2}$, $\beta a = (1-\beta)a \rightarrow \beta = \frac{1}{2}$).

25 For example, if farmers can remember the farmers with whom they had previous contact, then they may slowly build a database of farmers and base their choices on this database when they meet again. Or alternatively, every farmer may only observe a small portion of what happened in the market, and base their decisions on this information.

26 That is, Schotter assumes that only the last period matters.

27 In Schotter's presentation the updating rule is a little bit different. Although it does not make a difference for us, let us present an updating rule which is closer to Schotter's. Every period the players associate particular strategies with certain probabilities, creating a probability vector p. Assume that the probability vector is $p = (p(A), p(B))$, where $p(x)$ is the probability that each player expects others to play x ($x = A, B$). For period 0, $p = (0.5, 0.5)$. If in period 0 A is observed more often than B, then players update their strategies in the following way: they add a small constant ε ($0 < \varepsilon < 1$) to $p(A)$ and subtract ε from $p(B)$. This is useful if we were to talk about many strategies. Yet for this coordination game we have assumed that there are only two pure strategies, that is, play A no matter what the other does, and play B no matter what the other does. We could not have done this for another type of game (i.e. other than a coordination game of the type above) as the super-game strategies (i.e. strategies for the repeated game) can be much more complex in other types of games.

28 This is a highly simplified presentation of what Schotter is doing. But it is enough to make our point. The interested reader may examine Schotter (1981: Ch. 3).

29 Alternatively, it may be assumed that player i has access to the information about the actions of all other players for the last m periods, and she draws a sample size s from this set of actions. But it seems to be more plausible that agents observe the actions of some individuals who are close to them, for example, neighbours (see Young 1998: 42).

30 See Gintis (2000: 228–236) for an excellent presentation of Young's approach.

31 In other models ε would represent the mutation rate (e.g. Gintis 1997, 2000; Luo 1999).

32 Remember that this change is represented with the error rate and it is external to the model.

33 We may assume that the reason for this is that A (e.g. gold) is more convenient than B (e.g. silver), for A is more durable, or it is easier to store, etc. (Young 1998: 12).

34 It is not argued here that these models are merely 'conceptual explorations' (Hausman 1992). As it is argued in the following pages, exploration of the model worlds increases our chances to explain particular cases.

35 Note that Young argues, 'These examples [in our case the currency game] are illustrative and meant to suggest directions for future work; I do not pretend to give a definitive account of the history of any one institutional form' Young (1998: xi).

36 Remember 'alerting to new possibilities' is a characteristic of a good invisible-hand explanation (see Chapter 5).

37 Of course, a new explanation does not necessarily require a new model but we use this expression to relate with our examination of several models of the origin of money.

38 Assuming that if some results are theoretically rejected they are removed from the meta-model. On the other hand, since some results only hold under some specific conditions, even if these results are rejected under different conditions they may be kept. Because if similar conditions are observed in the real world, they would be better approximations.

39 It is not argued here that people should start working on a comprehensive meta-model. The argument is that we should conceive each of them as contributing partially to a meta-model that provides the resources for explaining particular cases. See the conclusion to this chapter for a more detailed discussion.

40 Iwai (1997) argues that 'there is a fundamental limit on the power of theory to explain the origin of money ex post facto. History matters essentially.' He tries to show this analytically.

7 Models and representation

1 It is in no way argued here that such models rest on common folk wisdom alone. As it will later become clear, 'familiar' implies that both common sense and scientific knowledge is utilised by these models.

2 Ruben (1990) argues that 'facts' rather than 'events' are explained. For the sake of easy presentation, we use the following phrases without distinguishing between them: 'explaining a fact', 'explaining an event', 'explaining a phenomenon'. The latter two phrases can always be expressed in the language of facts.

3 The most discussed account of scientific explanation in philosophy of science is the deductive–nomological (D–N) model of explanation, which was first developed by Hempel and Oppenheim (1948) (all page references are to Hempel (1965), where H-O (1948) is reprinted). The discussions of the D–N model and its variants (e.g. the deductive–statistical (D–S) model, the inductive–statistical (I–S) model) – or more generally, the 'received view' (Suppe 1977 and Salmon 1990) – constitute most of the body of thought developed from 1948 to 1990s (Salmon 1990). One may be surprised about this after reading Hempel's (1965: 412) remark that 'these models are not meant to describe how working scientists actually formulate their explanatory accounts'.

4 Here, Hempel is referring to another type of incompleteness, called *elliptical formulation*. If an explanation does not mention all the laws that are necessary to produce the phenomenon, then it is elliptically formulated. For example, if I explain the fact that the glass which was placed in a hot oven broke because this glass is not heat-resistant, I 'forgo mentioning certain laws' of physics. Yet the explanation is 'incomplete, but in a rather harmless sense' (Hempel 1965: 415).

5 Another type of incompleteness mentioned by Hempel is that actual explanations do not explain everything contained in the explanation. That is, there is no explanatory closure. 'But completeness in this sense obviously calls for an infinite regress in explanation and is therefore unachievable' (Hempel 1965: 423).

6 Yet theoretical explanations are not discussed in a satisfactory manner (see note 10 below).

7 The causal view is one of the most discussed alternatives to the D–N account (see note 3 above) of explanation – the other is the unification view (Friedman 1974; Kitcher 1981, 1985, 1989; Mäki 2001a). The main criticism of the causal view may be presented as follows: providing a valid deductive argument does not guarantee that a good explanation is provided. A good explanation should also explicate the causal relationships that are responsible for the explanandum. There are several counterexamples to the D–N model that indicate this problem. But it is not necessary for our ar-

gument to give an overview of these counterexamples, which are now a standard part of textbooks of philosophy of science (e.g. Rosenberg 2000). An excellent overview of these counterexamples can be found in Salmon (1990: 46–50) and Ruben (1990: 138–154, 181–205). Also see Bromberger (1966), McCarthy (1977) and Ylikoski (2001: 66–68).

8 Also see Scriven (1975) and Dowe (2000).

9 Consider the case of astrology. We may take it that astrology suggests that the position of the planets and stars at the exact time of our birth causally affects our personal traits. Yet under the causal view of explanation, this would not be a proper explanation of the formation of our personal characteristics. To provide a proper explanation, astrology has to inform us about the way in which the positions of the planets are connected to our personal traits.

10 Salmon (1998: 398) discusses the problem of explaining generalisations. He admits that the problem has not been adequately addressed by philosophers of science. The problem is first acknowledged in footnote 33 of Hempel and Oppenheim (1948: 273): 'The precise rational reconstruction of explanation as applied to general regularities presents peculiar problems for which we can offer no solution at present.'

11 Mäki (1992b: 46) uses Ellis's (1985) distinction between a model theory and a process theory to describe the difference between what we have called the end-state interpretation and process interpretation. According to this description, process interpretation (process theory) of the invisible hand describes the causal process that produces unintended social consequences, and the end-state interpretation describes certain ideal states that would serve as reference points in explaining the states of the real world. Nevertheless, it has been argued in Chapter 6 that a better way to understand these models is to see them as attempts to find out the conditions under which some hypotheses about the real world may hold.

12 It is wiser for a view of explanation not to assume too much about the way the world works, for it is the task of scientists to inform us about it. The causal view may be considered as assuming much about the world. For example, Thalos (2002) argues that causation is an unnecessarily limiting notion for explanation. Instead, she proposes the notion of *physical dependence*. One may argue that a theory of explanation does not necessarily require a theory of causation. One may be agnostic about the nature of causation, but a theory of explanation should not be ignorant about the way in which scientists themselves use notions of causality and dependence.

13 See Mäki (2001a) and Lloyd (1998) for two excellent overviews of the literature on models. For discussions of the use of models in science see: Achinstein (1965, 1968), Braithwaite (1962), Cartwright (1999), Duhem (1914), Campbell (1920), Churchland and Hooker (1985), Freudenthal (1961), Giere (1988a), Harre (1970), Hesse (1970), Horgan (1994), Hughes (1990), McMullin (1978), Redhead (1980), Morgan and Morrison (1999), Suppe (1989), van Fraassen (1980) and Wartofsky (1979). For discussions of the use of models in economics see: Arrow (1951b), Bicchieri (1988), Boumans and Morgan (2001), Cartwright (1999), Gibbard and Varian (1978), Hausman (1992), Krugman (1995), Morgan (2000, 2001a,b), Morgan and Morrison (1999), Rappaport (2001), Schotter (1996), Sugden (2000) and Varian (1993).

14 Here, we are basically concerned with the process interpretation of the invisible hand; representing how constitution of phenomena may be discussed separately under the title of structural models. Yet this is not pursued here. We are interested in the types of models that focus on the causal mechanisms that bring about a certain phenomenon, which we may call state–space models.

15 Holland *et al.* (1989) introduce a general framework for representation of knowledge. Their framework assumes that human beings, as well as animals, have 'mental models' of the world they are living in, and examine how these cognitive systems can generate mental models. Whether human beings have mental models, as Holland *et al.* characterise it, is a controversial issue. However, their framework can be useful

for us in characterising the *explicit* models that scientists use, without assuming the existence of 'mental models' of the type they put forward. Their framework is useful for studying scientific models, because it gives guidelines for what to look for in a model.

16 Note here that 'segregation' itself is an interpretation of the real world. It is an abstraction. The next section discusses similar issues.

17 Isomorphism is a characteristic of small-scale material models of real objects, for example, of a small-scale model of a particular train. Some area maps are isomorphic to the particular area they represent in a special way: they ignore some details but everything in the map represents one particular object, that is, they are partially isomorphic. Also see Chapter 3 for Menger's argument that 'isomorphism' is not feasible in science.

18 Quoted in Holyoak and Thagard (1996: 20) from Borges (1964: 65).

19 See Holyoak and Thagard (1996) and Holland *et al.* (1989) for a lengthy discussion on similar points.

20 Since we have seen that isolation is an important aspect of modelling in Chapters 3, 4 and 6, we do not discuss it further.

21 For a discussion of the relation between models, analogies and metaphors see Achinstein (1964), Bicchieri (1988), Black (1962), Hesse (1970, 2000), Horgan (1994), Kroes (1989) and Leatherdale (1974). Also see Morgan and Morrison (1999).

22 Usually, analogical thinking helps us form a q-morphic (mental) model of the phenomenon we would like to understand. In this way, it also facilitates creative thinking:

> Indeed an interesting trade-off emerges between the completeness of an analogy and its usefulness in generating inferences. The more complete the initial correspondences between the source and the target, the more confident you can be the two are in fact isomorphic. But, unless you know more about one analog than the other – in other words, unless the initial correspondences between source and target propositions are incomplete – the mapping will not allow any new inferences to be made. A complete isomorphism has nothing to be filled in, leaving no possibility for creative leaps. Incompleteness may well weaken confidence in the overall mapping, but it also provides the opportunity for using the source to generate a plausible (but fallible) inference about the target.
>
> (Holyoak and Thagard 1996: 30)

23 We have also argued that the term 're-description' is prone to such an interpretation. Moreover, many philosophers use 'maps' as examples when they discuss representation. The map analogy also suggests a similar interpretation.

24 Both Bobo and Zubrinsky (1996) and Farley *et al.* (1997) report that the tolerance levels of whites are inconsistent with the existing segregation, that is, they are tolerant to mixed neighbourhoods, and this contradicts the view that segregation is caused by strong discriminatory preferences. This inconsistency may also suggest the hypothesis that residential segregation may be an unintended consequence.

25 We may speculate that this is indeed what Schelling did, consciously or unconsciously. In fact, he told me (private interview, 6 March 2001) that he had an intuition that residential segregation may be an unintended consequence of human action, and that because he could not find anything like this in the literature he tried to develop a model of segregation along these lines (see Aydinonat 2005).

26 'There are two main ways in which the theoretical representations seem to deviate from commonsense representations. I call them modification and rearrangement' Mäki (1996: 434). Also see Rosenberg (1995), for a discussion of folk psychology in social sciences.

27 It should be noted here that I differ from Mäki in that what I call 'familiar elements'

include the existent body of scientific knowledge in addition to common sense elements. Also see Figure 7.3.

28 It should be noted, however, that in *Aspects of Scientific Explanation*, Hempel (1965: 433–447) discusses two types of models: analogical models and theoretical models. He (1965: 439) argues that 'analogical models can be dispensed with for the systematic purposes of scientific explanation'. Yet, he accepts that they may be important for *discovery* purposes. Concerning theoretical models (one of his examples is 'economic models'), he argues that they may be considered as a theory with a limited scope:

> However, a limited scope and only approximate validity within that scope may severely restrict the actual explanatory and predictive value of a theoretical model.
>
> Hempel (1965: 447)

29 See Suppe (1989: Chapter 2) for a general criticism of the syntactic view.

30 The semantic view is a highly stylised account of models and theories, and has many controversial features. In fact, Suppe (1989: 426) admits that the relation between theories and models should be studied further. But in terms of getting closer to what we have seen in the previous chapters, it is better than the syntactic account of theories. For a brief introduction to the origin of semantic view theories, see Suppe (1989: 5–20). Key representatives of the semantic conception may be listed as follows: Giere (1984, 1985a,b, 1988a,b,c), Lloyd (1988), Suppe (1977, 1989), Suppes (1961, 1967), Thompson (1983, 1989) and van Fraassen (1970, 1972, 1980, 1987). It should be noted, however, that Giere refrains from talking with the language of theories in his later work (e.g. Giere 1999).

31 Philosophers of biology have argued that the syntactic approach to theories do not make justice to the use of models in evolutionary biology. For a discussion of semantic conception of theories in biology, see Lloyd (1988), Plutynski (2001), Sloep and van der Steen (1987), Thompson (1983, 1989) and Wilkins (1998).

32 van Fraassen (1980: 47) argues that 'to believe in a theory is to believe that one of its models correctly represents the world'. But this belief need not hold that what the theory says about unobservable phenomena is true. Rather, 'the belief involved in accepting a scientific theory is only that it "saves the phenomena" that it correctly describes what is observable' (van Fraassen 1980: 4).

33 Note that van Fraassen (1980: 69) thinks that this condition is weaker than the truth condition. Yet one may argue that theories and models, or the theoretical hypothesis connected with these are close to truth, or approximately true. The requirement that models and theories have to be close to the truth does not necessarily require that models are empirically adequate in the sense that van Fraassen suggests. See Niiniluoto (1999) for a defence of the realist interpretation of theories and models. For a defence of realism in economics, see Mäki (1990a, 1992a, 1994, 1996).

34 Moreover, van Fraassen's observable/non-observable dichotomy does not seem to be adequate for most economic models. For the argument that the economic model worlds do not contain entities that go beyond our 'ordinary conceptualised experience' and that economic model worlds are created by way of 'modifying' and 'rearranging' folk generalisations, see Mäki (1996).

35 See Blaug (1992), Cartwright (1983, 1999) and Hausman (1992). For the argument that the use of *ceteris paribus* clauses render economic models unfalsifiable, see Hutchison (1938).

36 The discussion about the kind of assumptions is also relevant here. Yet it does not add much to our point. See Musgrave (1981) and Mäki (1994).

37 For example, Liu (1997: 162) argues that the concept of similarity has to be 'fleshed out'.

38 The reader should remember Smith's analogy between models/theories/systems and

machines, where he argues that a model/theory/system is an imaginary machine (see Chapter 5).

39 For the argument that *ceteris paribus* laws indicate tendencies in social sciences, see Kincaid (1990, 1996).

40 Note here that the similarities between what we have called 'modification' and 'rearrangement' (Mäki 1996) and Cartwright's suggestion.

41 Remember from Chapter 5 that the process interpretation of the invisible hand implies the following idea: by means of starting from what is thought to be the basic principles or facts about the object of inquiry, rational reconstruction, conjectural history (or a theoretical explanation) may give us some of the connecting principles of society or of the social phenomenon we wish to examine.

42 We have talked about logical plausibility in Chapters 3, 4 and 6. It amounts to explaining the transformation of M-World *(t)* to M-World *(t + n)* successfully. By definition, a logically coherent model will be logically plausible. The idea is that we should prefer a model with results following from its premises.

43 In the chequerboard model individuals who have mild discriminatory preferences do not strictly prefer a mixed neighbourhood. They are content as long as they do not have an extreme minority status, and they are equally happy with other compositions. That is, *A*s (or *B*s) do not care whether they have a reasonable minority status or a majority status, or whether the neighbourhood is integrated or not; as long as they do not have an extreme minority status. Strict preference for perfect integration implies that individuals prefer a mixed (integrated) neighbourhood to a segregated one.

44 The philosophical literature on thought experiments is full of controversies, and there is little consensus on the definition of 'thought experiments'. However, a working definition is enough for our purposes:

> [Thought experiment] is something functioning, or intended to function, as an experiment, in the following sense. It aspires to test some hypothesis or theory. It is performed in thought – and hence it is real – but need not thereby shun such prosthetic devices as pencil and paper, encyclopaedias, or computers.
>
> Häggqvist (1996: 15)

The interested reader is advised to read the following two excellent books: Sorensen (1992a) and Häggqvist (1996). These books introduce the controversial issues in the literature with their own solutions. They are also good sources for further references on thought experiments (also see Sorensen 1992b).

45 One may argue that this experiment cannot be called quasi-material. Yet this term at least conveys the idea that there is some materiality involved in the model (e.g. real individuals). This terminology is helpful for distinguishing between purely theoretical models and models that contain some real world entities.

46 Because of the complex relationship between models, theory and the real world, Morrison and Morgan argue that the representative function of the models should not be understood as some kind of mirroring of a phenomenon:

> the idea of representation used here is not the traditional one common in the philosophy of science; in other words, we have not used the notion of 'representing' to apply only to cases where there exists a certain kind of mirroring of a phenomenon, system, or theory by a model. Instead, a representation is seen as a kind of rendering – a partial representation that either abstracts from, or translates into another form, the real nature of the system or theory, or one that is capable of embodying only a portion of a system.
>
> (Morrison and Morgan 1999: 27)

In this chapter we have seen that Morrison and Morgan are correct. However, Morrison and Morgan seem to underestimate the importance of isolation in model building. This book, and particularly this chapter, has emphasised the role of isolation in model building and explanation. Models are mediators, but in the view developed here, they can only mediate if they utilise the tool of isolation.

8 Game theory and conventions

1 It is also argued that conventions may emerge out of games that involve partial conflict. For example, Sugden (1986) lists three types of conventions: coordination conventions, conventions of property and conventions of reciprocity (also see Ullmann-Margalit 1977). Under this broader definition, this chapter discusses coordination conventions.

2 There is another game where individuals may unintendedly coordinate their activities, which is called 'the minimal social condition'. In one version of this game players do not even know that the consequences of their action depend on other players' actions, in another version they are not informed about the existence of others. Nevertheless, they know the available strategies and observe the consequences of their actions. The basic idea behind this type of game is that individuals may coordinate their actions even if they do not know the rules and structure of the game. On 'the minimal social condition', see Sidowski, Wyckoff and Tabory (1956), Sidowski (1957) and Colman (1982a). For a brief overview of the literature, see Colman (1995: 40–50).

3 What Lewis calls 'proper coordination equilibrium' is a stronger (solution) concept than the Nash equilibrium. This difference does not change the nature of our argument. Note that Sugden (1998a: 4) argues that 'one consequence of Lewis's definition is that he is able to argue that conventions tend to become norms, while on the usual game-theoretic account, Nash equilibria are sustained simply by self-interest'.

4 Moreover, mixed strategy equilibrium is a problematic concept: 'it is hard to see how a mixed strategy equilibrium can be a solution' (Bicchieri 1993: 60; see also Fudenberg and Levine 1998: 19).

5 Note also that Schelling argues: 'the mathematical structure of the payoff function should not be permitted to dominate the analysis' and that 'there is a danger in too much abstractness: we change the character of the game when we drastically alter the amount of contextual detail. [. . .] *It is often contextual detail that can guide the players* to the discovery of a stable or, at least, mutually non-destructive outcome. [. . .] This corner of game theory is inherently *dependent on empirical evidence*' (Schelling 1958: 252, emphasis added).

6 The importance of history has been studied widely in different research areas in economics. For example, Tirole (1996) studies what happens if an individual's reputation is dependent on his past behaviour and on the behaviour of the group to which he belongs. He argues that dishonest behaviour in the past increases the time needed to establish a reputation of honesty in a way that a new generation of agents may suffer from the dishonesty of their predecessors. The literature on path-dependency studies how an economy may lock into an inefficient equilibrium because of historical accidents and feedbacks created by externalities (e.g. Arthur 1984, 1989; David 1985). Other interesting examples are Azariadis and Drazen (1990) and Krugman (1991).

7 Something is common knowledge if everyone knows it, if everyone knows that it is known by others, if everyone knows that the fact that everyone knows that is known by others is known by others, and so on. See Lewis (1969) and Aumann (1976).

8 However, leaving this problem aside, we may easily see that sometimes it may be useful to use a mixed strategy. For example, if we really do not know what to do, or better, if the outcome of our action is dependent on things we cannot control, we may want to choose one of the alternatives randomly to increase our chances of achieving

the 'right' result. Considering the above game, or the money game (see Chapter 6), we may argue that agents may use a mixed strategy when they have no clue about what to expect from the other player. Thus, the concept of mixed strategy captures the idea that 'clueless' individuals may randomise their choices in the context of a novel coordination problem.

9 For example, Bernheim argues:

> Nash hypothesis, far from being a consequence of rationality, arises from certain restrictions on agents' expectations which may or may not be plausible, depending on the game being played.
>
> (Bernheim 1984: 1007)

Concerning the rationality of individuals who are playing their part in the Nash equilibrium, Luce and Raiffa argue:

> Even if we are tempted at first to call a [Nash] non-conformist 'irrational', we would have to admit that [his opponent] might be 'irrational' in which case it would be 'rational' for [him] to be 'irrational' – to be a [Nash] non-conformist.
>
> (Luce and Raiffa 1957: 63; also quoted in Bernheim 1984: 1009)

Interestingly, this leads them to discuss the nature of game theory:

> We belabour this point because we feel that it is crucial that the social scientist recognise that game theory is not descriptive, but rather (conditionally) normative. It states neither how people do behave nor how they should behave in an absolute sense, but how they should behave if they wish to achieve certain ends. It prescribes for given assumptions courses of action for the attainment of outcomes having certain formal 'optimal' properties. These properties may or may not be deemed pertinent in any given real world conflict of interest. If they are, the theory prescribes the choices which must be made to get that optimum.
>
> (Luce and Raiffa 1957: 63)

Also see Bicchieri (1993), Crawford (1997: 210), Janssen (1998a), Mailath (1992: 250–259, 1998: 1351).

10 Jacobsen (1996: 68) argues that 'the problem of justifying Nash equilibrium has nothing in particular to do with multiple equilibria'. This is true in that the Nash solution concept is in need of a justification even in the absence of multiple equilibria. However, the solution of the problem of equilibrium selection necessitates a justification of the Nash equilibrium.

11 While orthodox justifications keep the standard structure of the game, non-orthodox justifications do not (see Janssen 1998a). Some of the orthodox justifications of Nash equilibrium are as follows: correlated equilibrium (Aumann 1987; also see Janssen 1998a and Sugden 1991 for a criticism), coordinated expectations (Aumann and Brandenburger 1995; Brandenburger 1992). Other well-known refinements of the Nash equilibrium concept are subgame-perfect-equilibrium (Selten 1975), trembling hand (Selten 1975) intuitive criterion (Cho and Kreps 1987).

12 According to Aumann (1987: 1), the puzzle is the following: 'why and under what conditions the players in an n-person game might be expected to play such an equilibrium', in an n-player game. Particularly, he asks: 'why should we expect players to play their part in the equilibrium?' To expect players to play their part for an unique equilibrium, player I has to expect player II to play his part, and player II would play his part if and only if he expects player I to play his part. Correspondingly, Aumann introduces the notion of 'correlated equilibrium'. In the correlated equilibrium play-

ers do not (need to) know what others are doing. Yet it is assumed to be common knowledge among players that players maximise their expected utility given their information.

13 'In the language of Schelling, Nash equilibrium may be "focal". If agents share the common belief that Nash equilibrium is normally realised, they no longer entertain the rationally admissible doubt that an opponent will fail to conform' (Bernheim 1984: 1009).

14 Remember that in Chapter 6 we have seen that Young's model of the emergence of the medium of exchange 'predicts' that the Pareto-dominant equilibrium will be selected. Also see the section on learning, below.

15 This relates to another refinement, known as the 'global games approach'.

16 However, it should be noted that evolutionary game theory is argued to provide a firmer basis for risk dominance (e.g. Kandori, Mailath and Rob 1993; Young 1998) and recent theories of focal points follow the idea that Pareto-dominant equilibrium will be selected (e.g. Bacharach and Bernasconi 1997; Janssen 1998a, 2001b; also see below).

17 For a good overview of 'focal points' see Janssen (1998b). For an early attempt to formalise the focal points see Gauthier (1975). Some of the recent work on focal points can be listed as follows: Bacharach (1993, 1994); Bacharach and Bernasconi (1997); Bacharach and Stahl (1997); Binmore and Samuelson (2002); Casajus (2001); Crawford and Haller (1990); Goyal and Janssen (1996); Janssen (2001a,b); Mehta *et al.* (1992, 1994a,b); and Stahl (1993).

18 The experimenters marked the red balls in a way that cannot be seen, for example, by placing pieces of papers marked with different numbers in the balls. After the selection is made, the balls are opened to see whether coordination has been achieved.

19 This approach is based on two principles. PIR (principle of insufficient reason): one cannot rationally discriminate two strategies if they have the same attributes. POC (principle of coordination): 'if in a class of strategy combinations that respect PIR there is a unique strategy combination that is Pareto-optimal then individual players should do their part of that strategy combination' (Gauthier 1975; Janssen 1998a, 2001a). If $p(i)(j)$ is the probability that player i chooses strategy j then $p(1)(\text{red})+p(1)(\text{red})+p(1)(\text{green})=1$. That is, $2p(1)(\text{red})+p(1)(\text{green})=1$. Similarly, $2p(2)(\text{red})+p(2)(\text{green})=1$. The class of mixed strategy combinations that respect PIR are $\{(p(1)(\text{red}), p(1)(\text{red}), p(1)(\text{green})), ((p(2)(\text{red}), p(2)(\text{red}), p(2)(\text{green}))\}$. According to PIR, the Pareto optimal strategy combination is: $\{(0, 0, 1), (0, 0, 1)\}$. This means that individuals should choose the green object, according to POC (Janssen 1998a, 2001a). (Note the similarity between POC and payoff dominance (or Pareto dominance) of Harsanyi and Selten (1988)).

20 Indeed, Sugden (1995) is conscious of the incompleteness of his attempt. He follows Schelling in that salience is an empirical matter and no theory of focal points can be complete. In another place he argues:

> They [game theorists] have been unable to integrate salience into the formal structure of game theory. [. . .] few game theorists have been interested in investigating the facts of salience. [. . .] Instead, they have continually been puzzled by their inability to fit salience into a theoretical structure based on a priori deduction from premises about rationality.
>
> (Sugden 2001: F220)

21 Similarly, focal alternatives are embedded in Games I, II and III, which are discussed in Appendix IV.5.

22 In fact, Mehta *et al.*'s (1994b) experiments confirm a similar result.

23 For example, Mailath argues:

The refinements literature still serves the useful role of providing a language to describe properties of different equilibria. Applied researchers find the refinements literature of value for this reason, even though they cannot depend on it mechanically to eliminate 'uninteresting' equilibria.

(Mailath 1998: 1372)

24 Bernheim (1984) and Pearce (1984) suggest an alternative concept of rationalisability:

an individual is rational [. . .] if he optimises subject to some probabilistic assessment of uncertain events, where his assessment is consistent with all of his information. [. . .] If it is possible to justify the choice of a particular strategy by constructing infinite sequences of self-justifying conjectured assessments in this way, then I call the strategy 'rationalizable'.

(Bernheim 1984: 1011)

However, rationalisability cannot easily be considered as a refinement or a justification of Nash equilibrium concept, because it is less strict than the Nash equlibrium concept. For example, Samuelson (1998) considers the concept of rationalisable strategy as being somewhat opposite to the refinements movement. Rationalisability criterion helps us justify players' choices, but it does not help us discriminate among several equilibria of the game (Bicchieri 1993: 51–52).

25 Another mechanism that may explain successful coordination is replicator dynamics. The interested reader may refer to the discussion on 'Evolutionary stability and replicator dynamics' in Appendix IV.6.

26 For an extended discussion on learning in game theory, see: Fudenberg and Levine (1998); Marimon (1997); and Milgrom and Roberts (1991).

27 Note that Chapter 6 discusses models with imitation (Luo 1999), artificially intelligent learning (Marimon *et al.* 1990), Bayesian learning (Schotter 1981) and fictitious play (Young 1998).

28 See Appendix IV.8 for a lengthier version of this discussion.

29 Fictitious play has been first employed as a tool to compute Nash equilibria (see Brown 1951; Robinson 1951; also see Young 1998: 31). For an extensive discussion of fictitious play, see Fudenberg and Levine (1995, 1998) and Krishna and Sjostrom (1995).

30 Note that in this case Pareto-dominant equilibrium is also the risk-dominant equilibrium (see Appendix IV.4).

31 For learning and stochastic dynamics see Foster and Young (1990); Kaniovski and Young (1995); Fudenberg and Harris (1992); Kandori, Mailath and Rob (1993); and Kandori and Rob (1995).

32 Note that in Kandori, Mailath and Rob (1993) randomness is at the individual level while in Foster and Young (1990) it is introduced at the aggregate level.

33 In fact, Kandori, Mailath and Rob (1993) argue that their results are more applicable to small populations than large populations. Ellison (1993) argues that the rate of convergence decreases as the number of players increase. Yet local interaction allows that the results hold for large populations. Similarly, Young (1998) argues that in large populations the process may stick to an inferior state, but in small groups it is more likely that optimum 'technology' is selected. An example of this may be the QWERTY case, where a large population of individuals stick to an inferior technology (Arthur 1984, 1989; David 1985).

34 See Appendix IV.7 for other issues concerning local interaction. Young's (1996, 2001) model of local interaction and Goyal and Janssen's (1997) discussion of coexistence of conventions are discussed in the appendix.

35 It can be argued that rationality is not a necessary condition either. However, note that end-state models generally rest on the rationality assumption.

36 The concept of evolutionarily stable strategies (ESSs) (see Appendix IV.6) may be considered as relating to the stability and persistence of certain equilibria. In the context of explanation of conventions, static evolutionary analysis examines the conditions under which a certain convention is stable and persistent in the model world. Although ESS is a static concept, it rests on an idea of evolutionary dynamics and the mechanisms behind evolutionarily stable equilibria have been studied with dynamic models.

37 If concordant mutual expectations are created by agreement then the convention is intentionally created. That is, if agents recognise the problem and agree on a solution (e.g. driving on the right) then they are explicitly intending to bring about a consequence at the social level (see Chapter 2). If agents solve the problem through salience or precedence then the emerging convention may be considered as an unintended consequence of their actions, given that they do not intend to bring about the convention.

38 It should be noted here that game-theoretic models have triggered a large number of experiments: for a good overview of the experimental literature see Crawford (1997). For some surveys on experimental games (especially prisoners' dilemma games), see: Apfelbaum (1974); Colman (1982a); Gallo and McClintock (1965); Good (1991); Nemeth (1972); Pruitt and Kimmel (1977); and Rappoport and Orwant (1962).

39 A good example of how game theory may be used in historical research is Greif (2006). See my review of Greif (2006) in *History of Economic Ideas* (Aydinonat 2006).

40 For example, Janssen argues:

> In the literature (e.g., industrial organisation) they do not seem to reflect rules of game in real world (neither rules of competition, nor legal and cultural constraints). The rules seem to be nothing but the invention of the theorist (who invents the game).
>
> (Janssen 1998a: 23)

41 Remember here that it has been argued that Schotter's and Young's models deal with a small part of the whole emergence story (see Chapter 6).

42 It should be noted here that game theory is sometimes considered as a toolbox or as a collection of techniques for analysing strategic interaction. While this is an acceptable interpretation, it does not tell us much about models that employ game theory.

9 Concluding remarks

1 Simulations have played an important role for analysing individual behaviour in prisoner's dilemma games. See Axelrod (1981, 1984, 1997); Axelrod and Dion (1988); Axelrod and Hamilton (1981); Bicchieri (1989); Binmore (1992); Boyd and Lorberbaum (1987); Donninger (1986); Nowak and Sigmund (1992, 1993); Poundstone (1993); and Vanderschraaf (1988).

2 For an overview see www.econ.iastate.edu/tesfatsi/ace.htm

3 For a general introduction see Waldrop (1992); Holland (1995, 1998); and Kauffman (1995).

Bibliography

Aaronson, D. (2001) 'Neighborhood dynamics', *Journal of Urban Economics*, 49: 1–31.

Aberg, Y. (2000) 'Individual social action and macro level dynamics: a formal theoretical model', *Acta Sociologica*, 43: 193–205.

Achinstein, P. (1964) 'Models, analogies and theories', *Philosophy of Science*, 31: 328–350.

Achinstein, P. (1965) 'Theoretical models', *British Journal for the Philosophy of Science*, 16: 102–119.

Achinstein, P. (1968) *Concepts of Science: A Philosophical Analysis*, Baltimore, MD: The Johns Hopkins University Press.

Aiyagari, S. R. and Wallace, N. (1991) 'Existence of steady states with positive consumption in the Kiyotaki-Wright Model', *Review of Economics Studies*, 58: 901–916.

Akerlof, G. A. (1970) 'Market for "lemons": quality uncertainty and the market mechanism', *Quarterly Journal of Economics*, 84: 488–500.

Alchian, A. A. (1977) 'Why money?', *Journal of Money, Credit, and Banking*, 9 (1): 133–140.

Alexander, R. D. (1974) 'The evolution of social behavior', *Annual Review of Ecology and Systematics*, 5: 325–383.

Alvarez, A. (2004) 'Learning to choose a commodity-money: Carl Menger's theory of imitation and the search monetary framework', *European Journal of the History of Economic Thought*, 11 (1): 53–78.

Anderson, P. W., Arrow, K. J. and Pines, D. (1988) *The Economy as an Evolving Complex System*, vol. 5, Santa Fe Institute Studies in the Sciences of Complexity, Redwood City, CA: Addison-Wesley.

Apfelbaum, E. (1974) 'On conflicts and bargaining', *Advances in Experimental Social Psychology*, 7: 103–156.

Armstrong, W. E. (1924) 'Rossel Island money: a unique monetary system', *Economic Journal*, 34: 423–429.

Armstrong, W. E. (1928) *Rossel Island*, Cambridge: Cambridge University Press.

Arnold, R. A. (1980) 'Hayek and institutional evolution', *The Journal of Libertarian Studies*, 4 (4): 341–352.

Arrow, K. J. (1951a) 'An extension of the basic theorems of classical welfare economics', *Proceedings of the Second Berkeley Symposium on Mathematical Statistics and Probability* (Berkeley and Los Angeles: University of California Press), pp. 507–532.

Arrow, K. J. (1951b) 'Mathematical models in the social sciences' in D. Lerner and H. D. Laswell (eds), *The Policy Sciences*, Stanford: Stanford University Press, pp. 129–54.

Arrow, K. J. and Debreu, G. (1954) 'Existence of an equilibrium for a competitive economy', *Econometrica*, 22 (3): 265–290.

Arthur, W. B. (1984) 'Competing technologies and economic prediction', *Options,* April: 10–13; reprinted in D. MacKenzie and J. Wajcman (eds) (1999), *The Social Shaping of Technology*, Philadelphia: Open University Press.

Arthur, W. B. (1989) 'Competing technologies, increasing returns, and lock-in by historical events', *The Economic Journal*, 99 (394): 116–131.

Arthur, W. B. *et al* (1989) *Emergent Structures: A Newsletter of the Economic Research Program,* March, Santa Fe: The Santa Fe Institute.

Arthur, W. B. *et al* (1990) *Emergent Structures: A Newsletter of the Economic Research Program,* August, Santa Fe: The Santa Fe Institute.

Aumann, R. J. (1976) 'Agreeing to disagree', *Annals of Statistics,* 4: 1236–1239.

Aumann, R. J. (1985) 'What is game theory trying to accomplish?' in K. J. Arrow and S. Honkapohja (eds), *Frontiers of Economics,* Oxford: Basil Blackwell, pp. 28–78.

Aumann, R. J. (1987) 'Correlated equilibrium as an expression of Bayesian rationality', *Econometrica,* 55 (1): 1–18.

Aumann, R. J. and Brandenburger, A. (1995) 'Epistemic conditions for Nash equilibrium', *Econometrica,* 63: 1161–1180.

Aydinonat, N. E. (2005) 'An interview with Thomas C. Schelling: interpretation of game theory and the checkerboard model', *Economics Bulletin,* 2 (2): 1–7.

Aydinonat, N. E. (2006) 'Institutions, theory, history and context-specific analysis', *History of Economic Ideas,* 14 (3): 143–156.

Axelrod, R. (1981) 'The emergence of cooperation among egoists,' *The American Political Science Review,* 75: 306–318.

Axelrod, R. (1984) *The Evolution of Cooperation*, New York: Basic Books.

Axelrod, R. (1997) *The Complexity of Cooperation: Agent-Based Models of Competition and Collaboration*, Princeton: Princeton University Press.

Axelrod, R. and Dion, D. (1988) 'The further evolution of cooperation', *Science,* 242: 1385–1390.

Axelrod, R. and Hamilton, W. (1981) 'The evolution of cooperation', *Science,* 211: 1390–1396.

Axtell, R. L. and Epstein, J. M. (1994) 'Agent-based modelling: understanding our creations', *The Bulletin of the Santa Fe Institute*, Winter.

Axtell, R. L., Epstein, J. M. and Young, H. P. (2001) 'The emergence of classes in a multi-agent bargaining model', in S. N. Durlauf and P. H. Young (eds), *Social Dynamics*, Cambridge: The MIT Press, pp. 191–211.

Azariadis, C. and Drazen, A. (1990) 'Threshold externalities in economic development', *Quarterly Journal of Economics*, 105 (2): 501–526.

Bacharach, M. (1991) 'Games with concept-sensitive strategy spaces', unpublished manuscript, Department of Economics, University of Oxford.

Bacharach, M. (1993) 'Variable universe games', in K. G. Binmore, A. Kirman and P. Tani (eds), *Frontiers of Game Theory*, Cambridge: The MIT Press, pp. 255–276.

Bacharach, M. (1994) 'Report on EEEN-related research April 1993–February 1994', in European Economics Experimental Research Network (EEEN) Project 1993–1994 Report.

Bacharach, M. and Bernasconi, M. (1997) 'The variable frame theory of focal points: an experimental study', *Games and Economic Behavior,* 19: 1–45.

Bacharach, M. and Stahl, D. (1997) 'The variable frame level-n theory of games', *FRAM working paper and working paper of Department of Economics*, University of Texas at Austin.

Backhouse, R. E., Hausman, D. M., Mäki, U. and Salanti, A. (eds) (1998) *Economics and Methodology: Crossing Boundaries*, Proceedings of the IEA Conference held in Bergamo, Italy, London: Macmillan.

Baird, C. W. (2000) 'Alchian and Menger on Money', *Review of Austrian Economics*, 13: 115–120.

Barchas, P. (1986), 'A sociophysiological orientation to small groups', in E. Lawler (ed.), *Advances in Group Processes*, Vol. 3, Greenwich, CT: JAI Press, pp. 209–246.

Barry, N. (1982) 'The tradition of spontaneous order; bibliographical essay', *Literature of Liberty*, 5: 7–58.

Barry, N. (1985) 'In defense of the invisible hand', *Cato Journal*, 5 (1): 133–148.

Bates, F. L. (1974) 'Alternative models for the future of society: from the invisible to the visible hand', *Social Forces*, 53 (1): 1–11.

Batten, D. F. (2001), 'Complex landscapes of spatial interaction', *The Annals of Regional Science*, 35: 81–111.

Baumeister, R. F. and Leary, M. R. (1995) 'The need to belong: desire for interpersonal attachments as a fundamental human motivation', *Psychological Bulletin*, 117: 497–529.

Bayer, P., McMillan, R. and Reuben, K. (2001) 'The causes and consequences of residential segregation: an equilibrium analysis of neighborhood sorting', PublicPolicy Institute of California.

Bell, S. (2001) 'The role of the state and the hierarchy of money', *Cambridge Journal of Economics*, 25: 149–163.

Benenson I. (1998) 'Multi-agent simulations of residential dynamics in the city', *Computers, Environment and Urban Systems*, 22: 25–42.

Benenson I. (1999) 'Modeling population dynamics in the city: from a regional to a multi-agent approach', *Discrete Dynamics in Nature and Society*, 3: 149–170.

Benenson, I. (2004) 'Agent-based modeling: from individual residential choice to urban residental dynamics', in D. Goodchild and D. G. Janelle (eds), *Spatially Integrated Social Science: Examples in Best Practice*, Oxford: Oxford University Press, pp. 67–94.

Benito, J. M. and Hernandez, P. (2004) 'Schelling's dynamic models of segregation: a cellular automata approach', paper presented at the XXIX Simposio de Análisis Económico, Universidad de Navara.

Bernheim, D. (1984) 'Rationalizable strategic behavior', *Econometrica*, 52: 1007–1028.

Berry, C. J. (1974) 'Adam Smith's considerations on language', *Journal of the History of Ideas*, 35: 130–138.

Berry, C. J. (1997) *Social Theory of the Scottish Enlightenment*, Edinburgh: Edinburgh University Press.

Berry, C. J. (2006) 'Smith and science' in K. Haakonssen (ed.), *The Cambridge Companion to Adam Smith*, Cambridge: Cambridge University Press, pp. 112–135.

Bianchi, M. (1994) 'Hayek's spontaneous order: the "correct" versus the "corrigible" society', in J. Birner and R. van-Zijp (eds), *Hayek, co-ordination and evolution: His legacy in philosophy, politics, economics and the history of ideas*, London and New York: Routledge, pp. 232–251.

Bicchieri, C. (1988) 'Should a scientist abstain from metaphor' in A. Klamer, D. McCloskey and R. M. Solow (eds), *The Consequences of Economic Rhetoric*, Cambridge: Cambridge University Press, pp. 100–114.

Bicchieri, C. (1989) 'Self-refuting theories of strategic interaction,' *Erkenntinis* 30: 69–85.

Bicchieri, C. (1993) *Rationality and Coordination*, Cambridge: Cambridge University Press.

Bierman, H. S. and Fernandez, L. (1998) *Game Theory with Economic Applications*, Reading: Addison-Wesley.

Binmore, K. G. (1990) *Essays on the Foundations of Game Theory*, Oxford: Basil Blackwell.

Binmore, K. G. (1992) *Fun and Games*, Lexington, MA: D.C. Heath and Company.

Binmore, K. G. and Samuelson, L. (1994) 'An economist's perspective on the evolution of norms', *Journal of Institutional and Theoretical Economics,* 150 (1): 45–63.

Binmore, K. G. and Samuelson, L. (1992) 'Evolutionary stability in repeated games played by finite automata', *Journal of Economic Theory,* 57: 278–305.

Binmore, K. G. and Samuelson, L. (2002) 'The evolution of focal points', unpublished manuscript.

Binmore, K. G., Kirman, A. and Tani, P. (1993a) 'Introduction: famous gamesters', in *Frontiers of Game Theory*, Cambridge: The MIT Press, pp. 1–25.

Binmore, K. G., Kirman, A. and Tani, P. (eds) (1993b) *Frontiers of Game Theory*, Cambridge: The MIT Press.

Birner, J. and van-Zijp, R. (eds) (1994) *Hayek, co-ordination and evolution: His legacy in philosophy, politics, economics and the history of ideas*, London and New York: Routledge.

Bishop, J. D. (1995) 'Adam Smith's invisible hand argument', *Journal of Business Ethics,* 14 (3): 165–180.

Black, M. (1962) *Models and Metaphors: Studies in Language and Philosophy*, Ithaca, NY: Cornell University Press.

Blaug, M. (1992) *The Methodology of Economics*, Cambridge: Cambridge University Press.

Blaug, M. (1997) *Economic Theory in Retrospect*, 5th edn, Cambridge: Cambridge University Press.

Blume, A., Kim, Y. G. and Sobel, J. (1993) 'Evolutionary stability in games of communication', *Games and Economic Behavior,* 5: 547–575.

Blume, L. E. and Durlauf, S. N. (2001) 'The interaction based approach to socioeconomic behaviour' in S. N. Durlauf and P. H. Young (eds), *Social Dynamics*, Cambridge: The MIT Press, pp. 15–44.

Bobo, L. and Zubrinsky, C. L. (1996) 'Attitudes on residential integration: perceived status differences, mere in-group preference, or racial prejudice?', *Social Forces,* 74 (3): 883–909.

Bohannan, P. and Dalton, G. (1971) 'Markets in Africa', in G. Dalton (ed.), *Economic Anthropology and Development: Essays on Tribal and Peasant Economies*, New York: Basic Books, pp. 143–166.

Boniolo, G. (1997) 'On a unified theory of models and thought experiments in natural sciences', *International Studies in the Philosophy of Science,* 11 (2): 12–42.

Borgers, T. and Sarin, R. (1995) 'Naïve reinforcement learning with endogenous aspirations', unpublished manuscript, University College London.

Borgers, T. and Sarin, R. (1996) 'Learning through reinforcement and replicator dynamics', unpublished manuscript, University College London.

Borges, J. L. (1964) 'Funes the Memorious', in *Labyrinths: Selected Stories and other Writings*, New York: New Directions, pp. 59–66.

Boudon, R. (1982) *The Unintended Consequences of Human Action*, London: Macmillan.

Boumans, M. (1999) 'Built-in justification', in M. Morgan and M. Morrison (eds), *Models as Mediators: Perspectives on Natural and Social Science*, Cambridge: Cambridge University Press, pp. 66–97.

Boumans, M. and Morgan, M. (2001) 'Ceteris paribus conditions: materiality and the application of economic theories', *Journal of Economic Methodology,* 8 (1): 11–26.

Bowles, S. and Sethi, R. (2006) 'Social segregation and the dynamics of group inequality', *SFI Working Paper*, 6 February.

Bowles, S., Boyd, R., Fehr, E. and Gintis, H. (1997) 'Homo reciprocans: a research initiative on the origins, dimensions, and policy implications of reciprocal fairness', research proposal, University of Massachusetts. Available at www.umass.edu/preferen/gintis/homo.pdf (accessed 10 June 2005).

Boyd, R. and Lorberbaum, J. (1987) 'No pure strategy is evolutionarily stable in the repeated Prisoner's Dilemma game,' *Nature*, 327: 58–59.

Braithwaite, R. B. (1962) 'Models in the empirical sciences', in E. Nagel, P. Suppes and A. Tarski (eds), *Logic Methodology and Philosophy of Science: Proceedings of the 1960 International Congress*, Stanford: Stanford University Press, pp. 224–231.

Brandenburger, A. (1992) 'Knowledge and equilibrium in games', *Journal of Economic Perspectives*, 6: 83–101.

Brennan, G. and Pettit, P. (1993) 'Hands invisible and intangible', *Synthese*, 94 (2): 191–225.

Brewer, A. (2006) 'On the other (invisible) hand. . .', University of Bristol discussion paper no. 06/594.

Bridel, P. and Salvat, C. (2004) 'Reason and sentiments: review of Emma Rothschild's economic sentiments: Adam Smith, Condorcet and the Enlightenment', *European Journal of the History of Economic Thought*, 1 (1): 131–145.

Bromberger, S. (1966) 'Why-Questions', in R. N. Colodny (ed.), *Mind and Cosmos*, Pittsburgh: University of Pittsburgh Press, pp. 86–111.

Brown, G. W. (1951) 'Iterative solutions of games by fictitious play', in T. Koopmans (ed.), *Activity Analysis of Production and Allocation*, New York: Wiley, pp. 374–376.

Brown, P. M. (1996) 'Experimental evidence on money as a medium of exchange', *Journal of Economic Dynamics and Control*, 20 (4): 583–600.

Brown, V. (1997) ' "Mere inventions of the imagination": a survey of recent literature on Adam Smith', *Economics and Philosophy*, 13: 281–312.

Bruch, E. E. and Mare, R. D. (2003) 'Neighborhood choice and neighborhood change', working paper CCPR-003-03, California Center for Population Research.

Bruch, E. E. and Mare, R. D. (2004) 'Neighborhood choice and neighborhood change', unpublished manuscript.

Brunner, K. and Meltzer, A. H. (1971) 'The uses of money: money in the theory of an exchange economy', *American Economic Review*, 61: 784–805.

Caldwell, B. (1982) *Beyond Positivism: Economic Methodology in the Twentieth Century*, London: George Allen & Unwin.

Caldwell, B. (ed.) (1990) *Carl Menger and his Legacy in Economics*, Durham and London: Duke University Press.

Caldwell, B. (1997) 'Hayek and Socialism', *Journal of Economic Literature,* 35: 1856–1890.

Camber, C. and Thaler, R. (1995) 'Ultimatums, dictators, and manners', *Journal of Economic Perspectives*, 9 (2): 209–219.

Cameron, L. (1995) 'Raising the stakes in the ultimatum game: experimental evidence from Indonesia', discussion paper no. 345, Department of Economics, Princeton University.

Campagnolo, G. (2005) 'Carl Menger's "Money as measure of value": an introduction', *History of Political Economy*, 37(2): 233–243.

Campbell, N. R. (1920 [1957]) *Foundations of Science*, New York: Dower.

Carey, T. V. (1998) 'The invisible hand of natural selection, and vice versa', *Biology and Philosophy,* 13 (3): 427–442.

Carlsson, H. and van Damme, E. (1993a) 'Equilibrium selection in stag-hunt games', in K.

G. Binmore, A. Kirman and P. Tani (eds), *Frontiers of Game Theory*, Cambridge: The MIT Press, pp. 237–254.

Carlsson, H. and van Damme, E. (1993b) 'Global games and equilibrium selection', *Econometrica*, 61: 989–1018.

Cartwright, N. (1983) *How the Laws of Physics Lie*, Oxford: Oxford University Press.

Cartwright, N. (1989) *Natures Capacities and their Measurement*, Oxford: Clarendon Press.

Cartwright, N. (1991) 'Fables and models', *The Arisotelian Society, Supplementary Volume* 65: 55–68; reprinted in J. Worall (ed.) (1994), *The Ontology of Science*, The International Research Library of Philosophy, Vol. 10, Aldershot: Dartmouth Publishing Co., pp. 191–204.

Cartwright, N. (1999) *The Dappled World: A Study of the Boundaries of Science*, Cambridge: Cambridge University Press.

Casajus, A. (2001) *Focal Points in Framed Games – Breaking the Symmetry*, Berlin: Springer.

Casti, J. L. (1989) *Alternate Realities: Mathematical Models of Nature and Man*, New York: John Wiley and Sons.

Casti, J. L. (2000) *Paradigms Regained: A Further Exploration of the Mysteries of Modern Science*, London: Little, Brown and Company.

Cheng, W. L. (1999) 'Division of labor, money, and economic progress', *Review of Development Economics*, 3 (3), 354–368.

Chipman, J. S. (2002) 'The fundamental theorems of welfare economics', unpublished manuscript, University of Minnesota.

Cho, I-K., and Kreps, D. M. (1987) 'Signalling games and stable equilibria', *Quarterly Journal of Economics*, 102: 179–221.

Churchland, P. M. and Hooker, C. W. (eds) (1985) *Images of Science*, Chicago: University of Chicago Press.

Clark, C. M. A. (1990) 'Adam Smith and society as an evolutionary process', *Journal of Economic Issues*, 24 (3): 825–844.

Clark, C. M. A. (1993) 'Spontaneous order versus instituted process: the market as cause and effect', *Journal of Economic Issues*, 27 (2): 373–385.

Clark, W. A. V. (1991) 'Residential preferences and neighbourhood racial segregation: a test of the Schelling segregation model', *Demography*, 28 (1): 1–19.

Coase, R. H. (1976) 'Adam Smith's view of man', Selected Papers No. 50, Graduate School of Business, The Chicago University; originally published in *Journal of Law and Economics,* 19, (3): 529–546.

Coase, R. H. (1992) 'The institutional structure of production', *The American Economic Review*, 82 (4): 713–719.

Collin, F. (1997) *Social Reality*, London: Routledge.

Colman, A. M. (1982a) 'Experimental games', in *Cooperation and Competition in Humans and Animals*, Workingham: Van Nostrand Reinhold, pp. 113–140.

Colman, A. M. (ed.) (1982b) *Cooperation and Competition in Humans and Animals*, Workingham: Van Nostrand Reinhold.

Colman, A. M. (1995) *Game Theory and its Applications in the Social and Biological Sciences*, Oxford: Butterworth & Heinemann.

Colman, A. M. and Bacharach, M. (1997) 'Payoff dominance and the Stackelberg heuristic', *Theory and Decision*, 43: 1–19.

Conte, R. and Gilbert, N. (1995) 'Introduction: computer simulation for social theory', in *Artificial Societies: The Computer Simulation of Social Life*, London: UCL Press, pp. 1–18.

Cosmides, L. and Tooby, J. (1994) 'Better than rational: evolutionary psychology and the invisible hand', *The American Economic View, Papers and Proceedings*, 84 (2): 327–332.

Cowen, T. (1998) 'Do economists use social mechanisms to explain?' in P. Hedström and R. Swedberg (eds), *Social Mechanisms: An Analytical Approach to Social Theory*, Cambridge: Cambridge University Press, pp. 125–146.

Crawford, V. P. (1997) 'Theory and experiment in the analysis of strategic interaction', in D. M. Kreps and K. F. Wallis (eds), *Advances in Economics and Econometrics: Theory and Applications*, Seventh World Congress, Vol. I., Cambridge: Cambridge University Press, pp. 206–242.

Crawford, V. and Haller, H. (1990) 'Learning how to cooperate: optimal play in repeated coordination games', *Econometrica*, 58: 571–596.

Curren R. (1987) 'Invisible-hand explanations reconsidered', *Humanities Working Paper 120*, California Institute of Technology.

Dalton, G. (1965) 'Primitive money', *American Anthropologist*, 67: 44–65.

Dalton, G. (ed.) (1971a) *Economic Anthropology and Development: Essays on Tribal and Peasant Economies*, New York: Basic Books.

Dalton, G. (1971b) 'Primitive money', in *Economic Anthropology and Development: Essays on Tribal and Peasant Economies*, New York: Basic Books, pp. 167–192.

David, P. (1985) 'Clio and the economics of QWERTY', *American Economic Review*, 75 (2): 332–337.

David, P. (1988) 'Path-dependence: putting the past into the future of economics', *IMSSS Technical Report* no. 5333, Stanford University.

David, P. A. (1994) 'Why are institutions the "carriers of history"?: path dependence and the evolution of conventions, organizations and institutions', *Structural Change and Economic Dynamics*, 5 (2): 205–220.

Davies, G. (1996) *A History of Money from Ancient Times to the Present Day*, Cardiff: University of Wales Press.

Davis, J. B. (1989) 'Smith's invisible hand and Hegel's cunning of reason', *International Journal of Social Economics,* 16 (6): 50–66.

Davis, J. R. (1990) 'Adam Smith on the provential reconciliation of individual and social interests: is man led by an invisible hand or misled by a sleight of hand', *History of Political Economy,* 22: 341–352.

Dawid, H. (2000) 'On the emergence of exchange and mediation in a production economy', *Journal of Economic Behaviour and Organisation*, 41 (1): 27–53.

Dawkins, R. (1976) *The Selfish Gene*, Oxford: Oxford University Press.

Debreu, G. (1954) 'Valuation equilibrium and Pareto optimum', *Proceedings of the National Academy of Sciences*, 40: 588–592.

Debreu, G. (1959) *Theory of Value*, New York: Wiley.

Demeny, P. (1986) 'Population and the invisible hand', *Demography,* 23 (4): 473–487.

Denis, A. (1999) 'Was Adam Smith an individualist?', *History of the Human Sciences,* 12 (3): 71–86.

Denton, N. A. and Massey, D. S. (1991) 'Patterns of neighborhood transition in a multiethnic world: U.S. metropolitan areas, 1970–80', *Demography*, 28 (1): 41–63.

Donninger, C. (1986) 'Is it always efficient to be nice?' in A. Deikmann and P. Mitter (eds), *Paradoxical Effects of Social Behavior: Essays in Honor of Antol Rapoport*, Heidelberg: Physica Verlag, pp. 123–134.

Dowd, K. (1999) 'The invisible hand and the evolution of the monetary system' in J. Smithin (ed.), *What is Money*, London: Routledge, pp. 139–156.

Dowd, K. (2001) 'The emergence of fiat money: a reconsideration', *Cato Journal*, 20 (3): 467–476.

Dowe, P. (2000) *Physical Causation*, Cambridge: Cambridge University Press.

Downs, A. (1981) *Neigborhoods and Urban Development*, Washington, DC: The Brookings Institution Press.

Drogoul, A. and Ferber, J. (1994) 'Multi-agent simulation as a tool for studying emergent processes in societies' in N. Gilbert and J. Doran (eds), *Simulating Societies: The Computer Simulation of Social Phenomena*, London: UCL Press, pp. 127–142.

Duffy, J. and Ochs, J. (1999) 'The emergence of money as a medium of exchange: an experimental study', *American Economic Review*, 89 (4): 847–877.

Duhem, P. (1914 [1954]) *The Aim and Structure of Physical Theory*, trans. P. P. Wiener, Princeton: Princeton University Press.

Durlauf, N. D. (1991) 'Multiple equilibria and persistence in aggregate fluctuations', *The American Economic Review, Papers and Proceedings*, 81 (2): 70–74.

Durlauf, S. N. and Young, P. H. (eds) (2001) *Social Dynamics*, Cambridge: The MIT Press.

Easterly, W. (2004) 'Empirics of strategic interdependence: the case of the racial tipping point', unpublished manuscript, New York University.

Ebeling, R. M. (ed.) (1991) *Austrian Economics: A Reader*. Vol. 18, Hillsdale, MI: Hillsdale College Press.

Edmonds, B. and Hales, D. (2005) 'Computational simulation as theoretical experiment', *Journal of Mathematical Sociology*, 29: 209–232.

Einzig, P. (1966) *Primitive Money*, Oxford: Pergamon Press.

Ellen, I. G. (2000) 'Race-based neighbourhood projection: a proposed framework for understanding new data on racial integration', *Urban Studies*, 37 (9): 1513–1533.

Ellis, B. (1985) 'What science aims to do' in P. M. Churchland and C. A. Hooker (eds), *Images of Science*, Chicago: Chicago University Press, pp. 48–74.

Ellison, G. (1993) 'Learning, local interaction and coordination', *Econometrica*, 61 (5): 1047–1071.

Elster, J. (1979) *The Ulysses and the Sirens*, Cambridge: Cambridge University Press.

Eltis, W. (2004) 'Emma Rothschild on economic sentiments: and the true Adam Smith', *European Journal of the History of Economic Thought*, 1 (1): 147–159.

Engineer, M. and Shi, S. (1998) 'Asymmetry, imperfectly transferable utility, and the role of money in improving terms of trade', *Journal of Monetary Economics*, 41: 153–183.

Engineer, M. and Shi, S. (2000) 'Bargains, barter and money', unpublished manuscript.

Epstein, J. M. (2005) 'Remarks on the foundations of agent-based generative social science', *CSED Working Paper* No. 41, Washington DC: The Brookings Institution.

Epstein, J. M. and Axtell, R. (1996). *Growing Artificial Societies: Social Science from the Bottom Up*, Washington DC: Brookings Institution Press.

Erev, I. and Roth, A. (1996) 'On the need for low rationality cognitive game theory: reinforcement learning in experimental games', unpublished manuscript, University of Pittsburgh.

Evensky, J. (1993) 'Retrospectives: ethics and the invisible hand', *The Journal of Economic Perspectives*, 7 (2): 197–205.

Evensky, J. (1998) 'Adam Smith's moral philosophy: the role of religion and its relationship to philosophy and ethics in the evolution of society', *History of Political Economy*, 30 (1): 17–42.

Evensky, J. (2005) *Adam Smith's Moral Philosophy: A Historical and Contemporary Perspective on Markets, Law, Ethics, and Culture*, Cambridge: Cambridge University Press.

Fagiolo, G., Valente, M. and Vriend, N. J. (2005) 'Segregation in networks', Working Paper no. 549, Department of Economics, Queen Mary University of London.

Farley, R., Fielding, E. L. and Krysian, M. (1997) 'The residential preferences of blacks and whites: a four-metropolis analysis', *Housing Policy Debate*, 8 (4): 763–800.

Fehr, E. and Tougareva, E. (1995) 'Do competitive markets with high stakes remove reciprocal fairness? Experimental evidence from Russia', working paper, Institute for Empirical Economic Research, University of Zurich.

Feiner, S. F. and Roberts, B. B. (1990) 'Hidden by the invisible hand: neoclassical economic theory and the textbook treatment of race and gender', *Gender and Society*, 4 (2): 159–181.

Ferguson, A. (1767 [1995]) *An Essay on the History of Civil Society*, ed. F. Oz-Salzberger, Cambridge: Cambridge University Press.

Fielding, E. L. (1997) 'How long can it go: applying survey data to Schelling's model of segregation', paper presented at the annual meetings of the Population Association of America, March, Washington DC.

Fiori, S. (2001) 'Visible and invisible order. The theoretical duality of Smith's political economy', *European Journal of the History of Economic Thought*, 8 (4): 429–448/

Flache, A. and Hegselmann, R. (2001) 'Do irregular grids make a difference? Relaxing the spatial regularity assumption in cellular models of social dynamics', *Journal of Artificial Societies and Social Simulation*, 4 (4). Available at www.soc.surrey.ac.uk/ JASSS/4/4/6.html (accessed 15 May 2007).

Fleischacker, S. (2004) *On Adam Smith's Wealth of Nations: A Philosophical Companion*, Princeton: Princeton University Press.

Flew, A. (1987). 'Social science: making visible the invisible hands', *Journal of Libertarian Studies*, 8: 197–211.

Foley, V. (1976) *The Social Physics of Adam Smith*, West Lafayette: Prude University Press.

Forsythe, R., Horowitz, J., Savin, N. E. and Sefton, M. (1994) 'Replicability, fairness and pay in experiments with simple bargaining games', *Games and Economic Behavior*, 6 (3): 347–369.

Fosset, M. and Waren, W. (2005) 'Overlooked implications of ethnic preferences for residential segregation in agent-based models', *Urban Studies*, 42 (11): 1893–1917.

Foster, D. and Young, H. P. (1990) 'Stochastic evolutionary game dynamics', *Theoretical Population Biology*, 38: 219–232.

Foster, D. P. and Young, H. P. (2001) 'On the impossibility of predicting the behaviour of rational agents', unpublished manuscript.

van Fraassen, B. C. (1970) 'On the extension of Beth's semantics of physical theories', *Philosophy of Science* 37: 325–339.

van Fraassen, B. C. (1972) 'A formal approach to the philosophy of science', in R. N. Colodny (ed.), *Paradigms and Paradoxes*, Pittsburgh: University of Pittsburgh Press, pp. 303–366.

van Fraassen, B. C. (1980) *The Scientific Image*, Oxford: Clarendon Press.

van Fraassen, B. C. (1987) 'The semantic approach to scientific theories', in N. J. Nersessian (ed.), *The Process of Science*, Dordrecht: Kluwer Academic Publishers, pp. 105–124.

van Fraassen, B. C. (1989) *Laws and Symmetry*, Oxford: Clarendon Press.

Freudenthal, H. (ed.) (1961) *The Concept and the Role of the Model in Mathematics and Natural and Social Sciences*, Dordrecht: Reidel.

Friedman, D. (1991) 'Evolutionary games in economics', *Econometrica*, 59: 637–666.

Friedman, M. (1974) 'Explanation and scientific understanding', *Journal Philosophy*, 71 (1): 5–19.

Friedrichs, J. (1998) 'Ethnic segregation in Cologne, Germany, 1984–94', *Urban Studies,* 35 (10): 1745–1763.

Frye, T. and Shleifer, A. (1996) 'The invisible hand and the grabbing hand', *NBER Working Paper Series* no. 5856, National Bureau of Economic Research.

Fudenberg, D. and Harris, C. (1992) 'Evolutionary dynamics with aggregate shocks', *Journal of Economic Theory,* 57: 420–441.

Fudenberg, D. and Levine, D. K. (1995) 'Consistency and cautious fictitious play', *Journal of Economic Dynamics and Control* 19: 1065–1090.

Fudenberg, D. and Levine, D. K. (1998) *The Theory of Learning in Games,* Cambridge: MIT Press.

Gallie, W. B. (1955) 'Explanations in history and the genetic sciences', *Mind, New Series,* 64 (254): 160–180; reprinted in P. Gardiner (ed.), *Theories of History,* New York and London: Free Press, pp. 386–402.

Gallo, P. S. and McClintock, C. G. (1965) 'Cooperative and competitive behavior in mixed-motive games', *Journal of Conflict Resolution,* 9: 68–78.

Gardiner, P. (ed.) (1959) *Theories of History,* New York and London: Free Press.

Gauthier, D. (1975) 'Coordination', *Dialogue,* 14: 195–221.

Gibbard, A. and Varian, H. R. (1978) 'Economic models', *Journal of Philosophy,* 664–677.

Giere, R. N. (1984) *Understanding Scientific Reasoning,* 2nd edn. New York: Holt, Rinehart, and Winston.

Giere, R. N. (1985a) 'Philosophy of science naturalized', *Philosophy of Science,* 52: 331–356.

Giere, R. N. (1985b) 'Constructive realism', in P. M. Churchland and C. W. Hooker (eds), *Images of Science,* Chicago, IL: University of Chicago Press, pp. 75–98.

Giere, R. N. (1988a) *Explaining Science: A Cognitive Approach,* Chicago IL: University of Chicago Press.

Giere, R. N. (1988b) 'Laws, theories, and generalizations', in A. Grünbaum and W. C. Salmon (eds), *The Limits of Deductivism,* Berkeley: University of California Press, pp. 37–46.

Giere, R. N. (1988c) 'The cognitive structure of scientific theories,' *Philosophy of Science* 61: 276–296.

Giere, R. N. (1999) 'Using models to represent reality', in L. Magnani, N. J. Nersessian and P. Thagard (eds), *Model Based Reasoning in Scientific Discovery,* New York: Kluwer/Plenum, pp. 41–57.

Giere, R. N. (2000) 'Theories', in W. H. Newton-Smith (ed.), *A Companion to the Philosophy of Science,* Oxford: Blackwell Publishers, pp. 515–524.

Gilbert, M. (1989) 'Rationality and salience', *Philosophical Studies,* 57: 61–77.

Gilbert, M. (1990) 'Rationality, coordination and convention', *Synthese,* 84: 1–21.

Gilbert, N. and Doran, J. (1994a) 'Simulating societies: an introduction' in N. Gilbert and J. Doran (eds), *Simulating Societies: The Computer Simulation of Social Phenomena,* London: UCL Press, pp. 1–18.

Gilbert, N. and Doran, J. (1994b) *Simulating Societies: The Computer Simulation of Social Phenomena,* London: UCL Press.

Gintis, H. (1997) 'A markov model of production, trade, and money: theory and artificial life simulation', *Computational and Mathematical Organisation Theory,* 3 (1): 19–41.

Gintis, H. (2000) *Game Theory Evolving,* Princeton: Princeton University Press.

Good, D. A. (1991) 'Cooperation in a microcosm: lessons from laboratory games', in R. A. Hinde and J. Groebel (eds), *Cooperation and Prosocial Behavior,* Cambridge: Cambridge University Press, pp. 224–237.

Goyal, S. and Janssen, M. C. W. (1996) 'Can we rationally learn to coordinate?', *Theory and Decision*, (40): 29–49.

Goyal, S. and Janssen, M. C. W. (1997) 'Non-exclusive conventions and social coordination', *Journal of Economic Theory*, 77 (1): 34–57.

Grampp, W. D. (2000) 'What did Smith mean by the invisible hand?', *The Journal of Political Economy*, 108 (3): 441–465.

Greif, A. (1998) 'Historical and comparative institutional economics', *The American Economic Review, Papers and Proceedings*, 88 (2): 80–84.

Greif, A. (2006) *Institutions and the Path to the Modern Economy: Lessons from Medieval Trade*, Cambridge, Cambridge University Press.

Grodzins, M. (1957) 'Metropolian segregation', *Scientific American*, 24: 33–41.

Grünbaum, A. and Salmon, W. C. (eds) (1988) *The Limits of Deductivism*, Berkeley: University of California Press.

Guala, F. (2001) 'Models, simulations, and experiments', unpublished manuscript.

Gürke, B. (2006) *Ethnic segregation in London*, Master thesis, abridged version (trans. B. Gürke), Department of Sociology, University of Cologne, Germany.

Güth, W. and Tietz, R. (1990) 'Ultimatum bargaining behavior: a survey and comparison of experimental results', *Journal of Economic Psychology*, 11: 417–449.

Güth, W., Schmittberger, R. and Schwarz, B. (1982) 'An experimental analysis of ultimatum bargaining', *Journal of Economic Behavior and Organisation*, 3: 367–388.

Häggqvist, S. (1996) *Thought Experiments in Philosophy*, Stockholm: Almqvist & Wiksell.

Hahn, F. H. (1970) 'Some adjustment problems', *Econometrica*, 38 (1): 1–17.

Hahn, F. H. (1973) 'The winter of our discontent', *Economica*, 40: 322–330.

Hahn, F. (1981) 'Reflections on invisible hand', text of the Fred Hirsch Memorial Lecture presented at the University of Warwick on 5 November.

Hamowy, R. (1987) *The Scottish Enlightenment ant the Theory of Spontaneous Order*, The Journal of the History of Philosophy Monograph Series, Carbondale and Edwardsville: Southern Illinois University Press.

Hands, D. W. (1995) 'Social epistemology meets the invisible hand: Kitcher on the advancement of science', *Dialogue*, 34 (3): 605–621.

Harre, R. (1970) *The Principles of Scientific Thinking*, Chicago: University of Chicago Press.

Harsanyi, J. (1973) 'Games with randomly distributed payoffs: a new rationale for mixed-strategy equilibrium points', *International Journal of Game Theory*, 2: 1–23.

Harsanyi, J. and Selten, R. (1988) *A General Theory of Equilibrium in Games*, Cambridge: MIT Press.

Hausman, D. M. (1992) *The Inexact and Separate Science of Economics*, Cambridge: Cambridge University Press.

Hayek, F. A. (1937) 'Economics and knowledge', *Economica*, 4: 33–54; reprinted in F. A. Hayek (1949), *Individualism and Economic Order*, London: Routledge and Kegan Paul, pp. 33–56.

Hayek, F. A. (1943) 'The facts of the social sciences', *Ethics*, 54 (1): 1–13; reprinted in F. A. Hayek (1949), *Individualism and Economic Order*, London: Routledge and Kegan Paul, pp. 57–76.

Hayek, F. A. (1945) 'The use of knowledge in society', *American Economic Review*, 35 (4): 519–530; reprinted in F. A. Hayek (1949), *Individualism and Economic Order*, London: Routledge and Kegan Paul, pp. 77–91.

Hayek, F. A. (1946a) 'Individualism: true and false', The twelfth Finlay Lecture, University College, Dublin, 17 December 1945; reprinted in F. A. Hayek (1949), *Individualism and Economic Order*, London: Routledge and Kegan Paul, pp. 1–32.

Hayek, F. A. (1946b) 'The meaning of competition', Stafford Little lecture, Princeton University; reprinted in F. A. Hayek (1949) *Individualism and Economic Order*, London: Routledge & Kegan Paul, pp. 92–106.

Hayek, F. A. (1949) *Individualism and Economic Order*, London: Routledge and Kegan Paul.

Hayek, F. A. (1967a) *Studies in Philosophy, Politics and Economics*, London: Routledge and Kegan Paul.

Hayek, F. A. (1967b) 'Notes on the evolution of systems of rules of conduct' in *Studies in Philosophy, Politics and Economics*, London: Routledge and Kegan Paul, pp. 66–81.

Hayek, F. A. (1967c) 'The results of human action but not human design' in *Studies in Philosophy, Politics and Economics*, London: Routledge and Kegan Paul, pp. 96–105.

Heath, E. (1992) 'Rules, function, and the invisible hand: "an interpretation of Hayek's social theory"', *Philosophy of the Social Sciences*, 22(1): 28–45.

Heath, E. (1998) 'Mandeville's bewitching engine of praise', *History of Philosophy Quarterly*, 15 (2): 205–226.

Hedström, P. and Swedberg, R. (eds) (1998) *Social Mechanisms: An Analytical Approach to Social Theory*, Cambridge: Cambridge University Press.

Hempel, C. G. (1965) *Aspects of Scientific Explanation and Other Essays in the Philosophy of Science*, New York: Free Press.

Hempel, C. G. and Oppenheim, P. (1948) 'Studies in the logic of explanation', *Philosophy of Science* 15: 135–75; reprinted in C. G. Hempel (1965) *Aspects of Scientific Explanation and Other Essays in the Philosophy of Science*, New York: Free Press, pp. 291–295.

Henrich, J. (2000) 'Does culture matter in economic behavior? Ultimatum game bargaining among the Machiguenga', *American Economic Review*, 90 (4): 973–979.

Henrich, J., Boyd, R. *et al.* (2001) 'In search of homo economicus: behavioral experiments in 15 small-scale societies', *American Economic Review*, 91 (2): 73–79.

Henrich, J., Boyd, R. *et al.* (2005) ' "Economic man" in cross-cultural perspective: behavioral experiments in 15 small-scale societies', *Behavioral and Brain Science*, 28: 795–855.

Henry, J. F. (2002) 'The social origins of money: the case of Egypt', unpublished manuscript.

Hesse, M. (1970) *Models and Analogies in Science*, Notre Dame: University of Notre Dame Press.

Hesse, M. (2000) 'Models and analogies' in W. H. Newton-Smith (ed.), *A Companion to the Philosophy of Science*, Oxford: Blackwell Publishers, pp. 299–307.

Hinde, R. A. and Groebel, J. (eds) (1991) *Cooperation and Prosocial Behavior*, Cambridge: Cambridge University Press.

Hirshleifer, J. (1985) 'Evolution, spontaneous order, and market exchange', working paper, UCLA Department of Economics: 358, January.

Hodgson, G. M. (1992) 'Carl Menger's theory of the evolution of money: some problems', *Review of Political Economy*, 4 (4): 396–412.

Hodgson, G. (1994) 'Hayek, evolution and spontaneous order', in P. Mirowski (ed.), *Natural Images in Economic Thought*, Cambridge: Cambridge University Press, pp. 408–447.

Hodgson, G. M. (2001) *How Economics Forgot History*, London: Routledge.

Hofbauer, J. and Sigmund, K. (1988) *The Theory of Evolution and Dynamical Systems*, Cambridge: Cambridge University Press.

Hofbauer, J., Schuster, P. and Sigmund, K. (1979) 'A note on evolutionary stable strategies and game dynamics', *Journal of Theoretical Biology*, 81: 609–612.

Hoffman, E., McCabe, K. and Smith, V. L. (1998) 'Behavioral foundations of reciprocity: experimental economics and evolutionary psychology', *Economic Inquiry*, 36 (3): 335–352.

Hoffman, E., McCabe, K., Shachat, K. and Smith, V. L. (1994) 'Preferences, property rights, and anonymity in bargaining games', *Games and Economic Behavior*, 7: 346–380.

Holcombe, R. G. (1999) 'Equilibrium versus the invisible hand', *Review of Austrian Economics*, 12: 227–243.

Holland, J. H. (1975) *Adaptation in Natural and Artificial Systems*, Cambridge: The MIT Press.

Holland, J. H. (1995) *Hidden Order: How Adaptation Builds Complexity*, Reading: Perseus.

Holland, J. H. (1998) *Emergence: from Chaos to Order*, Oxford: Oxford University Press.

Holland J. H., Holyoak, K. J., Nispet, R. E. and Thagard, P. R. (1989) *Induction: Processes of Inference, Learning, and Discovery*, Cambridge: The MIT Press.

Holmes, R. L. (1977) 'Nozick on anarchism', *Political Theory*, 5: 247–256.

Holyoak, K. J. and Thagard, P. (1996) *Mental Leaps: Analogy in Creative Thought*, Cambridge: The MIT Press.

Horgan, J. (1994) 'Icon and bild: a note on the analogical structure of models – the role of models in experiment and theory', *The British Journal for the Philosophy of Science*, 45: 599–604.

Horwitz, S. (1999) 'From Smith to Menger to Hayek: liberalism in the tradition of the Scottish enlightenment', paper presented at HES meetings, June 1999, Greensboro, NC.

Howitt, P. and Clower, R. (2000) 'The emergence of economic organization', *Journal of Economic Behavior and Organization*, 41: 55–84.

Hughes, R. I. G. (1990) 'The Bohr atom, models, and realism', *Philosophical Topics*, 18: 71–84.

Hull, D. (1988) *Science as a Process*, Chicago: The University of Chicago Press.

Hull, D. L. (1997) 'What's wrong with invisible-hand explanations?', *Philosophy of Science*, 64 (4): 117–126.

Hutchison, T. W. (1938) *The Significance and Basic Postulates of Economic Theory*, New York: Kelley.

Hutchison, T. W. (1962) *A Review of Economic Doctrines 1870–1929*, Oxford: Clarendon Press.

Hutchison, T. W. (1981) *The Politics and Philosophy of Economics*, Oxford: Basil Blackwell.

Hutter, M. (1994) 'Organism as a metaphor in German economic thought', in P. Mirowski (ed.), *Natural Images in Economic Thought*, Cambridge: Cambridge University Press, pp. 289–321.

Huttman, E. D., Blauw, W. and Saltman, J. (eds) (1991) *Urban Housing Segregation of Minorities in Western Europe and the United States*, Durham, NC: Duke University Press.

Iceland, J. (2002) 'Beyond black and white: metropolitan residential segregation in multiethnic America', paper presented at the American Sociological Association meetings, Chicago, Illinois, 16–19 August.

Iceland, J., Weinberg, D. H. and Steinmetz, E. (2002) 'Racial and ethnic residential segregation in the United States: 1980–2000', *Census 2000 Special Reports* (Issued August 2002) US Department of Commerce, Economics and Statistics Administration.

Ihlanfeldt, K. R. and Scafidi, B. (2002) 'Black self-segregation as a cause of housing segregation: evidence from the multi-city study of urban inequality', *Journal of Urban Economics*, 51: 366–390.

Ingham, G. (1996) 'Money is a social relation', *Review of Social Economy*, 54 (4): 507–529.

Ingham, G. (1998a) 'On the underdevelopment of the "sociology of money"', *Acta Sociologica*, 41 (1): 3–18.

Ingham, G. (1998b) 'Money is a social relation', in S. Fleetwood (ed.), *Critical Realism in Economics: Development and Debate*, London: Routledge, pp. 103–105.

Ingham, G. (1999) '"Babylonian madness": on the historical and sociological origins of money', in J. Smithin (ed.), *What is Money?*, London: Routledge, pp. 16–41.

Ingham, G. (2004) *The Nature of Money*, Cambridge: Polity Press.

Ingrao, B. (1998) 'Invisible hand', in H. Kurz and N. Salvadori (eds), *The Elgar Companion to Classical Economics*, Cheltenham: Edward Elgar, pp. 436–441.

Innes, A. M. (1913) 'What is money', *The Banking Law Journal*, 30: 377–408.

Iwai, K. (1997) 'Evolution of money', Working Paper, Faculty of Economics, The University of Tokyo; reprinted in A. Nicita and U. Pagano (eds) (2000), *The Evolution of Economic Diversity*, London: Routledge, pp. 396–431.

Iway, K. (1996) 'The bootstrap theory of money: a search-theoretic foundation of monetary economics', *Structural Change and Economic Dynamics*, 7: 451–477.

Jacobsen, H. J. (1996) 'On the foundations of Nash equilibrium', *Economics and Philosophy*, 12: 67–88.

Janssen, M. C. W. (1993) *Microfoundations: A Critical Inquiry*, London: Routledge.

Janssen, M. C. W. (1998a) 'Individualism and equilibrium coordination in games', in R. Backhouse, D. Hausman, U. Mäki and A. Salanti (eds), *Crossing Boundaries: Case studies in Economic Methodology*, London: MacMillan, pp. 1–35.

Janssen, M. C. W. (1998b) 'Focal points', in P. Newman (ed.) (1998) *The New Palgrave Dictionary of Economics and the Law*, London: Macmillan, pp. 150–155.

Janssen, M. C. W. (2001a) 'Towards a justification of the principle of coordination', *Economics and Philosophy*, 17: 221–234.

Janssen, M. C. W. (2001b) 'Rationalising focal points', *Theory and Decision*, 50 (1): 119–148.

Jevons, W. S. (1876) *Money and the Mechanism of Exchange*. New York: D. Appleton and Co.

Jones, R. A. (1976) 'The origin and development of media of exchange', *Journal of Political Economy*, 84 (4): 757–775.

Kalai, E. and Lehrer, E. (1993a) 'Rational learning leads to Nash equilibrium', *Econometrica*, 61: 1019–1045.

Kalai, E. and Lehrer, E. (1993b) 'Private-beliefs equilibrium', *Econometrica*, 61: 1231–1240.

Kalai, E. and Samet, D. (1984) 'Persistent equilibria in strategic games', *International Journal of Game Theory*, 13: 129–144.

Kandori, M. and Rob, R. (1995) 'Evolution of equilibria in the long run: a general theory and applications', *Journal of Economic Theory*, 65: 383–414.

Kandori, M., Mailath, G. J. and Rob, R. (1993) 'Learning, mutation and long run equilibria in games', *Econometrica*, 61: 29–56.

Kaniovski, Y. and Young, P. H. (1995) 'Learning dynamics in games with stochastic perturbations', *Games and Economic Behavior*, 11: 330–363.

Karlson (1993) *The State of the State*, PhD thesis, Uppsala University Press.

Kauder, E. (1965) *A History of Marginal Utility Theory*, New Jersey: Princeton University Press.

Kauffman, S. (1995) *At Home in the Universe: The Search for the Laws of Self-Organization and Complexity,* Oxford: Oxford University Press.

Keller, R. (1994) *On Language Change: The Invisible Hand in Language*, London: Routledge.

Khalil, E. L. (2000a) 'Beyond natural selection and divine intervention: the Lamarckian implication of Adam Smith's invisible hand', *Journal of Evolutionary Economics*, 10: 373–393.

Khalil, E. L. (2000b) 'Making sense of Adam Smith's invisible hand: beyond Pareto optimality and unintended consequences', *Journal of the History of Economic Thought*, 22 (1): 49–63.

Kincaid, H. (1990) 'Defending laws in the social sciences', *Philosophy of the Social Sciences*, 20: 56–83.

Kincaid, H. (1996) *Philosophical Foundations of the Social Sciences*, Cambridge: Cambridge University Press.

Kitcher, P. (1981) 'Explanatory unification', *Philosophy of Science*, 48: 507–531.

Kitcher, P. (1985) 'Two approaches to explanation', *Journal of Philosophy*, 82: 632–639.

Kitcher, P. (1989) 'Explanatory unification and the causal structure of the world', in P. Kitcher and W. C. Salmon (eds), *Scientific Explanation*, Minnesota Studies in the Philosophy of Science, vol. 13. Minneapolis: University of Minnesota Press, pp. 410–506.

Kitcher, P. (1993) *The Advancement of Science. Science without Legend, Objectivity without Illusions,* Oxford: Oxford University Press.

Kitcher, P. and Salmon, W. C. (eds) (1989) *Scientific Explanation*, Minnesota Studies in the Philosophy of Science, vol. 13. Minneapolis: University of Minnesota Press.

Kiyotaki, N. and Wright, R. (1989) 'On money as a medium of exchange', *Journal of Political Economy*, 97 (4): 927–954.

Kiyotaki, N. and Wright, R. (1991) 'A contribution to the pure theory of money', *Journal of Economic Theory*, 53: 215–235.

Kiyotaki, N. and Wright, R. (1992) 'Acceptability, means of payment, and media of exchange' *Federal Reserve Bank of Minneapolis Quarterly Review*, 16 (3): 2–10.

Kiyotaki, N. and Wright, R. (1993) 'A search theoretic approach to monetary economics', *American Economic Review*, 83 (1): 63–77.

Klein, D. B. (1997) 'Convention, social order, and two coordinations', *Constitutional Political Economy*, 8: 319–335.

Knapp, G. F. (1905 [1924]) *The State Theory of Money*, abridged edn (trans. H. M. Lucas and J. Bonar), London: Macmillan and Co.

Knudsen, C. (1993) 'Equilibrium, perfect rationality and the problem of self-reference in economics', in U. Mäki, B. Gustafsson and C. Knudsen (eds), *Rationality, Institutions and Economic Methodology*, London: Routledge, pp. 133–170.

Kohlberg, E. and Mertents, J. F. (1986) 'On the strategic stability of equilibria', *Econometrica*, 54: 1003–1037.

Koppl, R. (1992) 'Invisible-hand explanations and neoclassical economics: toward a post marginalist economics', *Journal of Institutional and Theoretical Economics,* 148 (2): 292–313.

Koppl, R. (1994) 'Invisible-hand explanations', in P. Boettke (ed.), *The Elgar Companion to Austrian Economics*, Cheltenham: Edward Elgar, pp. 192–197.

Kreps, D. and Wilson, R. (1982) 'Sequential equilibria', *Econometrica*, 50: 863–894.

Krishna, V. and Sjostrom, T. (1995) 'On the convergence of fictitious play', unpublished manuscript, Harvard University.

Kroes, P. (1989) 'Structural analogies between physical systems', *British Journal for Philosophy of Science*, 40: 145–154.

Krugman, P. R. (1991) 'History versus expectations', *Quarterly Journal of Economics*, 106: 65–67.

Krugman, P. R. (1995) *Development, Geography and Economic Theory*, Cambridge: The MIT Press.

Kühne, U. (1997) 'Wie erklärt man mit unsichtbaren Händen?' paper presented at GAP3 conference in Munich, 15–18 September.

Kuniński, M. (1992) 'Freidrich von Hayek's theory of spontaneous order: between "versthen" and "invisible hand explanation" in J. L. Auspitz *et al* (eds), *Praxiologies and the Philosophy of Economics*, London: Transaction Publishers, pp. 347–367.

Kurz, H. and Salvadori, N. (eds) (1998) *The Elgar Companion to Classical Economics*, Cheltenham: Edward Elgar.

Land, S. K. (1977) 'Adam Smith's "Considerations Concerning the First Formation of Languages"', *Journal of the History of Ideas*, 38: 677–690.

Langlois, R. (ed.) (1986a) *Economics as a Process, Essays in New Institutional Economics*, Cambridge: Cambridge University Press.

Langlois, R. (1986b) 'The new institutional economics: an introductory essay' in *Economics as a Process, Essays in New Institutional Economics*, Cambridge: Cambridge University Press, pp. 1–25.

Langlois, R. (1986c) 'Rationality, institutions and explanation' in *Economics as a Process, Essays in New Institutional Economics*, Cambridge: Cambridge University Press, pp. 225–255.

Lapavitsas, C. (2005) 'The emergence of money in commodity exchange, or as monopolist of the ability to buy', *Review of Political Economy*, 17 (4): 549–569.

Latzer, M. and Schmitz, S. W. (2002) *Carl Menger and the Evolution of Payment Systems: From Barter to Electronic Money*, Cheltenham and Northampton: Elgar.

Laurie, A. J. and Jaggi, N. K. (2003) 'Role of "vision" in neighbourhood racial segregation: a variant of the Schelling segregation model', *Urban Studies*, 40 (13): 2687–2704.

Leatherdale, W. H. (1974) *The Role of Analogy, Model and Metaphor in Science*, Amsterdam: North-Holland.

Leibenstein, H. (1982) 'The Prisoner's Dilemma in the invisible hand: an analysis of intrafirm productivity', *The American Economic Review Papers and Proceedings*, 72 (2): 92–97.

Leonard, T. C. (2002) 'Reflection on rules in science: an invisible hand perspective', *Journal of Economic Methodology*, 9 (2): 141–168.

Levy, D. (1985) 'David Hume's invisible hand in The Wealth of Nations: the public choice of moral information', *Hume Studies*, 10th anniversary issue: 110–149.

Lewis, D. (1969) *Convention: A Philosophical Study*, Cambridge: Harvard University Press.

Liebrand, W. B. G. (1998) 'Computer modelling and the analysis of complex human behaviour: retrospect and prospect' in W. B. G. Liebrand, A. Nowak and R. Hegselmann (eds), *Computer Modelling of Social Processes*, London: SAGE Publications, pp. 1–14.

Liebrand, W. B. G., Nowak, A. and Hegselmann, R. (eds) (1998) *Computer Modelling of Social Processes*, London: SAGE Publications.

Lind, D. (1989). 'The failure of Nozick's invisible-hand justification of the political state', *Auslegung*, 15: 57–68.

Lindgren, J. R. (1969) 'Adam Smith's theory of inquiry', *The Journal of Political Economy*, 77 (6): 897–915.

Liu, C. (1997) 'Models and theories I: the semantic view revisited', *International Studies in the Philosophy of Science*, 11 (2): 147–164.

Lloyd, E. (1988) *The Structure and Confirmation of Evolutionary Theory*, Greenwood Press: New York.

Lloyd, E. (1998) 'Models', in *Routledge Online Encyclopaedia of Philosophy*, version 1.0, London: Routledge.

Logan, J. R. and Molotch, H. L. (1987) *Urban Fortunes: The Political Economy of Place*, Berkeley, University of California Press.

Luce, R. D. and Raiffa, H. (1957 [1989]) *Games and Decisions: Introduction and Critical Survey*, New York: Dover Publications.

Luo, G. Y. (1999) 'The evolution of money as a medium of exchange', *Journal of Economic Dynamics and Control*, 23: 415–458.

McCarthy, T. (1977) 'On an Aristotelian model of scientific explanation', *Philosophy of Science*, 44: 159–166.

Macfie, A. (1971) 'The invisible hand of Jupiter', *Journal of the History of Ideas*, 32: 595–599.

Maclachlan, F. (2003) 'Max Weber and the state theory of money', paper presented at the 44th annual meeting of the International Studies Association , Portland, Oregon, 1 March. Available at http://home.manhattan.edu/~fiona.maclachlan//maclachlan-23feb03.htm (accessed 10 January 2005).

McMahon, C. (1981) 'Morality and the invisible hand', *Philosophy and Public Affairs*, 10: 247–277.

McMullin, E. (1978) 'What do physical models tell us?', in B. van Rootselaar and J. F. Staal (eds), *Logic, Methodology and Philosophy of Science,* vol. III, Amsterdam: North-Holland, pp. 385–396.

McPherson, M., Smith-Lovin, L. and Cook, J. (2001) 'Birds of a feather: homophily in social networks', *Annual Review of Sociology*, 27: 415–444.

Magnani, L., Nersessian, N. J. and Thagard, P. (1999) *Model Based Reasoning in Scientific Discovery*, New York: Kluwer/Plenum.

Mailath, G. J. (1992) 'Introduction: symposium on evolutionary game theory', *Journal of Economic Theory*, 57: 259–277.

Mailath, G. J. (1998) 'Do people play Nash equilibrium? Lessons from evolutionary game theory', *Journal of Economic Literature*, 36: 1347–1374.

Mäki, U. (1990a) 'Scientific Realism and Austrian Explanation', *Review of Political Economy*, 2: 310–344.

Mäki, U. (1990b) 'Mengerian economics in realist perspective', in B. Caldwell (ed.), *Carl Menger and His Legacy in Economics*, Durham and London: Duke University Press, pp. 289–310.

Mäki, U. (1991) 'Practical syllogism, entrepreneurship, and the invisible hand', in D. Lavoie (ed.), *Economics and Hermeneutics*, London: Routledge and Kegan Paul, pp. 149–176.

Mäki, U. (1992a) 'On the method of isolation in economics', *Poznan Studies in the Philosophy of the Sciences and the Humanities*, special issue (ed. C. Dilworth), Idealization IV: Intelligibility in Science, 26: 319–354.

Mäki, U. (1992b) 'Market as an isolated causal process: a metaphysical ground for realism', in B. Caldwell and S. Boehm (eds), *Austrian Economics: Tensions and New Developments*, Boston: Kluwer, pp. 35–59.

Mäki, U. (1993) 'Economics with institutions: agenda for methodological enquiry', in U. Mäki, B. Gustafsson and C. Knudsen (eds), *Rationality, Institutions and Economic Methodology,* London: Routledge, pp. 3–42.

Mäki, U. (1994) 'Isolation, idealization and truth in economics', *Poznan Studies in the Philosophy of the Sciences and the Humanities*, 38: 147–168.

Mäki, U. (1996) 'Scientific realism and some peculiarities of economics', *Boston Studies in the Philosophy of Science*, 169: 425–445.

Mäki, U. (1997) 'Universals and the *Methodenstreit*: a re-examination of Carl Menger's conception of economics as an exact science', *Studies in History and Philosophy of Science*, 28 (3): 475–495.

Mäki, U. (1998) 'Aspects of realism about economics', *Theoria*, 13 (2): 301–319.

Mäki, U. (2001a) 'Models', in N. J. Smelser *et al.* (eds), *The International Encyclopaedia of Social and Behavioral Sciences*, vol. 15, Amsterdam: Elsevier Sciences/Pergamon, pp. 9931–9937.

Mäki, U. (2001b) 'Explanatory unification: double and doubtful', *Philosophy of the Social Sciences*, 31 (4): 488–506.

Mäki, U. (2004) 'Theoretical isolation and explanatory progress: transaction cost economics and the dynamics of dispute', *Cambridge Journal of Economics*, 28 (3): 319–346.

Mäki, U., Davis, J. B. and Hands, D. W. (eds) (1998) *The Handbook of Economic Methodology*, Cheltenham: Edward Elgar.

Mäki, U., Gustafsson, B. and Knudsen, C. (eds) (1993) *Rationality, Institutions and Economic Methodology*, London: Routledge.

Marimon, M. and Miller, J. H. (1989) 'Money as a medium of exchange in an economy with genetically reproduced decision rules', unpublished manuscript.

Marimon, R. (1997) 'Learning from learning in economics', in D. M. Kreps and K. F. Wallis (eds), *Advances in Economics and Econometrics: Theory and Applications. Seventh World Congress*, vol. I, Cambridge: Cambridge University Press, pp. 278–315.

Marimon, R., McGrattan, E. and Sargent, T. J. (1990) 'Money as a medium of exchange in an economy with artificially intelligent agents', *Journal of Economic Dynamics and Control*, 14: 329–373.

Marris, R. and Mueller, D. C. (1980) 'The corporation, competition, and the invisible hand', *Journal of Economic Literature*, 18 (1): 32–63.

Marshall, A. (1923) *Money, Credit and Commerce*, London. MacMillan & Co. Limited.

Maskin, E. S. (1994) 'The invisible hand and externalities', *The American Economic Review Papers and Proceedings*, 84 (2): 333–337.

Massey, D. S. and Denton, N. A. (1993) *American Apartheid: Segregation and the Making of the Underclass*, Cambridge, MA: Harvard University Press.

Mauss, M. (1930 [1990]) *The Gift: The Form and Reason for Exchange in Archaic Societies* (trans. W. D. Halls), New York: Norton & Company.

Maynard Smith, J. (1974) 'The theory of games and the evolution of animal conflicts', *Journal of Theoretical Biology*, 47: 209–221.

Maynard Smith, J. (1982) *Evolution and the Theory of Games*, Cambridge: Cambridge University Press.

Maynard Smith, J. and Price, G. R. (1973) 'The logic of animal conflict', *Nature*, 246: 15–18.

Medema, S. G. and Samuels, W. J. (eds) (1996) *Foundations of Research in Economics: How Do Economists Do Economics?*, Aldershot: Edward Elgar Publishing.

Meen, D. and Meen, G. (2003) 'Social behaviour as a basis for modelling the urban housing market: a review', *Urban Studies*, 40 (5–6): 917–935.

Mehta, J., Starmer, C. and Sugden, R. (1992) 'An experimental investigation on focal points in coordination and bargaining' in J. Geweke (ed.), *Decision Making Under Risk and Uncertainty: New Models and Empirical Findings*, Dordrecht: Kluwer, pp. 211–220.

Mehta, J., Starmer, C. and Sugden, R. (1994a) 'The nature of salience: an experimental investigation of pure coordination games', *American Economic Review*, 84: 658–673.

Mehta, J., Starmer, C. and Sugden, R. (1994b) 'Focal points in pure coordination: an experimental investigation', *Theory and Decision*, 36: 163–185.

Menger, C. (1883 [1985]) *Investigations into the Method of the Social Sciences, with Special Reference to Economics*, New York and London: New York University Press.

Menger, C. (1892a) 'On the origins of money', *Economic Journal*, 2: 239–255; reprinted in R. M. Ebeling (ed.) (1991), *Austrian Economics: A Reader*, vol. 18, Hillsdale, MI: Hillsdale College Press, pp. 483–504.

Menger, C. (1892b [2005]) 'Money as measure of value' (trans. G. Campagnolo), *History of Political Economy*, 37(2): 245–261. (Original article: 'La monna esure de valeur', *Revue d'économie politique*, 6: 159–175.)

Menger, C. (1950) *Principles of Economics* (trans. J. Dingwall and B. F. Hoselitz), New York: The Free Press.

Merton, R. (1936) 'The unanticipated consequences of human action', *American Sociological Review*, 1 (6): 894–904.

Michihiro, K. (1997) 'Evolutionary game theory in economics', in D. M. Kreps and K. F. Wallis (eds), *Advances in Economics and Econometrics: Theory and Applications*. Seventh World Congress, vol. I., Cambridge: Cambridge University Press, pp. 243–277.

Milgrom, P. and Roberts, J. (1991) 'Adaptive and sophisticated learning in normal form games', *Games and Economic Behavior*, 3: 82–100.

Mill, J. S. (1843 [1858]) *A System of Logic, Ratiocinative and Inductive*, New York: Harper & Brothers Publishers.

Mirowski, P. (ed.) (1994) *Natural Images in Economic Thought*, Cambridge: Cambridge University Press.

Mirowski, P. (1997) 'On playing the economics trump card in the philosophy of science: why it did not work for Michael Polanyi', *Philosophy of Science*, 64 (4): 127–138.

Mises, L. von (1954 [1981]) *The Theory of Money and Credit* (trans. H. E. Batson), Indianapolis, IN: Liberty Fund Inc. Available at www.econlib.org/library/ (accessed 4 February 2005).

Montes, L. (2003) 'Smith and Newton: some methodological issues concerning general equilibrium theory', *Cambridge Journal of Economics*, 27: 723–747.

Montes, L. (2004) *Adam Smith in Context: A Critical Reassessment of Some Central Components of His Thought*, London: Palgrave-Macmillan.

Montes, L. (2006) 'On Adam Smith's Newtonianism and general economic equilibrium theory', in L. Montes and E. Schliesser (eds), *New Voices on Adam Smith*, London: Routledge, pp. 247–270.

Montes, L. and Schliesser, E. (eds) (2006), *New Voices on Adam Smith*. London: Routledge.

Morgan, M. (2000) 'Experiments without material intervention: model experiments, virtual experiments and virtually experiments', *Research Memoranda in History and Methodology of Economics*, no. 00-1 Universiteit van Amsterdam, Faculty of Economics and Econometrics and LSE Centre for Philosophy of Natural and Social Science.

Morgan, M. (2001a) 'Models, stories and the economic world', *Journal of Economic Methodology*, 8 (3): 361–384.

Morgan, M. (2001b) 'The curious case of Prisoner's Dilemma: model situation? exemplary narrative?', *Research Memoranda in History and Methodology of Economics*, no. 01-7 Universiteit van Amsterdam, Faculty of Economics and Econometrics and LSE Centre for Philosophy of Natural and Social Science.

Morgan, M. and Morrison, M. (eds) (1999) *Models as Mediators: Perspectives on Natural and Social Science*, Cambridge: Cambridge University Press.

Morrison, M. (1999) 'Models as autonomous agents', in M. Morgan and M. Morrison (eds), *Models as Mediators: Perspectives on Natural and Social Science*, Cambridge: Cambridge University Press, pp. 38–65.

Morrison, M. and Morgan, M. (1999) 'Models as mediating instruments', in M. Morgan, and M. Morrison (eds), *Models as Mediators: Perspectives on Natural and Social Science*, Cambridge: Cambridge University Press, pp. 10–37.

Morrow, G. R. (1923) 'The significance of the doctrine of sympathy in Hume and Adam Smith', *The Philosophical Review*, 32 (1): 60–78.

Musgrave, A. (1981) 'Unreal assumptions in economic theory: the f-twist untwisted', *Kyklos*, 34: 377–387.

Myerson, R. B. (1978) 'Refinement of the Nash equilibrium concept', *International Journal of Game Theory*, 7: 73–80.

Nadeau, R. (1998) 'Spontaneous order', in U. Mäki, J. B. Davis and D. W. Hands (eds), *The Handbook of Economic Methodology*, Cheltenham: Edward Elgar, pp. 477–484.

Nagel, E., Suppes, P. and Tarski, A. (eds), *Logic Methodology and Philosophy of Science: Proceedings of the 1960 International Congress*, Stanford: Stanford University Press.

Nash, J. F. (1951) 'Non-cooperative games', *Annals of Mathematics*, 54: 286–295.

Nelson, A. (1986) 'Explanation and justification in political philosophy', *Ethics*, 97 (1): 154–176.

Nemeth, C. (1972) 'A critical analysis of research utilizing Prisoner's Dilemma paradigm for the study of bargaining', *Advances in Experimental Psychology*, 6: 203–234.

Nersessian, N. J. (ed.) (1987) *The Process of Science*, Dordrecht: Kluwer Academic Publishers.

Newman, P. (ed.) (1998) *The New Palgrave Dictionary of Economics and the Law*, London: Macmillan.

Niiniluoto, Ilkka (1999) *Critical Scientific Realism*, Oxford: Oxford University Press.

North, D. (1990) *Institutions, Institutional Change and Economic Performance*, Cambridge: Cambridge University Press.

Norton, R. (2002) 'Unintended consequences', in D. R. Henderson (ed.), *The Concise Encyclopedia of Economics*, Library of Economics and Liberty, Liberty Fund, Inc. Available at www.econlib.org/library/Enc/UnintendedConsequences.html (accessed 12 July 2006).

Nowak, M. and Sigmund, K. (1992) 'Tit for tat in heterogeneous populations', *Nature*, 355: 250–253.

Nowak, M. and Sigmund, K. (1993) 'A strategy of win-stay, lose-shift that outperforms tit-for-tat in the Prisoner's Dilemma game,' *Nature*, 364: 56–58.

Nozick, R. (1974) *Anarchy, State and Utopia*, New York: Basic Books.

Nozick, R. (1994) 'Invisible-hand explanations', *American Economic Review Papers and Proceedings*, 84 (2): 314–318.

Ostroy, J. M. and Starr, R. M. (1990) 'The transactions role of money', in B. M. Friedman and F. K. Hahn (eds), *Handbook of Monetary Economics*, Amsterdam: North-Holland, pp. 3–62.

Otteson, J. (2002a) *Adam Smith's Market Place of Life*, Cambridge: Cambridge University Press.

Otteson, J. (2002b) 'Adam Smith's first market: development of language', *History of Philosophy Quarterly*, 19 (1): 65–86.

Page, S. E. (1999) 'Computational economics from A to Z', keynote lecture at SwarmFest

1999 held at the Anderson Business School at UCLA on 26 March. Available at www. econ.iastate.edu (accessed 10 July 2005).

Pancs, R. and Vriend, N. J. (2003) 'Schelling's spatial proximity model of residential segregation revisited', working paper no. 487, Queen Mary University of London, Department of Economics.

Pascal, A. H. (1972) *Racial Discrimination in Economic Life,* Lexington: Lexington Books.

Pearce, D. (1984) 'Rationalizable strategic behavior and the problem of perfection', *Econometrica,* 52: 1029–1050.

Peart, S. J. and Levy, D. M. (2005) *The 'Vanity of the Philosopher': From Equality to Hierarchy in Post-Classical Economics,* Michigan: The University of Michigan Press.

Persky, J. (1989) 'Retrospectives: Adam Smith's invisible hands', *The Journal of Economic Perspectives,* 3 (4): 195–201.

Pettit, P. (1996) *The Common Mind: An Essay on Psychology, Society, and Politics,* Oxford: Oxford University Press.

Pettit, P. (1998) 'The invisible hand', in U. Mäki, J. B. Davis and D. W. Hands (eds), *The Handbook of Economic Methodology,* Cheltenham: Edward Elgar, pp. 256–259.

Plutynski, A. (2001) 'Modeling evolution in theory and practice', *Philosophy of Science,* 68: 225–236.

Polanyi, K. (1944 [1957]) *The Great Transformation,* Boston: Beacon Press.

Popper, K. (1962) *Conjectures and Refutations: The Growth of Scientific Knowledge,* New York: Basic Books.

Popper, K. (1979) *Objective Knowledge: An Evolutionary Approach,* Oxford: Clarendon Press.

Portugali J. and Benenson, I. (1997) 'Human agents between local and global forces in a self-organizing city' in F. Schweitzer (ed.), *Self-Organization of Complex Structures: From Individual to Collective Dynamics,* London: Gordon and Breach, pp. 537–546.

Portugali J., Benenson, I. and Omer, I. (1994) 'Socio-spatial residential dynamics: stability and instability within a self-organized city', *Geographical Analysis,* 26 (4): 321–340.

Portugali J., Benenson, I. and Omer, I. (1997) 'Spatial cognitive dissonance and sociospatial emergence in a self-organizing city', *Environment and Planning B,* 24: 263–285.

Postema, G. J. (1980). 'Nozick on liberty, compensation, and the individual's right to punish', *Social Theory and Practice,* 6: 311–338.

Poundstone, W. (1993) *Prisoner's Dilemma,* Oxford: Oxford University Press.

Pruitt, D. G. and Kimmel, M. J. (1977) 'Twenty years of experimental gaming: critique, synthesis, and suggestions for the future', *Annual Review of Psychology,* 28: 363–392.

Quiggin, A. H. (1949) *A Survey of Primitive Money: The Beginnings of Currency,* Methuen: London.

Radnitzky, G. (1993) 'Knowledge, values and the social-order in Hayek oeuvre', *Journal of Social and Evolutionary Systems,* 16 (1): 9–24.

Rappaport, S. (2001) 'Economic models as mini-theories', *Journal of Economic Methodology,* 8 (2): 275–285.

Rappoport, A. and Orwant, C. (1962) 'Experimental games: a review', *Behavioral Science,* 7: 1–37.

Redhead, M. (1980) 'Models in physics', *British Journal for the Philosophy of Science,* 31: 145–163.

Ritter, J. A. (1995) 'The transition from barter to fiat money', *The American Economic Review,* 85 (1): 134–149.

Robinson, J. (1951) 'An iterative method of solving a game', *Annals of Mathematics,* 54: 296–301.

Roll, E. (1936) 'Menger on money', *Economica*, New Series, 3 (12): 455–460.

Rosenberg, A. (1989) 'Explanatory role of existence proofs', *Ethics*, 97 (1): 177–186.

Rosenberg, A. (1995) *Philosophy of Social Science*, Oxford: Westview Press.

Rosenberg, A. (2000) *Philosophy of Science: A Contemporary Introduction*, London: Routledge.

Rosenberg, N. (1960) 'Some institutional aspects of the Wealth of Nations', *The Journal of Political Economy*, 68 (6): 557–570.

Rosser, J. B. Jr. (1999) 'On the complexities of complex economic dynamics', *Journal of Economic Perspectives*, 13 (4): 169–192.

Roth, A. E., Prasnikar, V., Okuno-Fujiwara, M. and Zamir, S. (1991) 'Bargaining and market behavior in Jerusalem, Ljubljana, Pittsburgh, and Tokyo: an experimental study', *American Economic Review*, 81 (5): 1068–1095.

Rothschild, E. (1994) 'Adam Smith and the invisible hand', *American Economic Review Papers and Proceedings*, 84 (2): 319–322.

Rothschild, E. (2001) *Economic Sentiments: Adam Smith, Condorcet, and the Enlightenment*, Cambridge: Harvard University Press.

Ruben, D-H. (1990) *Explaining Explanation*, London: Routledge.

Rubinstein, A. (1991) 'Comments on the interpretation of game theory', *Econometrica*, 59 (4): 909–924.

Rutherford, M. (1994) *Institutions in Economics: The Old and the New Institutionalism*, New York: Cambridge University Press.

Sagoff, M. (1994). 'Four dogmas of environmental economics', *Environmental Values*, 3 (4): 285–310.

Salmon, W. C. (1984) *Scientific Explanation and the Causal Structure of the World*, Princeton: Princeton University Press.

Salmon, W. C. (1990) *Four Decades of Scientific Explanation*, Minneapolis: University of Minnesota Press.

Salmon, W. C. (1998) *Causality and Explanation*, Oxford: Oxford University Press.

Samuelson, L. (1998) *Evolutionary Games and Equilibrium Selection*, Cambridge: MIT Press.

Samuelson, L. and Zhang, J. (1992) 'Evolutionary stability in asymmetric games', *Journal of Economic Theory*, 57: 363–391.

Sander, R., D. Schreiber and J. Doherty (2000a) 'Empirically testing a computational model: the example of housing segregation', in D. Sallach and T. Wolsko (eds), *Proceedings of the workshop on simulation of socialagents: Architectures and institutions*, Chicago: ANL/DIS/TM-60, Argonne National Laboratory, pp. 109–116.

Sander, R., Schreiber, D. and Doherty, J. (2000b) 'A computational model of housing segregation' unpublished manuscript.

Sayigh, Y. A. (1961) 'Development: the visible or invisible hand?', *World Politics*, 13 (4): 561–583.

Schelling, T. C. (1958) 'The strategy of conflict prospectus for a reorientation of game theory', *The Journal of Conflict Resolution*, 2 (3): 203–264.

Schelling, T. C. (1960) *The Strategy of Conflict*, Cambridge, MA: Harvard University Press.

Schelling, T. C. (1969) 'Models of segregation', *American-Economic Review*, 59 (2): 488–493.

Schelling, T. C. (1971a) 'Dynamic models of segregation', *Journal of Mathematical Sociology*, 1: 143–186.

Schelling, T. C. (1971b) 'On the ecology of micromotives', *The Public Interest*, 25: 61–98.

Schelling, T. C. (1972) 'The process of residential segregation: neighbourhood tipping' in A. H. Pascal (ed.), *Racial Discrimination in Economic Life,* Lexington: Lexington Books, pp. 157–185.

Schelling, T. C. (1978) *Micromotives and Macrobehavior,* London and New York: W. W. Norton.

Schelling, T. C. (1984a) *Choice and Consequence,* Cambridge: Harvard University Press.

Schelling, T. C. (1984b) 'What is game theory?', in *Choice and Consequence,* Cambridge: Harvard University Press, pp. 213–242.

Schelling, T. C. (1995) 'What do economists know?', *American-Economist,* 39 (1): 20–22.

Schelling, T. C. (1998) 'Social mechanisms and social dynamics', in P. Hedström and R. Swedberg (eds), *Social Mechanisms: An Analytical Approach to Social Theory,* Cambridge: Cambridge University Press, pp. 32–44.

Schlefer, J. (1998) 'Today's most mischievous misquotation', *The Atlantic Monthly,* 281 (3): 16–19.

Schmitz, S. W. (2002) 'Carl Menger's "Money" and current neoclassical models of money', in M. Latzer and S. W. Schmitz, *Carl Menger and the Evolution of Payment Systems: From Barter to Electronic Money,* Cheltenham and Northampton: Elgar, pp. 159–183.

Schotter, A. (1981) *The Economic Theory of Social Institutions.* New York: Cambridge University Press.

Schotter, A. (1996) ' "You're not making sense, you're just being logical" ' in S. G. Medema and W. J. Samuels (eds), *Foundations of Research in Economics: How Do Economists Do Economics?,* Aldershot: Edward Elgar Publishing, pp. 204–215.

Scriven, M. (1975) 'Causation as Explanation', *Nous,* 9: 3–16.

Searle, J. (1980) 'Minds, brains, and programs', *The Behavioural and Brain Sciences,* 3: 417–424.

Selgin, G. A. and Klein, P. G. (2000) 'Menger's theory of money: some experimental evidence', in J. Smithin (ed.), *What is Money,* London: Routledge, pp. 217–234.

Selgin, G. A. and White, L. H. (1987) 'The evolution of free banking system', *Economic Inquiry,* 25: 439–457.

Selgin, G. A. and White, L. H. (1994) 'How would the invisible hand handle money', *Journal of Economic Literature,* 32: 1718–1749.

Selten, R. (1975) 'Reexamination of the perfectness concept for equilibrium points in extensive-form games', *International Journal of Game Theory,* 4: 25–55.

Selten, R. (1980) 'A note on evolutionary stable strategies in asymmetric animal contests', *Journal of Theoretical Biology,* 84: 93–101.

Sethi, R. (1999) 'Evolutionary stability and media of exchange', *Journal of Economic Behavior and Organization,* 40: 233–254.

Sethi, R. and Somanathan, R. (2004) 'Inequality and segregation', *Journal of Political Economy,* 112: 1296–1321.

Shepard, J. M. (1995) 'The place of ethics in business: shifting paradigms?', *Business Ethics Quarterly,* 5 (3): 577–601.

Sidowski, J. B. (1957) 'Reward and punishment in a minimal social situation', *Journal of Experimental Psychology,* 54: 318–326.

Sidowski, J. B., Wyckoff, L. B. and Tabory, L. (1956) 'The influence of reinforcement and punishment in a minimal social situation', *Journal of Abnormal and Social Psychology,* 54: 318–326.

Sloep, P. and van der Steen, W. J. (1987) 'The nature of evolutionary theory: the semantic challenge', *Biology and Philosophy,* 2: 1–15.

Smith, A. (1762 [1985]) 'Considerations concerning the first formation of languages', in *Lectures On Rhetoric and Belles Lettres* (ed. J. C. Bryce), vol. IV of the Glasgow Edition of the Works and Correspondence of Adam Smith, Indianapolis: Liberty Fund, pp. 201–226. Available at http://oll.libertyfund.org//files/202/0141-05_Bk.pdf (accessed 13 June 2007).

Smith, A. (1789 [1904]) *An Inquiry into the Nature and Causes of the Wealth of Nations*, London: Methuen and Co. Ltd. (ed. Edwin Cannan). (First published in 1776, Cannan edition is based on the fifth edition which was published in 1789). Available at www.econlib.org/library/Smith/smWN.html (accessed 10 June 2002).

Smith, A. (1790) *The Theory of Moral Sentiments*, 6th edn, London: A. Millar. (First published in 1759. Printed for A. Millar in the Strand; and A. Kincaid and J. Bell in Edinburgh.) Available at www.econlib.org/library/Smith/smMS1.html (accessed 10 June 2002).

Smith, A. (1795 [1980]) 'The principles which lead and direct philosophical enquiries: illustrated by the history of astronomy', in *Essays on Philosophical Subjects* (ed. I. S. Ross), Oxford: Oxford University Press, pp. 1–105.

Smith, C. (2006a) *Adam Smith's Political Philosophy: The Invisible Hand and Spontaneous Order*, London: Routledge.

Smith, C. (2006b) 'Adam Smith on progress and knowledge', in L. Montes and E. Schliesser (eds), *New Voices on Adam Smith*, London: Routledge, pp. 293–312.

Smith, T. (1995a) 'Response to Jan Narveson', *Journal of Agricultural and Environmental Ethics*, 8 (2): 157–158.

Smith, T. (1995b) 'The case against free market environmentalism', *Journal of Agricultural and Environmental Ethics*, 8 (2): 126–144.

Smithin, J. (ed.) (2000) *What is Money?*, London: Routledge.

Solomon, M. (1994) 'Multivariate models of scientific change', *Proceedings of the Biennial Meetings of the Philosophy of Science Association*, 2: 287–297.

Somanathan, R. and Sethi, R. (2004) 'Inequality and segregation', discussion paper 04-03, Delhi Planning Unit, Indian Statistical Institute.

Sorensen, R. A. (1992a) *Thought Experiments*, Oxford: Oxford University Press.

Sorensen, R. A. (1992b) 'Thought experiments and the epistemology of laws', *Canadian Journal of Philosophy*, 22: 15–44.

Stahl, D. (1993) 'Evolution of smart players', *Games and Economic Behavior*, 5: 607–617.

Stark, W. (1962) *Fundamental Forms of Social Thought*, London: Routledge and Keegan Paul.

Starr, R. M. (1999) 'Why is there money? Convergence to monetary equilibrium in a general equilibrium model with transaction costs', discussion paper 99-23, University of California, San Diego.

Steele, D. R. (1987) 'Hayek's theory of cultural and group selection', *The Journal of Libertarian Studies*, 8 (2): 171–195.

Stenkula, M. (2003) 'Carl Menger and the network theory of money', *European Journal of the History of Economic Thought* 10 (4): 587–606.

Stewart, D. (1793 [1858]) 'Account of the life and writings of Adam Smith' in W. Hamilton (ed.), *The Collected Works of Dugald Stewart*, vol. 10, Edinburgh: Thomas Constable and Co.

Stiglitz, J. E. (1991) 'The invisible hand and modern welfare economics', *NBER Working Paper Series*, working paper no. 3641, National Bureau of Economic Research.

Stinchcombe, A. (1975) 'Merton's theory of social structure', in L. A. Coser (ed.), *The Idea of Social Structure: Papers in Honor of Robert K. Merton*, New York: Harcourt Brace Jovanovich, pp. 11–33.

Sugden, R. (1986) *The Evolution of Rights, Cooperation and Welfare*, New York: Basil Blackwell.

Sugden, R. (1991) 'Rational choice: a survey of contributions from economics and philosophy', *Economic Journal*, 101: 751–786.

Sugden, R. (1993) 'Spontaneous order', in U. Witt (ed.) *Evolutionary Economics*, Elgar Reference Collection series, International Library of Critical Writings in Economics, vol. 25. Aldershot, UK: Elgar, pp. 508–520.

Sugden, R. (1995) 'A theory of focal points', *Economic Journal*, 105: 533–550.

Sugden, R. (1998a) 'The role of inductive reasoning in the evolution of conventions', unpublished manuscript.

Sugden, R. (1998b) 'Normative expectations: the simultaneous evolution of institutions and norms' in A. Benner and L. Putterman (eds), *Economics, Values and Organisation*, Cambridge: Cambridge University Press, pp. 73–100.

Sugden, R. (2000) 'Credible worlds: the status of theoretical models in economics', *Journal of Economic Methodology*, 7 (1): 1–31.

Sugden, R. (2001) 'Ken Binmore's evolutionary social theory', *The Economic Journal*, 111: F213–43.

Sugden, R. (2002) 'Beyond sympathy and empathy: Adam Smith's concept of fellow feeling', *Economics and Philosophy*, 18: 63–87.

Suppe, F. (1977) *The Structure of Scientific Theories*, 2nd edn, Urbana, IL: University of Illinois Press.

Suppe, F. (1989) *The Semantic Conception of Theories and Scientific Realism*, Urbana, IL: University of Illinois Press.

Suppes, P. (1961) 'A comparison of the meaning and use of models in mathematics and the empirical sciences', in H. Freudenthal (ed.), *The Concept and the Role of the Model in Mathematics and Natural and Social Sciences*, Dordrecht: Reidel, pp. 163–177.

Suppes, P. (1967) 'What is a scientific theory?', in S. Morgenbesser (ed.), *Philosophy of Science Today*, New York: Basic Books, pp. 55–67.

Tajfel, H., Billig, M., Bundy, R. P. and Flament, C. (1971) 'Social categorization and intergroup behavior', *European Journal of Social Psychology*, 1: 149–177.

Taub, R. P., Taylor, D. G. and Dunham, J. D. (1984) *Paths of Neighborhood Change: Race and Crime in Urban America*, Chicago: Chicago University Press.

Taylor, P. and Jonker, L. (1978) 'Evolutionary stable strategies and game dynamics', *Mathematical Biosciences*, 16: 76–83.

Tesfatsion, L. (2000) 'Agent-based computational economics: a brief guide to the literature', unpublished manuscript. Available at www.econ.iastate.edu/tesfatsi (accessed 2 September 2004).

Tesfatsion, L. (2002) 'Agent-based computational economics: growing economies from the bottom up', *ISU Working Paper* no. 1. Available at www.econ.iastate.edu/tesfatsi/ (accessed 2 September 2004).

Thagard, P. (1992) *Conceptual Revelations*, Princeton: Princeton University Press.

Thalos, M. (2002) 'Explanation is a genus: an essay on the varieties of scientific explanation', *Synthese* 130 (3): 317–354.

Thompson, P. (1983) 'The structure of evolutionary theory: a semantic approach', *Studies in the History and Philosophy of Science*, 14: 215–229.

Thompson, P. (1989) *The Structure of Biological Theories*, Albany: SUNY Press.

Thomson, H. F. (1965) 'Adam Smith's philosophy of science', *Quarterly Journal of Economics*, 79 (2): 212–233.

Thornton, M. (2006) 'The mystery of Adam Smith's invisible hand resolved', *Mises Institute Working Paper,* 31 August. Available at www.mises.org/workingpapers.asp (accessed 10 June 2007).

Thurnwald, R. C. (1932) *Economics in Primitive Communities*, Oxford: Oxford University Press.

Tirole, J. (1996) 'A theory of collective reputations (with applications to the persistence of corruption and to firm quality)', *Review of Economic Studies*, 63 (1): 1–22.

Tobin, J. (1991) 'The invisible hand in modern economics', *Cowles Foundation Discussion Paper* no. 966, Cowles Foundation for Research in Economics at Yale University.

Torrens, P. M. and Benenson, I. (2005) 'Geographic automata systems', *International Journal of Geographical Information Science,* 19 (4): 385–412.

Townsend, R. M. (1980) 'Models of money with spatially separated agents', in J. H. Kareken and N. Wallace (eds), *Models of Monetary Economics*, Minneapolis: Federal Reserve Bank of Minneapolis, pp. 265–313.

Trejos, A. and Wright, R. (1995) 'Search, bargaining, money and prices', *Journal of Political Economy*, 103: 118–141.

Ullmann-Margalit, E. (1977) *The Emergence of Norms*, Oxford: Oxford University Press.

Ullmann-Margalit, E. (1978) 'Invisible hand explanations', *Synthese*, 39 (20): 263–291.

Ullmann-Margalit, E. (1997) 'The invisible hand and the cunning of reason', *Social Research*, 64 (2): 181–198.

Ullmann-Margalit, E. (1998) 'Invisible hand explanations', in P. Newman (ed.), *The New Palgrave Dictionary of Economics and the Law*, London: Macmillan, pp. 365–370.

Van Damme, E. (1987) *Stability and Perfection of Nash Equilibria*, Berlin: Springer.

Van Damme, E. (1998) 'On the state of the art in game theory: an interview with Robert Aumann', *Games and Economic Behavior*, 24: 181–210.

Vanberg, V. J. (1986) 'Spontaneous market order and social rules: a critical examination of a Hayek's theory of cultural evolution', *Economics and Philosophy*, 2: 75–100.

Vanberg, V. J. (1994) *Rules and Choice in Economics*, London: Routledge.

Vanderschraaf, P. (1998) 'The informal game theory in Hume's account of convention', *Economics and Philosophy*, 14: 215–247.

Varian, H. R. (1993) 'What use is economic theory?', unpublished manuscript.

Vaughn, K. (1987) 'Invisible hand', in J. Eatwell, M. Milgate and P. Newman (eds), *The New Palgrave: A Dictionary of Economics*, vol. 2, London and New York: Macmillan and Stockton, pp. 997–999.

Veblen, T. (1899) 'The preconceptions of economic science', *The Quarterly Journal of Economics*, 13 (4): 396–426.

Vromen, J. J. (1995) *Economic Evolution: An Enquiry into the Foundations of New Institutional Economics*, London: Routledge.

Vromen, J. J. (2000) 'What has evolutionary economics brought us so far?', unpublished manuscript, paper presented at the ESHET Conference, 'Is there a progress in economics?', Graz, 24–26 February.

Vromen, J. J. (2001) 'If conventions are solutions what are the problems?', unpublished manuscript.

Waldrop, M. M. (1992) *Complexity: The Emerging Science at the Edge of Order and Chaos*, New York: Simon and Schuster.

Wallace, N. (1998) 'A dictum for monetary theory', *Federal Reserve Bank of Minneapolis Quarterly Review* 22: 20–26.

Wärneryd, K. (1991) 'Evolutionary stability in unanimity games with cheap talk', *Economic Letters*, 36: 375–378.

Wärneryd, K. (1994) 'An economist's perspective on the evolution of norms: comment', *Journal of Institutional and Theoretical Economics*, 150 (1): 68–71.

Wartofsky, M. W. (1979) *Models: Representation and Scientific Understanding*, Dordrecht: D. Reidel.

Weibull, J. (1995) *Evolutionary Game Theory*, Cambridge: MIT Press.

Werhane, P. H. (1989) 'The role of self-interest in Adam Smith's "Wealth of Nations"', *Journal of Philosophy*, 86: 669–680.

Werhane, P. H. (1991) *Adam Smith and his Legacy for Modern Capitalism*, New York: Oxford University Press.

Wilkins, J.S. (1998) 'The evolutionary structure of scientific theories', *Biology and Philosophy*, 13: 479–504.

Williamson, S. and Wright, R. (1994) 'Barter and monetary exchange under private information', *American Economic Review*, 84 (1): 104–123.

Winsberg, E. (2001) 'Simulations, models, and theories: complex physical systems and their representations', *Philosophy of Science*, 68: S442–454.

Wray, K. B. (2000) 'Invisible hands and the success of science', *Philosophy of Science* 67(1): 163–175.

Wright, R. (1995) 'Search, evolution and money', *Journal of Economic Dynamics and Control*, 19: 181–206.

Ylikoski, P. (1995) 'The invisible hand and science', *Science Studies*, 8: 32–43.

Ylikoski, P. (2001) *Understanding Interests and Causal Explanation*, PhD thesis, Department of Moral and Social Philosophy, University of Helsinki.

Young, H. P. (1993a) 'The evolution of conventions', *Econometrica*, 61: 57–84.

Young, H. P. (1993b) 'An evolutionary model of bargaining', *Journal of Economic Theory*, 59: 145–168.

Young, H. P. (1996) 'Economics of convention', *Journal of Economic Perspectives*, 10 (2): 105–122.

Young, H. P. (1998) *Individual Strategy and Social Structure: An Evolutionary Theory of Institutions*, Princeton: Princeton University Press.

Young, P. H. (2001) 'Dynamics of conformity', in S. N. Durlauf and H. P. Young (eds), *Social Dynamics*, Cambridge: The MIT Press, pp. 133–154.

Zappia, C. (1995) 'Private information, contractual arrangements, and Hayek's knowledge problem', unpublished manuscript, revised version of the paper presented at the Annual Meetings of the History of Economics Society, 10–13 June 1994, Babson College, Babson Park, MA, USA.

Zhang, J. (2000) 'A dynamic model of residential segregation', *Journal of Mathematical Sociology*, 28: 147–170.

Zhang, J. (2004a) 'Residential segregation in an all-integrationist world', *Journal of Economic Behavior and Organization*, 54: 533–550.

Zhang, J. (2004b) 'Revisiting residential segregation by income: a Monte Carlo test', *International Journal of Business and Economics*, 2(1): 27–37.

Zuidema, J. R. (2001) 'Carl Menger, author of a research programme', *Journal of Economic Issues*, 15 (3/4): 13–35.

Index

Aaronson, D. 178
Aberg, Y. 178
about this book: argument 7; definition
  of 'unintended consequences' 3;
  focus 3, 4, 6; game theory 7, 10; plan
  and summary 7–10; significance
  of unintended consequences 3;
  spontaneous order, theory of 3;
  unintended social consequences,
  invisible-hand explanations of
  processes leading to 4–6
abstractions 45, 82–3, 119–20, 127–8,
  133, 135, 166, 197n22, 197n25;
  q-morphisms and 127–8
Achinstein, P. 207n13, 208n21
aggregate mechanisms (processes) 44,
  55–7, 63, 66, 89, 90, 91, 94, 138, 144,
  198n7
Aiyagari, S.R. and Wallace, N. 98, 174–5,
  204n11
Akerlof, G.A. 204n13
Alchain, A.A. 203n4, 204n13
Alexander, R.D. 63
Alvarez, A. 205n21
analogical thinking 128–30; creative
  thinking and 208n22; Smith's views on
  200n11
Anderson, P.W., Arrow, K.J. and Pines, D.
  167
Apfelbaum, E. 215n38
Aristotle 195n5
Arrow, K.J. 207n13
Arrow, K.J. and Debreu, G. 82
Arthur, W.B. 214n33
Arthur, W.B. et al. 167
astrology 207n9
astronomy, history of: invisible hand in
  72–7; Smith's perspective on 69, 70–1,
  80

Aumann, R.J. and Brandenburger, A. 151,
  212n11
Aumann, Robert J. 7, 151, 211n7,
  212–13n12, 212n11
Axelrod, R. 4, 215n1
Axelrod, R. and Dion, D. 215n1
Axelrod, R. and Hamilton, W. 215n1
Axtell, R.L. et al. 4
Axtell, Robert L. 4
Aydinonat, N. Emrah 198n9, 208n25,
  215n39
Azariadis, C. and Drazen, A. 211n6

Bacharach, M. 153, 213n17
Bacharach, M. and Bernasconi, M. 153,
  186–8, 213n16, 213n17
Bacharach, M. and Stahl, D. 213n17
Baird, C.W. 203n4
Barchas, P. 63
Barry, N. 83, 84, 201n25, 201–2n26,
  202n30, 202n33, 202n35
Bates, F.L. 199n1
Batten, D.F. 198n13
Baumeister, R.F. and Leary, M.R. 63
Bayer, P., McMillan, R. and Rueben, K.
  178
Benenson, I. 179, 180
Benenson, I. and Omer, I. 179
Benito, J.M. and Hernandez, P. 179
Bernheim, D. 151, 152, 154, 155, 212n9,
  213n13, 214n24
Berry, C.J. 86, 200n10, 202n39
Bicchieri, C. 4, 207n13, 208n21, 211n4,
  212n9, 214n24, 215n1
Bierman, H.S. and Fernandez, L. 107
Binmore, K.G. 215n1
Binmore, K.G. and Samuelson, L. 189,
  213n17

Binmore, K.G. *et al.* 163
Biographical Memoir of Adam Smith
    (Stewart, D.) 86
Black, M. 208n21
Blaug, M. 82, 83, 84, 209n35
blind action 71, 77–8, 200n15, 200n17
Blume, A., Kim, Y.G. and Sobel, J. 189
Blume, L.E. and Durlauf, S.M. 8, 178
Bobo, L. and Zubrinsky, C.L. 140, 180,
    208n24
Bohannan, P. and Dalton, G. 37–8
Borges, Jorge Luis 127, 208n18
Boudon, Raymond 2, 15–16
Boumans, M. 141, 143
Boumans, M. and Morgan, M. 141,
    207n13
bounded rationality and learning 156–8
Bowles, S. and Sethi, R. 63
Boyd, R. and Lorberbaum, J. 215n1
Braithwaite, R.B. 207n13
Brandenburger, A. 151, 212n11
Brennan, G. and Pettit, P. 21, 202n30,
    203n44
Brewer, A. 199n3
Bridel, P. and Salvat, C. 199n6
Bromberger, S. 206–7n7
Brown, G.W. 214n29
Brown, P.M. 103, 205n20
Brown, V. 199n4
Bruch, E.E. and Mare, R.D. 180
Brunner, K. and Meltzer, A.H. 203n4

Caldwell, B. 202n34
Cameron, L. 184
Campagnolo, G. 195–6n10
Campbell, N.R. 207n13
Carey, T.V. 203n43, 203n44
Carlsson, H. and van Damme, E. 152
Cartwright, N. 134, 136–7, 207n13,
    209n35
Casajus, A. 213n17
Casti, John L. 8, 50, 168
Cheng, W.L. 203n4
chequerboard model: causal mechanisms,
    structural relationships and 59;
    coherence with existing knowledge 65;
    collective segregation, examination of
    61–2; complexity theory 61; conditions
    for near complete segregation 65;
    conjecture as starting point 60–1;
    constituent mechanisms 55–7, 57–61,
    61–4; credibility 63–4; critical-mass
    model 58–9; expectations, preferences
    and 62; experimenting with different
    mechanisms 60–1; explanandum
    phenomenon 59; explanatory value
    57–66; explorations in the world of
    the 178–81; group organization 61–2;
    individual and aggregate properties,
    distinction between 59–60; individual
    mechanisms in isolation 63–4; initial
    state 57–8; inputs 55–7; interactions
    among agents, complexity of 61;
    isolation, mechanisms in 61–4; meta-
    model, explanatory breadth of 64–6,
    66–7; neighbourhood dynamics 62;
    non-linear nature 66; outputs 55–7;
    partial (theoretical) explanation
    for segregation 50–1; postulating
    mechanisms 57–61; preferences,
    expectations and 62; real world
    and model world, recognizable
    similarities between 63; residential
    segregation and the 50–67; separate
    effects of active causes, distinction
    between 59–60; successive stages in
    phenomena generation, account of 64;
    thought experimentation 60–1, 63–4;
    tipping model 58–9; unorganized
    discriminatory behaviour 62
Chipman, J.S. 81
Cho, I.-K. and Kreps, D.M. 212n11
Churchland, P.M. and Hooker, C.W.
    207n13
Clark, C.M.A. 36, 38, 203n43
Clark, W.A.V. 9, 141, 178, 180
Coase, Ronald H. 82, 199n1, 201n24
collective behaviour 19–20, 24–5
collective segregation 61–2
Collin, F. 23, 194–5n10
Colman, A.M. 107, 211n2, 215n38
Colman, A.M. and Bacharach, M. 151
commodity money: commodities as money
    31, 34, 35, 37–8; equilibrium of 95–8,
    96, 97–8; fundamental equilibrium of
    97, 104–5; origin of 93; speculative
    equilibrium of 97, 104–5
common knowledge 148, 150–1, 153–4,
    155, 156, 158, 159, 192, 211n7,
    212–13n13
complexity 41, 42, 197n22; assumption
    in invisible-hand explanations
    89; complexity theory 61, 167; of
    exchanges 171; of interactions among
    agents 61; in relationship between
    model and real worlds 119–20, 142–3,
    144–5; of representation 125; of social
    world 87–8, 128

computer simulations 6, 100, 111, 146, 167
concrete phenomena 39–41, 132, 196n20
conjecture 2, 4, 28, 47–8, 49, 50, 64, 69,
   80, 82, 91–2; conceptual construction
   86–7, 89–91, 167, 178–9, 203n41;
   conjectural process 6, 56–7, 60–1;
   conjectural scenarios 141, 162–3,
   180; evaluation of plausibility of 118,
   119, 122, 125, 129–30, 132–4, 138,
   143; reality and imagination 74–6; as
   starting point in chequerboard model
   60–1
consequences: attribution of, problem of
   16–17; beneficial consequences 21–2;
   collective or aggregate consequences
   of individual actions 22, 23–5;
   concept of 12–13, 13–16; concrete
   consequences 15; desirability of 18,
   21–2; disadvantageous consequences
   21–2; of individual actions 15–16,
   17–20, 20–1; microlevel consequences
   22; peculiarities of invisible-hand
   consequences 22–5; of social action
   15–16, 17–20; sum-total consequences
   15; unanticipated and unintended
   consequences, distinction between 25;
   see also invisible-hand consequences;
   unintended social consequences
*Considerations concerning the first
   formation of languages* (Smith, A.) 86
constituent mechanisms 55–7, 57–61, 61–4
context: importance of 37–8, 184–5,
   211n5; of Smith's use of term 'invisible
   hand' 69–70, 71
coordination: conditions for 154;
   conventions for 147–50, 163–4;
   framework for understanding 161–2;
   principle of 153
Cosmides, L. and Tooby, J. 203n43
Cowen, Tyler 50, 66
Crawford, V.P. 151, 212n9, 215n38
Crawford, V.P. and Haller, H. 156, 193,
   213n17
credibility 7, 63–4, 172
critical-mass models 58–9
Curren, R. 203n44

Darwin, Charles (and Darwinian notions)
   177, 203n43
David, P.A. 214n33
Davis, J.B. 73, 200n13
Davis, J.R. 199n7
Dawid, H. 177, 203n4, 205n19
Dawkins, Richard 189

Demeny, P. 202n30
Denis, A. 70, 73
Denton, N.A. and Massey, D.S. 178
Descartes, René 200n12
design view of money 36, 37, 43, 48,
   196n12
direct exchange 32, 33, 35, 36
discriminatory preferences 4, 5, 6, 50,
   51–3, 57, 58, 62–3, 65–6, 130–3, 137,
   138–41, 178–81, 198n5, 198n12,
   208n24, 210n43; isomorphism and
   126; mild discriminatory preferences
   50, 51, 52–5, 57, 58, 63, 66, 132, 138,
   139, 178, 180, 198n5, 210n43; ranges
   of 62–3; representation and 130–2;
   residential segregation and 51–3, 57,
   58, 62–3, 65–6; strong discriminatory
   preferences 5, 6, 50, 51–2, 62–3, 65,
   130, 131, 133, 140, 180, 208n24
Donniger, C. 215n1
Dowd, K. 203n4, 204n10
Dowe, P. 207n8
Downs, A. 178
driving game 148–9, 154
Drogoul, A. and Ferber, J. 6
Duffy, J. and Ochs, J. 9, 103–4, 115, 142
Duhem, P. 207n13
Durlauf, N.D. 82, 199n1

Easterley, W. 181
Economic Sentiments (Rothschild, E.) 68
Edmonds, B. and Hales, D. 179
Ellen, I.G. 179
Ellis, B. 207n11
Ellison, G 157–8, 214n33
Elster, Jon 66
Eltis, W. 199n6
empirical forms, concrete phenomena and
   39–41
empirical research, necessity of 161
end-state: game theory and models 154,
   158; interpretation of invisible hand 3,
   69, 81–4; models of money, origin of
   94, 100, 117
Engineer, M. and Shi, S. 203n4
Epstein, J. and Axtell, R. 4, 5, 8, 50, 55,
   140, 178
Epstein, Joshua 4, 204n17
equilibrium points in game theory 149,
   150–1
Erasmus University Rotterdam Campus
   25–6
error 14, 15; error rates 110, 111, 205n32
ethnic segregation *see* residential
   segregation

Evensky, J. 199n1, 202n28, 202n30

exact orientation, theoretical knowledge and limitations of 41–3, 44–6

exact understanding 39–44, 48–9

exchange media: coordination models of emergence of 106–11; dependency on 32, 34; direct exchange 32, 33, 35, 36; emergence of 100–11; existence of 95–100; experiments in emergence of 103–5; hypothetical explanations of emergence of money as 93–5, 116–17; imitation models of emergence of 105–6; introduction of money through external actions 36; mechanisms in development of 48; money, exchange and the nature of trade 195–6n10; perspectives on emergence of 196n12; reciprocity in exchange 37–8; Schotter's model of emergence of 106–9, 114–15; simulations of emergence of 100–3; theoretical models of emergence of 105–11; universal acceptance of money as 31, 34–5, 35–9; Young's model of emergence of 110–11, 114–15

expectations 96, 155, 158, 179, 186, 192, 204n5, 212n9; correlated expectations 154, 156, 158, 212n11; individual expectations 104, 107–8, 110, 155, 192; knowledge and 85; motives and 59; mutual expectations, concordant 149, 150, 151, 158, 160, 193, 215n37; preferences and 62; strategies and 100

explanandum phenomena: chequerboard model 59; invisible-hand explanations 89, 90; models and representation 120, 121–2, 123, 143, 144, 206–7n7; origin of money 31, 49, 115–16; residential segregation 59

explanatory value: chequerboard model 57–66; Menger's story of origin of money 44–8, 49; models and explanations 123–5, 144–5; value of explanation 196n18

exploration with models 119–20, 139–43, 144–5

Fagiolo, G., Valente, M. and Vriend, N.J. 179

familiar elements 133, 138, 208–9n27

Farley, R. 140, 180

Farley, R., Fielding, E.L. and Krysian, M. 178, 208n24

feedback, effects of 34, 56, 211n6

Fehr, E. and Tougareva, E. 184

Feiner, S.F. and Roberts, B.B. 199n1

Ferguson, Adam 84, 86, 202–3n40

fiat money 9, 93, 95, 101–2, 109, 112, 117, 176, 177, 203n2, 204n11; fiat money equilibrium 98–9, 174–5

fictitious play 157, 158

Fielding, E.L. 178

Fiori, S. 199n8, 202n35

Flache, A. and Hegselmann, R. 179

Fleischacker, S. 200n10

Flew, A. 70

focal points: Bacharach and Bernasconi on 186–8; formal models of 153; in game theory 149, 152–8, 182–93

Forsythe, R. *et al.* 184

Fosset, M. and Waren, W. 179

Foster, D. and Young, H.P. 157, 193, 214n31, 214n32

van Fraassen, B.C. 134–5, 136, 200n10, 207n13, 209n30, 209n32

framing in game theory 153

Freudenthal, H. 207n13

Friedman, D. 189, 206–7n7

Friedrichs, J. 178

Frye, T. and Shleifer, A. 199n1

Fudenberg, D. and Harris, C. 214n31

Fudenberg, D. and Levine, D.K. 158, 190–1, 211n4, 214n26, 214n29

functions of money 27–8, 29–30

Funes the Memorious (Borges, J.L.) 127

Gallie, W.B. 64

Gallo, P.S. and McClintock, C.G. 215n38

Gambeta, Diego 66

game theory 146–64, 215n42; alternative conventions, equilibrium points and 150–1; bounded rationality and learning 156–8; coordination, conditions for 154; coordination, conventions for 147–50, 163–4; coordination, framework for understanding 161–2; coordination, principle of 153; driving game 148–9, 154; empirical research, necessity of 161; end-state models 154, 158; equilibrium selection 149, 150–1; fictitious play 157, 158; focal points 149, 152–8, 182–93; formal models of focal points 153; framing 153; game theory interpretation 162–3; games as informal experiments 183; insufficient reason, principle of 152–3; interpretation 159–63; labelling 153; mixed strategies, use of 211–12n8;

models, coordination in world of 150–2, 163–4; Nash equilibria and 147, 149, 151–2, 153, 155, 157, 159, 211n3, 212n9, 212n10, 213n13, 214n24; prediction, actual behaviour and 149–50, 183–5; process models 158–9, 165–6; rationality and learning 154–6, 192–3; refinements literature 151; select-a-ball game 153–4; unintended consequences, conventions as 160
Gauthier, D. 213n17, 213n19
Gibbard, A. and Varian, H.R. 207n13
Giere, R.N. 134, 135, 136, 207n13, 209n30
gift exchange 36, 38
Gilbert, N. and Doran, J. 6
Gintis, Herbert 9, 103, 107, 111, 205n30; models of emergence of money 177
Good, D.A. 215n38
Goyal, S. and Janssen, M.C.W. 155–6, 192–3, 213n17, 214n34
Grampp, W.D. 199n4
Greif, A. 150, 215n39
Grodzins, Morton 58–9
group organization 61–2
Gürke, B. 178
Güth, W. *et al.* 149, 184

Häggqvist, S. 210n44
Hahn, F.H. 82, 83, 84, 199n1
Hamowy, R. 3, 84, 202n30
Hands, D.W. 199n1
Harre, R. 207n13
Harsanyi, J. and Selten, R. 151, 152, 185, 213n19
Hausman, D.M. 64, 205n34, 207n13, 209n35
Hayek, Friedrich A. von 69, 84–6, 87, 202n32, 202n36
Heath, E. 202n30
Hedström, P. and Swedberg, R. 8, 66
Hempel, C.G. 121–2, 206n4, 209n28
Hempel, C.G. and Oppenheim, P. 122, 206n3, 207n10
Henrich, J. *et al.* 149, 184
Henry, J.F. 36–7
Hernes, Gudmund 66
Hesse, M. 207n13, 208n21
history: historical (singular) explanations of origin of money 45–6; knowledge of social phenomena 4–5, 164, 196n20; theoretical explanations and, relationship between 196n20; theoretical 'knowledge' and historical understanding 38, 40, 43, 49, 64

History of Economic Ideas (Aydinonat) 215n39
Hobbes, Thomas 86
Hodgson, G.M. 197n29, 204n14
Hofbauer, J. and Sigmund, K. 189
Hofbauer, J., Schuster, P. and Sigmund, K. 189
Hoffmann, E. *et al.* 184
Holcombe, R.G. 83, 202n28
Holland, J.H. *et al.* 126, 127, 128, 145, 196n11, 207–8n15, 208n19
Holland, John H. 56, 61, 100, 167, 204n15, 215n3
Holmes, R.L. 199n1
Holyoak, K.J. and Thagard, P. 127, 129, 208n18, 208n22
homomorphism 128
homophily 63, 198n12
Horgan, J. 207n13, 208n21
Horwitz, S. 195n8
Hughes, R.I.G. 207n13
Hull, D.L. 2, 199n1
Hume, David 4, 146
Hutchison, T.W. 197n32, 209n35
Hutter, M. 195n3
Huttman, E.D. *et al.* 178

Iceland, J. 178
ignorance 14, 15, 73
Ihlanfeldt, K.R. and Scafidi, B. 178
immediate neighbours 52–5
individual actions: aggregate and individual properties, distinction between 59–60; collective action and 80; consequences of 15–16, 17–20, 20–1; economizing actions of individuals 33, 34–5, 35, 39; independently pursued by many individuals 25; intended consequences of 208n25; intentions of 17–20, 21; interaction causing unintended consequences 76–7; organized actions 15, 16, 65–6, 67, 131–3, 144; separate effects of active causes, distinction between 59–60; social and individual level interests, distinction between 78–9
Ingham, G. 36
Ingrao, B. 199n7
initial state: of chequerboard model 57–8; within which money is developed 36–7, 39
Innes, A.M. 37
inputs to chequerboard model 55–7

insufficient reason, principle of 152–3
intentions: concept of 12; mixed
    intentions, possibility of 18; of social
    action 17–19; target of 16, 17
interests: coincidence of 111; economic
    interests 32, 34, 35, 39, 41–2, 201n21;
    imperious immediacy of 14–15, 194n7;
    individual pursuit of personal 21, 27,
    75–7, 78, 81, 82, 84, 85, 88, 93, 117,
    200n13; of merchants 79; of societies
    71, 78; *see also* self-interest
interpretation in game theory 159–63
invisible backhand 21
invisible hand 68–92, 81; astronomy,
    history of 72–7; blind action and
    71, 77–8; conjecture, reality and
    imagination 74–6; contexts of Smith's
    use of term 69–70, 71; domestic
    security, public interest and 70, 71;
    end-state interpretation 3, 69, 81–4;
    explanatory strategy for philosophers
    73–4; importance of concept 81, 91–2;
    individual actions, collective action
    and 80; individual actions, interaction
    causing unintended consequences
    76–7; individual and social level
    interests, distinction between 78–9;
    invisible-hand explanations 88–91;
    irregular events and power of gods
    69, 71, 72–3, 80; of Jupiter 69,
    72–4, 76–7, 80, 201n20; machines,
    philosophical systems and 76; modern
    conceptions 81–8; nature, connecting
    principles of 73–4, 89; privileged
    universal knowledge, foundation in
    notion of 71; process interpretation
    3, 4–7, 69, 84–8; providential order,
    implication of 71–2, 80–1; relation
    between Smith's invisible hands 70–1;
    religious connotations 69, 71, 72–3,
    80; Rothschild's perception of irony
    in Smith's 70, 80; self-interest 71,
    76, 78, 79, 84, 88; selfish behaviour
    and species multiplication 69–70,
    71; Smith's concept of the invisible
    hand 69–72, 77–81, 82, 83–4, 91;
    specialisation and knowledge 79–80;
    superstition 72, 73, 76; unanticipated
    and unintended consequences 80;
    unintended consequences and 68,
    77–81, 91
invisible-hand consequences 20–5,
    25–6; beneficial consequences 21–2;
    characteristics of 11–12, 23–5;

collective or aggregate consequences
    of individual actions 22, 23–5;
    disadvantageous consequences 21–2;
    individual action, independently
    pursued by many individuals 25;
    invisible backhand 21; location
    of 20–1; macro-social structures
    22; microlevel consequences 22;
    money, social phenomenon of 22–3;
    peculiarities of 22–5; properties of
    23–5; residential segregation 21–2, 23;
    social phenomena 22–3; unanticipated
    and unintended consequences,
    distinction between 25; *see also*
    unintended social consequences
invisible-hand explanations 88–91;
    complexity, assumption of 89;
    explanandum phenomena 89, 90;
    literature on 6; mental horizons and
    90–1; process models of 3, 4–7, 69,
    84–8; properties of 90; relationship
    with unintended social consequences
    11–26; *see also* money, origin of;
    residential segregation
irony 70, 71, 72, 74, 77, 80, 199n4,
    201n20, 201n24
irregular events and power of gods 69, 71,
    72–3, 80
isolation 128, 138–9, 166, 198n11;
    mechanisms in 61–4; *see also* models
    and representation
isomorphisms 126–7, 134–5, 208n17,
    208n22
Iwai, K. 36, 196n12, 206n40
Iway, K. 203n4

Jacobsen, H.J. 212n10
Janssen, M.C.W. 151, 153, 187, 202n27,
    212n9, 212n11, 213n16, 213n17,
    213n19, 215n40
Jevons, William S. 37, 195n7; on money
    170–1
Jupiter, invisible hand of 69, 72–4, 76–7,
    80, 201n20

Kalai, E. and Lehrer, E. 156, 193
Kandori, M. and Rob, R. 214n31
Kandori, M., Mailath, G.J. and Rob, R.
    151, 157, 191, 213n16, 214n31, 214n33
Kaniovski, Y. and Young, P.H. 214n31
Karlson, Nils 50, 203n44
Kauder, E. 197n32
Kauffman, S. 198n13, 215n3
Keller, R. 2, 12, 50, 202n39, 203n44

Khalil, E.L. 70, 202n28
Kincaid, H. 210n39
Kitcher, P. 199n1, 206–7n7
Kiyotaki, N. and Wright, R. 9, 95–8, 98–9, 99, 100, 101, 102, 103, 104, 105, 111, 112–13, 135, 203n4, 204n5; models of emergence of money 173–4, 174–5, 176, 177
Klein, D.B. 203n42
Knapp, G.F. 196n14
knowledge: chequerboard model, coherence with existing knowledge 65; exact orientation, theoretical knowledge and limitations of 41–3, 44–6; expectations and 85; historical understanding, theoretical 'knowledge' and 40; lack of 14, 15; privileged universal knowledge, notion of 71; of social phenomena 4–5, 164, 196n20; specialisation and 79–80; *see also* common knowledge
Knudsen, C. 83, 202n28
Koppl, R. 203n44
Krishna, V. and Sjostrom, T. 214n29
Kroes, P. 208n21
Krugman, P.R. 207n13, 211n6
Kühne, U. 203n44
Kuniński, M. 195n8, 202n30

labelling 153
Land, S.K. 86, 202n39
Langlois, R. 8
Lapavitsas, C. 204n9
Lauri, A.J. and Jaggi, N.K. 179
Leatherdale, W.H. 208n21
Leibenstein, H. 199n1
Leonard, T.C. 199n1
Lewis, David 2, 4, 146–7, 148–9, 150, 152, 156, 160, 182–3, 211n3, 211n7
Liebrand, W.B.G. 6
Liebrand, W.B.G. *et al.* 6
Lind, D. 199n1
Liu, C. 209n37
Lloyd, E. 207n13, 209n30, 209n31
local interaction 191–2, 214n33, 214n34
Luce, R.D. and Raiffa, H. 212n9
Luo, G.Y. 105, 111

McCarthy, T. 206–7n7
Macfie, A. 70
McMahon, C. 199n1
McMullin, E. 207n13
McPherson, M., Smith-Lovin, L. and Cook, J. 63

macro-social structures 2, 3, 4, 7, 8, 9, 11, 20, 22, 60, 166, 168
Mailath, G.J. 189, 212n9, 213–14n23
Mäki, U. 3, 8, 11, 91, 124, 126, 128, 132–3, 141, 197n32, 198n11, 202n31, 203n44, 204n7, 206–7n7, 207n11, 207n13, 208n26, 209n34, 209n36, 210n40
Marimon, M. and Miller, J.H. 176
Marimon, R. 156, 214n26
Marimon, R., McGrattan, E. *et al.* 4, 9, 100–3, 104, 111, 112–13, 204–5n18, 214n27; models of emergence of money 175–6, 177
market dependence 32, 33, 39
market traffic 32–3, 34, 39
marketability, saleableness and 204n5
Marris, R. and Mueller, D.C. 82, 199n1
Maskin, E.S. 82, 199n1
Massey, D.S. and Denton, N.A. 178
materialisation, spaces of 16, 17, 18
materiality 141; quasi-material 210n45
Maynard Smith, J. 189
Maynard Smith, J. and Price, G.R. 189
mechanisms: causal mechanisms, structural relationships and 59; constituent mechanisms 55–7, 57–61, 61–4; experimenting with range of different 60–1; general mechanisms in emergence of money 38–9, 40; individual mechanisms, interaction between 46–8; individual mechanisms in isolation 63–4; in isolation 61–4; machines, philosophical systems and 76; postulation of mechanisms 57–61; for residential segregation 51–2, 55–7, 57–61, 61–4; of segregation, Schelling's postulation of 57–61, 66–7
mediators, models as 143–4
Meen, D. and Meen, G. 178
Mehta, J. *et al.* 149, 213n17, 213n22
Menger, Carl 3, 4–5, 7–9, 20, 21, 50, 51, 58, 61, 67, 68, 84, 87, 90–2, 96–7, 99, 105, 109, 111, 113–15, 116, 120, 144, 161, 195n1; emergence of generally acceptable medium of exchange 31–5, 49; exact understanding 39–44, 48–9; explanation of origin of money 27, 28–35, 48–9, 93–5; explanation of origin of money, constraints on 46; explanation of origin of money, explanatory value of 44–8, 49; explanation of origin of money, partial success of 47–8, 49; Mengerian

question 29–30, 31; money, exchange
and the nature of trade 195–6n10;
'money', meaning of 30–1; partial
nature of explanation 121–2, 124;
realism, abstraction and 197n22,
197n25; reconstruction of Menger's
explanation 35–9; theory and history,
relationship between 196n20; value of
explanation 196n18
mental horizons and invisible-hand
explanations 90–1
Merton, Robert K. 2, 13–16, 16–17,
194n2, 194n3
meta-models: chequerboard model,
explanatory breadth of 64–6, 66–7;
of emergence of money 116, 117,
206n38, 206n39; of residential
segregation 64–6, 66–7, 144, 181; of
segregation, Schelling's chequerboard
as contribution to 64–6
Michihiro, K. 189
microlevel consequences 22
Micromotives and Macrobehavior
(Schelling, T.) 50
mild discriminatory preferences 50, 51,
52–5, 57, 58, 63, 66, 132, 138, 139,
178, 180, 198n5, 210n43
Milgrom, P. and Roberts, J. 214n26
Mill, John Stuart 59–60
minimal social condition 211n2
Mirowski, P. 199n1
Mises, Ludwig von 172
Models and Mediators (Morgan, M. and
Morrison, M.) 143
models and representation: abstractions
and q-morphisms 127–8; analogical
thinking 128–30; complexity in
relationship between model and
real worlds 119–20, 142–3, 144–5;
construction of models 119–20, 130–4,
142–3; coordination in world of game
theory 150–2, 163–4; example of
representation 130–4; explanandum
phenomena 120, 121–2, 123, 143,
144, 206–7n7; explanations and
models 123–5, 144–5; exploration
and models 119–20, 139–43, 144–5;
isolation 128, 138–9; isomorphisms
126–7; mediators, models as 143–4;
'nomological machines', models as
136–8; partial potential explanations
120–2; representation and models
126–34; similarity, notion of 38–9,
136–7; tendencies, information from

137–9; theories and models 134–6
modern conceptions of invisible hand 81–8
modern money 30–1, 36, 38
money, origin of: characterizations and
explanations of 27–8; commodities as
money 31, 34, 35, 37–8; commodity
money 93; commodity money
equilibrium 95–8; concrete phenomena
39–41; context, importance of 37–8;
dependency on exchange 32, 34;
design view of 36; direct exchange 32,
33, 35, 36; empirical forms, concrete
phenomena and 39–41; end-state
models 94, 100, 117; exact orientation,
theoretical knowledge and limitations
of 41–3, 44–6; exact understanding
39–44; explanandum phenomenon
31, 49, 115–16; explanatory progress,
explorations in model worlds 111–16,
117; explanatory value of Menger's
story 44–8, 49; feedback, effect of 34;
fiat money 9, 93, 95, 98–9, 101–2,
109, 112, 117, 176, 177, 203n2,
204n11; fiat money equilibrium 98–9,
174–5; functions and explanations
of 27–8, 29–30; fundamental
equilibrium of commodity money 97,
104–5; general mechanisms 38–9,
40; gift exchange 36, 38; historical
understanding, theoretical 'knowledge'
and 40; individual mechanisms,
interaction between 46–8; individuals,
economising actions of 33, 34–5, 35,
39; initial state within which money
is developed 36–7, 39; laws of nature,
exact laws and 40–1, 42; market
dependence 32, 33, 39; market traffic
32–3, 34, 39; meaning of term 'money'
30–1; Mengerian question 29–30, 31;
Menger's account, constraints on 46;
Menger's account, partial success of
47–8, 49; Menger's explanation 4–5,
27, 28–35, 48–9; models of emergence
of money, explorations in 111–16, 117;
modern money 30–1, 36, 38; objections
to Menger's account 36–9; obligatory
payments 37; organic analogy,
limitations of 29; organic explanation
28–9; organic phenomena 28–9;
organic phenomenon, money as 43–4;
potential explanatory power, model
worlds and 116, 117–18; primitive
monies 30–1, 37–8; process models
27–8, 33–4, 44–5, 47–8, 49, 95, 96, 98,

100, 102–3, 104–5, 105–6, 108–10, 115, 117–18; rationality, tendency towards 33–4; real and model worlds, relationship between 45–6; realism and realist-empirical orientation 41, 42–4, 46–8; reciprocity in exchange 37–8; reconstruction of Menger's story 35–9; saleableness 31–2, 34–5, 35, 39; singular (historical) explanations 45–6; social phenomena 28; social phenomenon, money as 22–3; speculative equilibrium of commodity money 97, 104–5; traditional description, inaccuracy of 37–8; unit of account, development as 36–7; *see also* exchange media
Money and the Mechanism of Exchange (Jevons, W.S.) 170–1
Montes, L. 199–200n9, 200n10, 202n29
moral philosophy 75–6
Morgan, M. 103, 141–2, 207n13
Morgan, M. and Morrison, M. 143, 207n13, 208n21
Morrison, M. and Morgan, M. 210–11n46
multiple individuals, consequences of actions of 20
Musgrave, A. 209n36

Nadeau, R. 202n30
Nash, John F. (and Nash equilibria) 107, 110; evolutionary stability and 189; game theory and 147, 149, 151–2, 153, 155, 157, 159, 211n3, 212n9, 212n10, 213n13, 214n24; Pareto dominance and 86, 185; replicator dynamics and 190; *see also* Pareto
nature: connecting principles of 73–4, 89; laws of, exact laws and 40–1, 42
neglected causal factors 13–14
neighbourhood dynamics 62; immediate neighbours 52–5
Nemeth, C. 215n38
Newton, Sir Isaac 74–6, 199–200n9, 202n29; Smith and Newton's system of astronomy 74–5
Niiniluoto, Ilkka 209n33
North, D. 150
Norton, R. 194n8
Nowak, M. and Sigmund, K. 215n1
Nozick, R. 2, 4, 8, 50, 88, 199n1, 202n37, 203n44

obligatory payments 37, 44
organic analogy, limitations of 29

organic explanation 28–9
organic phenomenon, money as 43–4
organized actions 15, 16, 65–6, 67, 131–3, 144
On the Origins of Money (Menger, C.) 31–4
Ostroy, J.M. and Starr, R.M. 203n4
Otteson, J. 202n39, 203n42
outputs of chequerboard model 55–7

Pancs, R. and Vriend, N.J. 8, 140, 178, 198n5
Pareto, Vilfredo 83, 152; Pareto-dominance 152, 157, 185, 190, 192, 213n14, 213n16, 213n19, 214n30; Pareto optima 81, 82, 83, 153, 157, 186, 192, 213n19; *see also* Nash, John F. (and Nash equilibria)
partial potential explanation 9, 10, 165; chequerboard model of segregation 50–1; game theory 147, 160, 164; invisible hand 68, 90, 91–2; Menger's explanation of origin of money 121–2, 124; models and representation 119, 120–2, 123, 125, 138, 143, 144; money, origin of 28, 39, 44, 49, 50, 64, 67, 68, 90, 91–2, 93, 115; potential explanatory power, model worlds and 116, 117–18; residential segregation 50, 64, 67; Schelling's explanation of segregation 121–2, 124
Pearce, D. 214n24
Peart, S.J. and Levy, D. 79
Persky, J. 70
Pettit, P. 8, 203n44
phenomena: concrete phenomena 39–41, 132, 196n20; economic phenomena 32, 40, 150; natural phenomena 73, 200n13, 201n20; organic phenomena 3, 20, 28, 29; pragmatic phenomena 28, 29; social phenomena 5, 22–3, 28–30, 43, 68, 119–20, 126, 133–9, 142–6, 164–6, 171, 199n8, 203n42; successive stages in phenomena generation 64; *see also* explanandum phenomena
philosophy of science 73–7, 165
physical dependence 207n12
Plato 195n5
Plutynski, A. 209n31
Polanyi, K. 37, 196n16
Popper, Karl 3
Portugali, J. and Benenson, I. 179
Portugali, J., Benenson, I. and Omer, I. 179
Postema, G.J. 199n1

Poundstone, W. 215n1
precedence 149, 155, 158, 160, 182–3, 215n37
prediction, actual behaviour and 149–50, 183–5
preferences 52, 56, 59, 60, 66, 128, 132, 135, 198n6; expectations and 62; individual preferences 55, 61, 62, 180; motivations and 57; segregationist preferences 65; *see also* discriminatory preferences
primitive monies 30–1, 37–8
process interpretation: interpretation in game theory 159–63; of invisible hand 3, 4–7, 69, 84–8; of origin of money 27–8, 33–4, 44–5, 47–8, 49, 95, 96, 98, 100, 102–3, 104–5, 105–6, 108–10, 115, 117–18; process models 27–8, 33–4, 44–5, 47–8, 49, 95, 96, 98, 100, 102–3, 104–5, 105–6, 108–10, 115, 117–18, 158–9, 165–6; of residential segregation 52, 53, 55, 56, 58, 61, 65, 66
providential order 71–2, 80–1
Pruitt, D.G. and Kimmel, M.J. 215n38

quasi-morphisms 127–8
Quiggin, A.H. 196n17
Quine, Willard van Orman 146

Raphael, D.D. 200n10
Rappaport, A. and Orwant, C. 215n38
Rappaport, S. 207n13
rationality 85, 103, 114, 151, 154, 156, 161, 177, 212n9, 215n35; bounded rationality 156, 158; learning and 154–6, 192–3; tendency towards 33–4; utility and 159
re-description 208n23
real and model worlds: abstraction and 45, 82–3, 119–20, 127–8, 133, 135, 166, 197n22, 197n25; complexity in relationship between 119–20, 144–5, 163–4; realist–empirical orientation 41, 42–4, 46–8; recognisable similarities between 63; relationship between 45–6, 166–7; *see also* models and representation
reciprocity in exchange 37–8
Redhead, M. 207n13
refinements literature 151
religious connotations of invisible hand 69, 71, 72–3, 80
replicator dynamics 162, 189–91, 214n25;

evolutionary stability and 188–9; Nash equilibria and 190
representation 130–4; mirroring 210–11n46; *see also* models and representation
reputation, past behaviour and 211n6
residential segregation: chequerboard model 50; discriminatory preferences and 4, 6, 50, 51–3, 57, 58, 62–3, 65–6, 130–3 138–40, 178, 180, 198n5, 208n24, 210n43; economic factors and 50, 51; explanandum phenomenon 59; immediate neighbours and 52–5; intended consequence of human action 208n25; interpretation of real world 208n16; invisible-hand consequences 21–2, 23; mechanisms for 51–2, 55–7, 57–61, 61–4; mild discriminatory preferences 50, 51, 52–5, 57, 58, 63, 66, 132, 138, 139, 178, 180, 198n5, 210n43; near complete segregation, conditions for 65; random distribution of residents 53–4; Schelling's observations and insights 4–5, 51–7, 62, 66–7, 191, 194n9; strong discriminatory preferences 5, 6, 50, 51–2, 62–3, 65, 130, 131, 133, 140, 180, 208n24; unintended consequence of human action 50, 51; welfare differences 50, 51; *see also* chequerboard model
risk dominance 152, 182, 185–6, 192, 213n16
Robinson, J. 214n29
Rosenberg, A. 8, 194n1, 204n12, 206–7n7, 208n26
Rosser, J.B., Jr. 8, 50, 167
Roth, A.E. *et al.* 184
Rothschild, E. 68, 70, 71–2, 77, 78–9, 80–1, 199n5, 201n20; perception of irony in Smith's invisible hand 70, 80
Rousseau, Jean-Jacques 86
Ruben, D.-H. 121, 123, 125, 206–7n7, 206n2
Rubinstein, A. 162
rules of the road, Young's model of emergence of 4–5, 26, 147–8
Rutherford, M. 8

Sagoff, M. 199n1
saleableness 31–2, 34–5, 35, 39; marketability and 204n5
salience 149, 152, 155, 160, 182–3, 213n20, 215n37

Salmon, W.C. 123, 206–7n7, 206n3, 207n10
Samuelson, L. 151, 190, 214n24
Samuelson, L. and Zhang, J. 189
Sander, R. *et al.* 9, 140–1, 180
Sayigh, Y.A. 199n1
Schelling, Thomas C. 6, 8, 21, 48, 49, 50, 68, 90, 91, 93, 118, 132, 140, 142, 144, 146–7, 149, 152, 153, 156, 163, 187; chequerboard model 178–81; contextual detail, importance of 184–5, 211n5; explanatory value of chequerboard model 57–66; games, informal experiments 183; mechanisms behind segregation, postulation of 57–61, 66–7; meta-model of segregation, Schelling's chequerboard model as contribution to 64–6; observations and insights on residential segregation 4–5, 51–7, 62, 66–7, 191, 194n9; partial nature of explanation 121–2, 124; residential segregation 51–7, 66–7; residual segregation model 191, 194n9; Schelling games 183
Schlefer, J. 79, 201n22
Schmitz, S.W. 204n9
Schotter, A. 8, 9, 106, 107–9, 110, 111, 114–15, 156, 205n27, 207n13, 214n27, 215n41
Scientific Explanation, Aspects of (Hempel, C.G.) 209n28
Scriven, M. 207n8
segregation *see* residential segregation
select-a-ball game 153–4
self-interest 71, 76, 78, 79, 84, 88; selfishness and 201n21; species multiplication and 69–70, 71
Selgin, G.A. and Klein, P.G. 9, 203n4
Selgin, G.A. and White, L.H. 203n4
Selten, R. 189, 212n11
Sethi, R. 104–5
Sethi, R. and Somanathan, R. 179
Shepard, J.M. 199n1
shortcuts, emergence of 1–2, 25–6
Sidowski, J.B. 211n2
Sidowski, J.B., Wyckoff, L.B. and Tabory, L. 211n2
similarity, notion of 38–9, 136–7
Skinner, A.S. 200n10
Sloep, P. and van der Steen, W.J. 209n31
Smith, Adam 3, 4, 11, 20–1, 37, 68, 69, 84, 85–8, 91; analogical thinking, views on 200n11; astronomy, history of 69, 70–1, 72–7, 80; general framework

and idea of invisible hand 71; invisible hand, concept of 69–72, 77–81, 82, 83–4, 91; on money 169–70; moral philosophy 75–6; Newton's system of astronomy 74–5; philosophy of science 73–7; relation between invisible hands of 70–1; unintended consequences and invisible hand 77–81; unintended social consequences, individual interactions and 200n13
Smith, C. 202n30
Smith, T. 199n1
social action: consequences of 15–16, 17–20; intentions of 17–19
Social Mechanisms, An Analytical Approach to Social Theory (Hedström, P. and Swedberg, R., eds) 66
social phenomena *see* phenomena
Solomon, M. 199n1
Somanathan, R. and Sethi, R. 179
Sorensen, R.A. 210n44
spontaneous order, theory of 3, 20, 84, 202n30, 203n41, 203n42
Stahl, D. 213n17
Stark, W. 195n3
Starr, R.M. 97
Stenkula, M. 195n9
Stewart, Dugald 61, 86–7
Stiglitz, J.E. 82, 199n1
strong discriminatory preferences 5, 6, 50, 51–2, 62–3, 65, 130, 131, 133, 140, 180, 208n24
Sugarscape (Epstein, J. and Axtell, R.) 4
Sugden, Robert 4, 6, 50, 63, 64, 65, 124–5, 147, 150, 153, 163, 182, 207n13, 211n1, 212n11, 213n20
superstition 72, 73, 76, 201n24
Suppe, F. 134, 135, 136, 206n3, 207n13, 209n29, 209n30
Suppes, P. 209n30

Tajfel, H. *et al.* 63
Taylor, P. and Jonker, L. 189
tendencies, information from 137–9
Tesfatsion, Leigh 4, 167
Thagard, P. 65
theoretical explanation *see* partial potential
The Theory of Money and Credit (Mises, L. von) 172
Theory of Moral Sentiments (Smith, A.) 69–70, 71, 72, 73, 74, 76–7, 80–1, 87
Thompson, P. 209n30, 209n31
Thomson, H.F. 75
Thornton, M. 199n2

thought experimentation 60–1, 63–4
Thurnwald, R.C. 196n16
tipping model 58–9
Tirole, J. 211n6
Tobin, J. 82
Torrens, P.M. and Benenson, I. 178
Townsend, R.M. 9, 203n4
Trejos, A. and Wright, R. 203n4

Ullmann-Margalit, Edna 4, 6, 8, 50, 56,
    88–9, 201n23, 202n30, 203n44, 211n1
ultimatum games 149, 183–5
*The Unanticipated Consequences of
    Purposive Social Action* (Merton, R.K.)
    13
unintended social consequences 2–7;
    actions-intentions, examples of pairs
    of 17; attribution of consequences,
    problem of 16–17; categorization
    of 16–20; classification of 13–16;
    collective behaviour 19–20; computer
    simulations 6; concept of 13, 14;
    concept of, elements of confusion
    on 12–13; concrete consequences
    15; consequences, concept of 12–13,
    13–16; conventions of game theory
    as 160; desirability of consequences
    18, 21–2; end-state explanations of
    invisible hand 3; error 14, 15; game
    theory 7; of human action in residential
    segregation 50, 51; ignorance 14, 15;
    individual action, consequences of
    15–16, 17–20, 20–1; individual action,
    intentions of 17–20, 21; individual
    interactions and 200n13; intention,
    concept of 12; intentions, target of 16,
    17; interests, imperious immediacy of
    14–15; invisible-hand explanations,
    literature on 6; invisible-hand
    explanations, process models of 3, 4–7,
    69, 84–8; invisible-hand explanations,
    relationship with 11–26; knowledge,
    lack of 14, 15; materialisation, spaces
    of 16, 17, 18; Merton's perspective
    on 2, 13–16, 16–17; methodological
    problems in categorisation of 16–17;
    mixed intentions, possibility of 18;
    money, Menger's model of emergence
    as medium of exchange 4–5; multiple
    individuals, consequences of actions
    of 20; neglected causal factors
    13–14; organised actions 15; process

explanations of invisible hand 3, 4–7,
    69, 84–8; relationship with invisible
    hand 68, 77–81, 91; residential
    segregation, Schelling's models of
    4–5; rules of the road, Young's model
    of emergence of 4–5; significance of
    3; Smith's invisible hand and 77–81;
    social action, consequences of 15–16,
    17–20; social action, intentions of
    17–19; spontaneous order, theory of 3;
    sum-total consequences 15; types of
    12–13; unanticipated consequences and
    14–16, 25, 80; unorganised actions 15;
    unseen causal factors 13–14; *see also*
    invisible-hand consequences
unorganized discriminatory behaviour 62

value of explanation 196n18
Van Damme, E. 7
Vanberg, V.J. 8, 202n30, 202n34, 202n38
Vanderschraaf, P. 215n1
Varian, H.R. 207n13
Vaughn, K. 71
Veblen, Thorstein 82
Vromen, J.J. 202n38

Waldrop, M.M. 215n3
Walras, Léon 83
Wärneryd, K. 189
Wartofsky, M.W. 207n13
The Wealth of Nations (Smith, A.) 70,
    71, 72, 73, 74, 76–7, 78–9, 80–1, 87,
    169–70
Weibull, J. 189, 190
welfare 28–9, 99, 174, 175; welfare
    differences 50, 51, 52, 61, 65, 67, 130,
    131, 132, 133; welfare economics 81,
    82, 83
Werhane, P.H. 201n21
Wilkins, J.S. 209n31
Williamson, S. and Wright, R. 99, 203n4
Wray, K.B. 199n1

Ylikoski, P. 121, 199n1, 203n44, 206–7n7
Young, H. Peyton 4, 5, 7, 9, 110–11,
    111–12, 113, 114–15, 156–7, 159–60,
    179, 191–2, 205n33, 205n35, 213n14,
    213n16, 214n27, 214n29, 214n33,
    214n34, 215n41

Zappia, C. 202n34
Zhang, J. 9, 140, 178, 179, 180
Zuidema, J.R. 197n32

Printed in the United States
by Baker & Taylor Publisher Services